Springer Water

The book series Springer Water comprises a broad portfolio of multi- and interdisciplinary scientific books, aiming at researchers, students, and everyone interested in water-related science. The series includes peer-reviewed monographs, edited volumes, textbooks, and conference proceedings. Its volumes combine all kinds of water-related research areas, such as: the movement, distribution and quality of freshwater; water resources; the quality and pollution of water and its influence on health; the water industry including drinking water, wastewater, and desalination services and technologies; water history; as well as water management and the governmental, political, developmental, and ethical aspects of water.

More information about this series at http://www.springer.com/series/13419

Florence Metz

From Network Structure to Policy Design in Water Protection

A Comparative Perspective
on Micropollutants in the Rhine River
Riparian Countries

 Springer

Florence Metz
Institute of Political Science
University of Bern
Bern
Switzerland

ISSN 2364-6934   ISSN 2364-8198   (electronic)
Springer Water
ISBN 978-3-319-85729-9   ISBN 978-3-319-55693-2   (eBook)
DOI 10.1007/978-3-319-55693-2

Printed on acid-free paper

This Springer imprint is published by Springer Nature
The registered company is Springer International Publishing AG
The registered company address is: Gewerbestrasse 11, 6330 Cham, Switzerland

Foreword

The first question I asked myself when writing this foreword was: Ideally, who should read this book? At least, four valid answers came to mind.

First, this book should be read by policy studies scholars who are interested in the broad puzzle of how public policies come about, how they are designed, and how they are introduced and implemented. Florence Metz nicely shows how a multi-level set of factors embracing networks (meso) and actor preferences (micro) can impact final policy output. When studying the policy output, she mainly focuses on policy instruments, and the variety of measures integrating a whole policy mix. She goes further than most policy process theories, not only studying policy change, but also considering the quality of the change and output, e.g., how comprehensive solutions for an identified problem come about.

Second, this is a "must read" for all researchers interested in the development and application of social network analysis in policy studies. Florence Metz conceptually and methodologically introduces networks, and thus stable relations among public and private actors involved in policymaking, as important drivers that shape final policy outputs. However, she goes beyond that. First, her book is a nice illustration of how robust data for a comparative case study design can be gathered through standardized surveys. In her analysis, she then goes beyond the current state of the art by linking insights from several streams of the network literature, in order to generate additional insights into relational structures between actors in policy networks, which is an important precondition to answer the final research question, i.e., What impact policy networks have on the design of comprehensive policy outputs?

Third, people interested in environmental issues, such as citizens, representatives of public environmental agencies, or practitioners in water supply and protection, would have great pleasure in reading this book. This book focuses on the highly topical issue of micropollutants in waters. These are chemical compounds that occur in waters in very small concentrations, but have a potentially toxic effect and have an equal impact upon aquatic ecosystems and human health. We are all involved in the production of and diffusion of micropollutants: They enter waters in a variety of ways. For example, through the application of pesticides on agricultural fields, but also via households and the consumption of health care and the application of cleaning products. Additionally, a variety of pharmaceuticals privately consumed or industrially produced enter our surface waters as micropollutants. Reducing micropollutants in waters is thus one of those very complex environmental problems that call for a non-negligible behavioral change if we want to eliminate them from water courses. Florence Metz manages to communicate the complex geophysical and chemical particularities of this issue not only to experts, but also to non-expert audiences.

Finally, this book should be read by Ph.D. candidates, young researchers, and postdocs as it is an ideal example in how to nicely and convincingly present a whole research endeavor. It is extremely well structured and written, with a rich but understandable vocabulary. Florence Metz demonstrates an excellent writing style and impressive skills in outlining scientific argumentation and research. Furthermore, the text is written on two levels, comprised of the main text and the footnotes, which makes the development of the arguments very sophisticated.

Addressing the history, the beginning, the scope, and the conclusions of the research project, the point of departure of Florence Metz's work is that she very specifically asked what is the relative contribution that the study of policy networks provides in answering the question of how policies, and thus programs, strategies or instruments, come about. Speaking more generally, she is also very interested in the puzzle of how to create comprehensive, effective measures to tackle complex problems, such as micropollutants in waters.

To address the puzzle and answer the research question, Florence Metz studied four different countries, such as Switzerland, Germany, France, and the Netherlands, in the transboundary setting of the Rhine river basin. More concretely, she adopted a comparative research design and assessed water quality management within the four countries. Another highlight of this book are the empirical sections and the quality of the data introduced. To achieve this, Florence Metz managed to gather data in 4 different countries, about 4 very diverse political processes embedded within different political systems, and embracing a variety of political stiles. Additionally, the survey had to be developed in a systematic way, in three different languages, taking into consideration the specificities of each country and process, without losing the rigor needed for the production of comparable results. Besides the written surveys in the three countries, Florence Metz interviewed a large majority of the organizations presented in the survey. The result is a product of endurance, patience, focus, and courage that Florence Metz exercised during this complex data gathering process.

This book concludes that several network types and structures embrace the potential for the adoption of a comprehensive policy design in water quality management. Most importantly, the combination of both ideological (actors' beliefs) and structural (relations in the network) aspects seems to drive policy outputs. Still, most importantly, a structure with cohesive beliefs is found to be beneficial for the adoption of a comprehensive policy design. To achieve such comprehensiveness, belief cohesion should, in the best cases, benefit from one of the following additional circumstances: some, but not too much interconnectedness among political actors; coalition structures with brokerage rather than diffuse across-coalition collaboration; or the presence of a dominant coalition or one entrepreneur, both in favor of the comprehensive index.

In my general and scientific appraisal of this research and book, I come to the conclusion that Florence Metz convincingly manages to make at least a threefold contribution to the literature addressing policy networks and policy design, and fills one major gap in the literature. First, she presents reflections about how problems arriving on the political agenda can be addressed in a comprehensive way. Florence Metz therefore focusses on policy instruments and instrument mixes and creates a new framework, via a policy comprehensiveness index, for evaluating, in an ex-ante manner, the quality of a policy design. Second, she makes an important contribution to the network literature and the conceptualization of politics as meso-level phenomena and as networks. The author succeeds in linking network structures to policy outputs. This is not only done in a discursive or metaphorical manner, but also through an empirical test within a comparative empirical setting.

Finally, the research fills an important gap not only by conceptually reflecting and empirically testing assumptions related to network effectiveness, but also by combining insights from network analysis, policy network theory, and the policy design literature. By so doing, Florence Metz enhances the understanding of specific types of network structures and their ability to promote given types of policy design.

In my personal appraisal, look back to 2010. It is when Florence arrived in Switzerland and came to join the same research group that I formed. She did so after having gathered an impressive set of experiences in various European universities, as well as international organizations. By that point already, Florence's research interests mainly lay in water-related issues and conflicts. During the past seven years, I have had the pleasure and honor of witnessing Florence's academic career, her personal development, and I have observed development into a true expert in policy studies, transboundary water management, and European water quality politics. Because of this, Florence is not only one of today's most promising young scholars when it comes to studying complex (environmental) problems and the related and potential policy solutions, but also she is an indispensable partner for international committees and platforms, such as the International Commission for the Protection of the Rhine, or the OECD. Florence is the person to call if one wishes to answer complex questions about how to design and successfully adopt new policies, especially related to the regulation of micropollutants and water quality issues, but also in the broader context of environmental, climate or societal change.

Bern   Karin Ingold
February 2017

Acknowledgements

Working on a research project can be compared to a journey that is exciting and enriching, and also exhausting and overwhelming at times. Without the support of many people who accompanied me on this journey, the final product of this research would not have been possible.

First of all, I would like to thank my supervisor Prof. Karin Ingold for her incredible and continuous support during all stages of this research process; for believing in me, for reading numerous versions of this work, and for her valuable inputs and critical mind, which substantially helped improving this work; for always being available, and aiding to find solutions; and for her great sense of humor, which makes work much more enjoyable.

I wish to express my particular thanks to my co-supervisor Prof. Hans Bressers, who enabled me to visit his research group at University of Twente; who provided me with great support to carry out the Dutch survey; who found motivating words at the appropriate moment; and whose work was a great source of inspiration. Thanks to Hans Bressers, I was able to experience the Netherlands, along with the cheerful international working environment at UTwente that I very much enjoyed.

A very special word of thanks goes to my co-supervisor Prof. Mark Lubell, who welcomed me into his research group at UC Davis, and thanks to whom I was granted the opportunity to live in California. The exchanges with him on topics of policy studies, and theoretical as well as methodological aspects of policy networks in particular, were of great help and provided valuable inputs to this research. From the research group at UC Davis, I would like to thank Gwen Arnold for her insightful comments on the very first draft of the policy comprehensiveness index; Michael Levy, Carlos Barahona, and Matthew Hamilton for their various inputs; and Susie Pike, Chantelise Pells, and Kasey Harding for welcoming me so warmly to Davis. Special thanks also go to my Davis lunch buddy Paul Wagner for all the entertaining conversations.

Of particular importance to me was the exchange with my colleagues at University of Berne and EAWAG. Special thanks go to Manuel Fischer, Philip Leifeld, Laurence Brandenberger, Laura Herzog, Mario Angst, and Victor Garcia for providing me with the literature and advice, and for reading and commenting on parts of my research, which greatly helped in improving it. I would also like to thank Alexa Jackson for her clear explanations of chemicals and micropollutants, which assisted me—as a social scientist—in enriching my understanding of the natural sciences.

I wish to thank my proofreader Rosalie Lipfert for her great work. Special thanks also go to Benedikt Wagner for his graphical work on the network diagrams. Moreover, I would like to thank the Swiss National Science Foundation for the support of this research project, research visits, and methods schools, which considerably improved the quality of this research.

This research would not exist without the information and expertise of the policy experts in Switzerland, Germany, France, and the Netherlands, who invested time to take my survey or respond to my interview questions. The exchanges with policy experts of diverse organizations enabled me to develop a better understanding of water protection policy in the twenty-first century, which largely informed this research.

Finally, I would like to thank my parents for their emotional support and sound advice on my research, and my friends for reminding me about the world outside of academia. Last but not least, I am very grateful to my partner Michael, who never got tired from motivating me to continue my research journey—in the words of Oliver Kahn: "weiter, immer weiter" (in English: go on, always go on).

Contents

Abbreviations

ACF	Advocacy Coalition Framework
CH	Switzerland
DV	Dependent variable
EQN	Environmental quality norms
EQSD	Environmental Quality Standards Directive
EU	European Union
F	France
G	Germany
ICPR	International Commission for the Protection of the Rhine
IPPC	Directive concerning Integrated Pollution Prevention and Control
IV	Independent variable
NL	The Netherlands
REACH	Directive concerning the Regulation on Registration, Evaluation, Authorization and Restriction of Chemicals
SAGE	Schémas d'Aménagement et de Gestion des Eaux
SDAGE	Schémas Directeurs d'Aménagement et de Gestion des Eaux
UN	United Nations
WFD	Water Framework Directive
WPA	Swiss Waters Protection Act
WPO	Swiss Waters Protection Ordinance

List of Figures

List of Tables

List of Annexes

Abstract

This work deals with the research question assessing whether structural characteristics of policy networks may help us to understand part of the variance of policy designs. While some policies are designed so as to comprehensively reduce a societal issue, others address a problem without being able to reduce or resolve it. In order to comprehend such variance, the present research specifically looks into the social structures of policymaking processes. Policy networks are considered snapshots representing the aggregated result of multi-actor interactions in the policymaking process. The present study seeks to explore whether the fabric of policy networks provides an explanation for the performance of policy designs in solving an underlying policy issue. If we understand which structural network characteristics promote comprehensive problem solving, we may also be able to improve policy designs.

This book concentrates on a new emerging policy field in water protection, namely aquatic micropollutants. Micropollutants are chemical substances present in very small concentrations in waters. The significance of reducing emissions into waters can be attributed to the fact that even very low concentrations of micropollutants can cause severe environmental impacts, and further impacts on humans can be expected. Finding ways to reduce micropollutants in waters is a relevant—but also a complex—political task because of the diversity of substances, uses, discharges, and effects. Due to the complexity of the issue, finding political solutions that comprehensively reduce micropollutants in waters is challenging. In order to understand how different countries respond to this challenge, this study takes the case of the Rhine River. This book analyzes water protection policies aiming at the reduction of aquatic micropollutants of the Rhine riparian countries, such as Switzerland, Germany, France, and the Netherlands.

Moreover, the study evaluates the performance of the studied policy designs in reducing aquatic micropollutants by means of a policy comprehensiveness index, a new tool introduced in this work, which renders measurable how comprehensively policy designs contribute to the reduction of a defined policy problem. With the help of the index, different hypotheses are formulated relating structural network properties to the comprehensiveness of policy designs. The analysis of policy networks is embedded into the literature on policy process theories, which discerned as crucial features of policy networks the combination of interconnected actors exhibiting belief cohesion and uniting into coalitions, brokerage, and entrepreneurship. To study these network properties, the study relies on network data from policy actors participating in policymaking processes on micropollutants that were gathered for this research. Social network analysis was employed to analyze network data. Results reveal significant differences in the ways water policy networks are structured in the four countries under study. Findings provide first indications that the relational structures between policy actors are an important element for comprehending how to achieve comprehensive policy designs and promote problem solving. A combination of several structural network properties impacts networks' ability to design comprehensive policies. A belief-cohesive network, in which policy actors collaborate and ally within coalitions, and where strong brokers mediate across coalitions, seems to promote comprehensive policy designs. Further conditions that are conducive to problem solving are belief-cohesive and well-interconnected dominant coalitions favorable to comprehensive policy designs, together with particularly active entrepreneurs pushing for a comprehensive design. Such insights about the social foundation of policy design may not only assist science, but also assist decision-makers in their search for optimal policy results.

Keywords Public Policy Analysis · Policy Networks · Network Structure · Social Network Analysis · Policy Design · Water Protection Policy · Micropollutants · Rhine River

Chapter 1
Introduction

In 2010, the United Nations (UN) recognized the human right to water and sanitation (Resolution 64/292). This UN decision fortifies that water is a public good and that nobody can be excluded from its use. While it is in every individual's interest to use water for drinking water purposes, for irrigation, as a sink for wastewater, a means of transport, or for leisure activities, its overuse is an inherent collective action problem that affects us all. Public policymaking takes a particularly important role regarding the protection of water quality, as it can help to overcome problems of collective action (Hardin 1968). The present study therefore analyzes the way in which the political realm handles a new challenge of water protection policy, namely aquatic micropollutants.

Micropollutants are chemical substances detected in very small concentration in waters (Schwarzenbach et al. 2006). Among others, the sources of pollution include our consumption of pharmaceuticals, cosmetics, or cleaning agents and reflect the increasing reliance of society on such chemicals (Hollender et al. 2008). Micropollutants challenge traditional water protection policies, as many substances are not vulnerable to conventional wastewater treatment (Miao et al. 2005). The International Commission for the Protection of the Rhine is among the first river basin organizations to take up this challenge and to place the issue of micropollutants on its political agenda since 2008.[1] Hence, the present work analyzes the policies that the four Rhine riparian states, i.e., Switzerland, Germany, France, and the Netherlands adopted to reduce micropollutants in waters.

In order to compare the four policy designs with regard to their prospective ability to effectively reduce aquatic micropollutants in the future, an index of policy comprehensiveness is introduced in this work. The index involves a problem-solving perspective by estimating how comprehensively policy designs

---

[1]Mandate for the MIKRO project group of the ICPR, see: http://www.iksr.org/index.php?id=317&L=3 (last access 13.9.2015).

© Springer International Publishing AG 2017
F. Metz, *From Network Structure to Policy Design in Water Protection*,
Springer Water, DOI 10.1007/978-3-319-55693-2_1

may contribute to the reduction of a defined policy problem. While existing indices (e.g., Environmental Sustainability Index, State of the Nation's Ecosystems, Ecological Indicators for the Nation) tend to measure environmental performance, i.e., the state of the environment, the proposed index seeks to evaluate *policy* performance. The purpose is to *ex-ante* evaluate policy designs' prospective impact on alleviating a policy problem by looking at the way policies are designed when they are adopted. Policy designs can vary, for example, in their policy goals, defined target groups, incentives for behavior changes, or arrangements regarding implementation, control, or sanctions. The purpose of the proposed index is to evaluate whether such arrangements are designed so as to comprehensively address an underlying policy issue. Indicators introduced here constitute an inaugural index proposal, which serve to rank and compare the comprehensiveness of policies across cases for the purpose of this study.

Policy designs are considered outputs of policymaking processes where multiple actors are involved and interacted with one another. To explain variation of policy design, the present study pays particular attention to the social mechanisms behind policymaking. Understanding the social setting of policymaking is a key to explaining policy design, because public policies are adopted by policy actors who are embedded in their own web of relations. These networked relations may socially influence actors' attitudes, perceptions, behavior, and policy preferences. Social interactions are said to promote social capital, trust, and mutual understandings (Henry et al. 2010), which are powerful factors in enabling social systems to overcome problems of collective action. While this claim exists within the research of collaborative governance (Koontz and Thomas 2006), it remains unclear which structural characteristics of policy networks promote (or hamper) effective environmental policies. Adopting a network approach enables one to analyze the relational aspect of policymaking, and to understand the social foundation of policy design. The present research seeks to assess the influence of policy networks on the adoption of certain policy designs and puts forward the following research question:

Do structural characteristics of policy networks help us to understand some part of the variance of policy designs' prospective problem-solving ability? The question represents a first step in the process of understanding whether certain structural network properties promote the emergence of particularly comprehensive policy designs. And which network structures, on the contrary, hamper policy solutions that could comprehensively solve an underlying policy problem?

In contrast to a pure actor perspective, the present study emphasizes the relational interdependencies between policy actors through its network approach. Actors' ability to formulate comprehensive policy designs is considered dependent upon the web of relations in which they are embedded. To identify those structural properties that are of relevance in policy networks, the present study borrows network concepts from policy process research such as belief cohesion, interconnectedness, coalition structure, brokerage, and entrepreneurship (Sabatier and Jenkins-Smith 1993; Kingdon 1984; Bressers and O'Toole 1998). These concepts are operationalized by adopting a formal social network analysis. To gather network data, a mixed-mode survey was conducted between 2011 and 2014 by the author.

A total of 199 Swiss, German, French, and Dutch policy actors, who participated in the respective policy processes on micropollutants, were either surveyed through a standardized mail questionnaire or through in-person interviews. Relational data was analyzed by means of network statistics.

The analytical strength of the network approach is its incorporation and transcendence of a pure micro- and macro-level explanation for policy design. It builds on policy actors (microlevel variable) as building blocks of networks, but goes beyond the mere aggregation of policy actors' attributes by taking into consideration actors' interdependencies. The network approach also incorporates the impact of a country's political system (macro-level variable) on actors' interactions. A country's political system provides certain actors with decision-making power and institutionalized access to policymaking processes, thereby constraining network interactions in the policymaking process. The present study employs the network approach as a meso-level explanation for policy design and focuses on studying the impact of structural network properties on policy design. Macro- and microlevel variables are assumed to be reflected in the structure of policy networks. They are taken into consideration here by analyzing the impact of actors' attributes on network structure.

1.1 Micropollutants in Surface Water—A New Policy Field

Micropollutants are chemical organic substances detected at very low concentration levels (ng/L to µg/L) in surface waters (Schwarzenbach et al. 2006). Their detection is possible today due to technological progress in chemical analysis throughout the last decades. While we are aware of the presence of micropollutants in surface waters today, there remain great uncertainties about their effects on humans and the environment. Due to these uncertainties, policy makers face the challenge of deciding whether or not to take action regarding micropollutants reduction measures. Following the precautionary principle, political measures are necessary because negative consequences for humans cannot be excluded today (Peel 2005; Auberson-Huang 2002). Even without certainty about potential risks, political action can be vindicated in order to minimize potential health risks associated with micropollutants, to ensure the environmental protection of surface waters, or drinking water safety. Following the risk-based principle, by contrast, one could consider political measures as premature as long as negative consequences cannot be proven with absolute certainty for all existing substances. Some might argue that political efforts for the reduction in micropollutants follow exaggerated environmental quality standards, and that resources should be invested in more urgent policy issues. This discussion highlights that the search for political answers regarding aquatic micropollutants represents a challenge to the political realm. In order to fully comprehend the complexities that decision-makers face, a brief overview is provided below about sources of micropollutants, entry-paths into the aquatic environment, and expected consequences.

Emission sources of micropollutants and entry-paths into the aquatic environment

Micropollutants represent a multifaceted problem characterized by an immense number and diversity of chemical substances, as well as various entry-paths into the aquatic environment (Metz and Ingold 2014). In Europe alone, there exist about 100,000 synthetic substances in use, and an additional 1000 new chemicals enter the market every year (Götz et al. 2010; Metz and Ingold 2014). Potential sources of emissions of micropollutants[2] into waters include the consumption in households of detergents, cleaning agents, or personal care products (Schwarzenbach et al. 2006; Bendz et al. 2005). Households, as well as hospitals or retirement homes, also emit pharmaceuticals, especially given the longer life span of populations today. Additional sources of pollution include diverse plant protection products used in agriculture, as well as corrosion inhibitors applied as protective agents to public places, roofs, and exteriors of buildings (Hollender et al. 2008; Wittmer et al. 2010). Moreover, industrial production processes apply diverse chemical substances such as plasticizers, solvents, dyes, and lubricants that can only be partly detected in waters (ICPR 2012). This broad range of pollution sources shows that the use of chemicals is deeply rooted in varied parts of society and the economy. While chemicals are developed in order to advance areas of human life, they are also accompanied by negative consequences and risks for the environment as well for humans. Those unintended consequences represent the beginning of a broader societal and political debate about how to manage an increasing chemicalization of society, including questions regarding a healthy usage of chemicals or about desired drinking water quality standards, or environmental ones.

The search for suitable political solutions is complicated not only by a large quantity of substances and their diverse fields of use, but also by their multiple entry-paths into the aquatic environment. The two main categories are *point* and *diffuse entry-paths*. Diffuse refers to surface runoff from fields, roads, or urban areas due to rainfall, and also to percolation, i.e., the slow movement of liquids through soil or permeable rock into waters (Wittmer et al. 2010). Point-source pollution, on the contrary, is a concentrated form of discharge, mostly from wastewater effluents containing chemicals not vulnerable to conventional treatment technologies (Miao et al. 2005). Point sources of pollution also include incompletely treated wastewater discharges after heavy rainfalls, or other forms of spills and incorrect disposals (Wittmer et al. 2010).

The primary source of point-source pollution is municipal and industrial wastewater treatment plants, which process pollution emitted by households,

---

[2]While in some single cases naturally occurring substances may impact water quality, the broader phenomenon of anthropogenic and synthetic pollution remains in the foreground of political regulation. In this book, the term *micropollutants* therefore refers to anthropogenic rather than naturally occurring pollution.

industry, and sometimes farms, if connected to the public sewerage system (Wittmer et al. 2010). Diffuse pollution is mainly caused by agriculture and urban areas after the application of plant protection products, biocides, or corrosion inhibitors.

Usage in society also affects input dynamics (Hollender et al. 2008). While wastewater treatment effluents are constantly discharged into surface waters, plant protection products are only seasonally applied and prone to sudden runoff from changing weather patterns. In the latter case, concentration patterns of micropollutants in waters seasonally differ, which pose another challenge for the search of appropriate policy solutions.

In summary, the large amount of substances, diverse entry-paths into the aquatic environment, and time-sensitive dynamics underline the complexity of designing policies on aquatic micropollutants. The challenge comes with developing political solutions that meet the various aspects of the underlying issue. Solutions that are tailored for the reduction in point-source pollution, for example, generally do not address the issue of diffuse pollution, and vice versa. Likewise, policies that deal with seasonal inputs do not solve the issue of constant discharges. The complexity of the issue suggests that there is no single best solution for this problem. Only a carefully fashioned policy design has the potential to comprehensively reduce micropollutants in waters (Metz and Ingold 2014).

Consequences of micropollutants for human and ecological health

To assess which substances are potentially harmful for humans and the environment, risk assessment is necessary. Such an evaluation builds on information about (a) exposure and (b) toxicity (Schwarzenbach et al. 2006). Exposure refers to measuring concentrations of single compounds in water or in certain aquatic species[3] and considers the environmental behavior of the compound in the aquatic system, including degradation, transport, effects of stressors such as temperature or UV light, volatilization into gas, sorption to solid surfaces, or bioaccumulation in living tissues. Toxicity can only be determined through toxicology testing, which involves an assessment of the harm to aquatic organisms and humans who encounter the substance. If the risk assessment reveals that substances are present in waters at higher levels than would be safe for the environment,[4] the general approach is to establish an environmental quality norm and to propose the

---

[3]An alternative to measuring concentration levels in waters is to analyze the frequency and level of entry into the aquatic system of compounds.

[4]To assess whether exposure levels are safe, researchers calculate the ratio of the predicted environmental concentration (PEC; based on statistical averages) to the predicted no-effect concentration (PNEC; the level at which there are no negative consequences for the ecosystem). If PEC:PNEC ratios are greater than one, an environmental quality norm value is established, and the substance is proposed for inclusion in future monitoring (Schwarzenbach et al. 2006).

substance for inclusion in future monitoring (Schwarzenbach et al. 2006).[5] An inherent drawback to this monitoring approach is that compounds are not analyzed for which there is no information about their exposure and toxicity. Even if a substance was quite dangerous and frequently present in surface waters, it would not be monitored. Without monitoring, no data can be obtained that would justify including the substance in future testing, and so it remains unanalyzed (Von der Ohe et al. 2011). This void concerns micropollutants in particular, since many compounds are so-called *emerging pollutants* that have only recently been detected and deemed a concern (Richardson and Ternes 2011). Representative examples of emerging pollutants, for which only limited ecotoxic evidence is available, include pharmaceutical residues, personal care products, the fuel additive MBTE, biocides, and metabolites of plant protection products (Hollender et al. 2008). Even if a given substance is at a concentration too low to be harmful, when mixed in water with other chemicals, research has shown that the combined effect can be devastating (Kortenkamp et al. 2007). In many countries, regulations such as the ones concerning the placing on the market of plant protection products (e.g., EC No 1107/2009 in the European Union) or biocides (e.g., EU No 528/2012 in the European Union) authorize marketing only for substances that pose no risk (PEC: PNEC < 1). However, many substances are used in several areas, e.g., as pesticides *and* as biocides, and their risk assessment does not take into consideration the exposure resulting from these combined uses. This discussion highlights the challenge involved in identifying potential risks of aquatic micropollutants for the environment and humans. Considering the high volume of substances in use, the ongoing development of new compounds, and potential interaction effects of substances and their metabolites, attaining full certainty regarding associated risks remains a challenge. Uncertainties are hence an inherent part of the decision-making process within the realm of micropollutants.

Well-investigated are the characteristics that lead substances to pose a significant risk, namely persistence, bioaccumulation, and ecotoxicity (Schwarzenbach et al. 2006; Buser et al. 2006). Persistence refers to chemical substances that withstand natural decomposition, evaporation, hydrolysis, or phytolysis. Therefore, persistent substances remain in their original form in aquatic systems for long periods of time, sometimes affecting waters hundreds or thousands of kilometers away from the contaminant source (Schwarzenbach et al. 2006). When persistent micropollutants transgress national borders, they may quickly turn into an international problem (Metz and Ingold 2014). For example, downstream locations that rely on surface water for drinking purposes may be particularly sensitive to contaminations in countries located upstream. In such cases, there is a need for internationally coordinated policies, which can be difficult to achieve where several national interests clash with one another.

---

[5]To overcome this monitoring issue, a few monitoring stations, e.g. Weil am Rhine, introduced screening methods, which search for all substances in waters, rather than only those that are known (Müller 2011).

Micropollutants can also be of high concern to living organisms if they bioaccumulate, i.e., incorporate into living tissue without being properly excreted or degraded within that living system. Hence, bioaccumulating substances remain in the organism and are found in progressively greater amounts in higher ranks among the food chain (Schwarzenbach et al. 2006; Schwaiger et al. 2004).[6] Due to this process, chemicals can quickly turn into a large-scale problem for which political solutions are needed.

Finally, toxicity, which refers to substances that have been proven to pose a significant risk to humans or the aquatic ecosystem, qualifies substances for political regulation. Thanks to research in the field, there is growing evidence of the negative impacts of micropollutants on aquatic ecosystems (Brodin et al. 2013; Kidd et al. 2007; Mostafa and Helling 2002) and human health (Touraud et al. 2011; Cunningham et al. 2009; Johnson et al. 2008; Rowney et al. 2009; Bercu et al. 2008). For example, researchers have found that micropollutants originating from estrogens in contraceptive pills cause the feminization of fish (Sedlak et al. 2000). Other studies show that micropollutants from psychiatric drugs reduce fish sociality and lead to higher feeding rates and increased activity (Brodin et al. 2013). Researchers expect further impacts on humans, such as genotoxic, immunotoxic, carcinogenic, and fertility-impairing effects. To date, however, there remains a great deal of uncertainty regarding adverse effects on human health (Touraud et al. 2011; Johnson et al. 2008; Rowney et al. 2009; Bercu et al. 2008).

We have seen that substances of concern are persistent, bioaccumulating, and toxic. These characteristics challenge the political realm in finding a simple answer to the question of designing appropriate policies that effectively reduce emissions at reasonable costs and administrative efforts. The question is further complicated since not all chemical substances share these mentioned characteristics of concern to the same degree. Some chemical compounds are persistent, but not of toxic danger, e.g., X-ray contrast agents (ICPR 2010). The question societies need to address then is whether, and to what level, such compounds are desired in waters, and whether a political solution is needed.

Micropollutants—a challenge for traditional water policy approaches

Traditional water protection policies, which built on wastewater treatment or environmental quality norms, have faced a new challenge with the emergence of micropollutants. First, conventional wastewater treatment technologies have not been designed to eliminate emerging pollutants, and therefore some substances are steadily transported into the aquatic environment (Wittmer et al. 2010). New wastewater treatment technologies, such as ozonation, membrane filtering, or activated carbon—which are able to eliminate emerging pollutants—are in the process of being developed, but many open questions regarding toxicity levels of

---

[6]Many micropollutants are also organic and/or polar, the latter being water soluble. Organic- or carbon-based compounds are generally persistent (or only degrade into harmful chemicals). They are prone to bioaccumulation because living tissue is also organic (Hollender et al. 2008).

transformation products, costs, or energy efficiency remain (Altmann et al. 2012). Secondly, the environmental quality norms approach consists of a compound-by-compound approach, where toxicology tests and fact sheets are needed for each and every substance in order to justify their inclusion in a regulation. The EU Water Framework Directive, for example, builds on environmental quality norms for securing water quality, which can only be established where detection and toxicity tests exist. Today's tests, however, are ill-adapted to detect emerging water quality issues considering the high number and diversity of pollutants, their low concentrations, metabolites, or interaction effects between substances mixed in water bodies (Daughton 2004). Regulation is constrained not only by the ability to detect substances, but also by the capacity to monitor them. Whether or not a substance is regulated depends decisively on the availability of off-the-shelf chemical analysis technology for monitoring (Daugthon 2004). Due to such practical constraints, many substances are not listed independently from their potential risks. As only those substances are regulated where chemical analysis possesses tools for detection and monitoring, there remains a void in anticipating risk in terms of the total sum of exposure to all contaminants. Existing governmental monitoring programs can only search for regulated substances in water. In such target-based environmental monitoring, unknown and potentially substantial portions of constituents in the aquatic environment go unnoticed. Monitoring programs are therefore only able to produce a filtered and limited view on water quality (Daughton 2004).

New developments in monitoring, screening, and modeling methods

The compound-by-compound approach for regulating and monitoring the occurrence of pollutants in the environment represents an ongoing, resource-intensive task, especially given the large number and constant engineering of new substances. By virtue of the limitations of today's compound-by-compound approach, alternatives to conventional monitoring of single stressors are under discussion. Single compound monitoring is already complemented by the analysis of sediments, which offers the opportunity to gain insights into long-term behavior of substances, their stability, and their fate (Chapman and Wang 2001; Förstner et al. 2004). Likewise, existing biological monitoring methods are able to detect potential harmful effects on living organisms such as fish, which are used as biomarkers and bioassays (Sanchez and Porcher 2009; Hogenboom et al. 2009; Leusch et al. 2010). Currently, being proposed is effect-based monitoring, which examines the effects of the entire effluent and not just its individual constituents. This monitoring technique additionally integrates other factors, unrelated to effluents, that may contribute to or diminish its effects. Effect-based tools are particularly suitable for investigative monitoring programs, for which the regulatory requirements are less formal (Wernersson et al. 2015). However, it remains a matter of debate whether effect-based monitoring is adequate for identifying effects not only in a responsive manner, i.e., once effects have occurred, but in a predictive way. There is great value in moving toward approaches which aim at identifying new environmental pollutants as early as possible, i.e., before they become pervasive in the aquatic environment (Daughton 2004). Alternatives to regulating and monitoring single compounds exist in the form of early warning systems, which are designed around the idea of 'detecting change in from of any sort of perturbation in a water's normal chemical fingerprint' (distribution pattern of types and quantities of solutes) (Daughton 2004).

Additionally, advances made in screening methods are highly appropriate for handling the specificities of micropollutants. Screening methods comprise systematic analytical tests used to recognize organic molecules, to characterize them physicochemically, and to determine their quantity in the water system. Research has developed new methods for nontarget screening and suspected-target screening in order to gain knowledge about the occurrence of so-called known unknowns and unknown unknowns in water bodies (Müller 2011). Both types of 'unknowns' refer to a class of substances that cannot be categorized into known molecules or identified by standard evaluation methods (Cleven et al. 2013). Their determination is dependent on exploratory research and comparisons with similar, known molecules (Krauss et al. 2010).

Developments also expand to holistic system modeling approaches, which consider the wide range of potential environmental pollutants, along with their interactions, in order to highlight pollutant scenarios with the highest effects potential. Holistic assessments also work toward considering multifactorial complexities of exposure to micropollutants, including frequency and timing, duration, cumulative exposure, or factors such as delayed-onset toxicity. Ultimately, it would be advantageous to move toward developing modeling approaches of real-world exposure where organisms continually face combinations of stressors, which vary over time in composition and concentrations (Daughton 2004). The traditional monitoring approach, by contrast, is limited to 'conventional' pollutants, i.e., lists of regulated pollutants, which typically comprise industrial chemicals and pesticides representing only a very small portion of the chemicals to which organisms experience real exposure.

In summary, while research has produced a number of advances pertaining to water quality, the relevant policy question today is: How do we integrate new detection and risk assessment methods into standard procedures of water policy?

Micropollutants—a complex policy problem

In conclusion, micropollutants are a complex policy problem, where each compound is associated with a unique combination of factors determining its usage, entry-path into waters, behavior in the environment, and impact on the ecosystem or human health. Regulating micropollutants becomes even more intricate considering uncertainties and the challenges involved in monitoring and risk assessment; their transboundary effects for certain compounds, and their local effects for others, which reflect the multi-level governance aspect of the issue. Moreover, several policy fields, such as environmental protection, chemical and agricultural policy, and also consumer health and workplace safety, are involved in the issue and must participate in order to find solutions. All in all, acquiring appropriate policy responses to the problem of micropollutants in an inter-sectoral, transboundary, and multi-level governance setting is a complex task. Understanding and explaining how countries along the Rhine respond to this challenge are subject of this research. Since the regulation of micropollutants is a fairly new issue on policy agendas, it has been largely neglected in social sciences. However, the search for appropriate solutions is of high political relevance at both the national and international levels, with many open questions arising that concern the most adequate governance structures and steering mechanisms. Solutions suitable for classical, macro-pollutants, such as nutrients, do not necessarily apply to micropollutants because of the diversity of compounds and sources, and for technical, financial, and societal reasons. To address this gap, the present book investigates the steering mechanisms at hand and their prospect for problem solving. In this regard, the research provides a systematic depiction and comparison of policy designs in place for the reduction in micropollutants in the Rhine basin (Sect. 2.3). Moreover, the study yields insights into the governance structures in place by examining water policy networks in the Rhine riparian states that deal with the issue of micropollutants (Sect. 3.3).

1.2   Do Policy Networks Matter to Explain Policy Design?

This research is a contribution to policy analysis that aims to achieve more optimal policy results by providing for a better understanding of the nature of policy designs and the social mechanisms behind the choice of them. With this goal, the present book distinguishes itself from previous work in policy analysis and policy process research that has so far largely focused on analyzing policy change (Sabatier 2007; Sabatier and Jenkins-Smith 1993; Mintrom and Norman 2009; Mintrom and Vergari 1996; Howlett 2002; Fischer 2013) or compromise (Fischer 2014). Policy change involves an alteration of policy between two points in time, but change is not necessarily equivalent to an optimal policy outcome. In fact, the word *change* is neutral with regard to a problem-solving perspective. At the same time, it is an inherent aim of policy analysis to understand policymaking in order to contribute to more optimal policy outcomes. For example, policy scholars' research about policy instruments relates to their interest in addressing public problems. Howlett (2005), for example, argues that policy instrument classifications (such as by Lowi 1972, 1964; Wilson 1986, 1974; Doern and Phidd 1983; Linder and Peters 1989; Salamon 2002) do not only aim at better descriptions but also at providing 'better prescriptions' or recommendations to decision-makers on how to best address a policy problem. Moreover, policy scholars studied complementarities and conflicts within policy instrument mixes in order to propose coherent, mutually reinforcing policy instrument mixes, and to successfully achieve defined policy goals (Salamon 2002; Grabosky 1995; Gunningham et al. 1998; Gunningham and Sinclair 1991; Gunningham and Young 1997; Howlett and Rayner 2007). Environmental policy scholars are interested in the question: How must policy outcomes be designed in order to ensure environmental protection (Carter 2007; Newig and Fritsch 2009;

Stavins 1989; Jänicke and Weidner 1995)? Even policy change scholars have often associated change with innovation, i.e., a positive direction of change (Mintrom and Vergari 1998; Jordan et al. 2013; Ingram and Fraser 2006; Doern and Niosi 1996). Despite the intrinsic aim of policy analysis at contributing to more optimal policy outcomes, there remains a lack of research regarding the quantification of policy designs' prospective performance in solving an underlying policy issue. A policy design's *prospective performance* here refers to its ability to comprehensively alleviate an underlying policy problem. The most thoughtful policy design, however, can still fail in addressing an issue if it is not implemented. Hence, a policy's prospective performance refers to a design that ensures (a) problem solving and (b) implementation.

Previous research has defined a successful policy in terms of its *effectiveness, efficiency, or legitimacy* (Howlett 2004, p. 6). *Legitimacy* refers to the acceptance and support of a policy design by actors involved in policymaking (Howlett 2004), the idea being that accepted policies are also more likely to be implemented (Thalmann 2004). In the present study, however, legitimacy is considered a characteristic of the policymaking process, rather than of policy design. In this line of thought, legitimacy is dependent on the ability of concerned actors to participate in the policymaking process and to make their voices heard. Depending on the openness of policymaking processes, policy designs can be more or less legitimate. However, legitimacy is not the key criterion when the analyst seeks to evaluate a policy design's problem-solving capacity. For example, a policy design could evolve out of an open policymaking process and, thus, be considered legitimate. The policy might then also be successfully implemented due to its high level of acceptance. However, the policy design might still fail at reducing an underlying problem, because certain target groups opposed a policy design that would effectively reduce the policy problem in the first place. Hence, the policy design would be deficient from a problem-solving perspective and, even if implemented, would not be able to considerably alleviate the underlying issue.

To disentangle questions of legitimacy from a problem-solving perspective, legitimacy is not taken into consideration for the evaluation of policy designs' performance here. Instead, legitimacy is considered a feature of the policymaking process. Actors' involvement in policymaking processes is taken into consideration in this work through the analysis of policy networks, which form the independent variable of this study. Defining legitimacy as a feature of the policymaking process is in accordance with the work of Howlett, who defines legitimacy as the 'ability of an instrument to attract the support of [...] those directly involved in policymaking in the issue area or sub-system' (2004, p. 6). While legitimacy is linked to the independent variable of this study, the idea of policy effectiveness is in congruence with the dependent variable, namely policy designs' prospective problem-solving capacity.

Policy *effectiveness* can be understood as congruence between the impact of an adopted policy and defined political goals (Pape 2009). While effectiveness relates to the benefits of political action, efficiency concerns the cost-benefit relation (Pape 2009, p. 7). Efficiency establishes a relation between the financial and personnel costs of a policy design and its impact on problem solving. The term *cost-effec-tiveness* combines both concepts and refers to achieving defined objectives at the lowest possible costs (Hahn 1989, p. 9). In environmental policy, one can define an effective policy as a substantive reduction in a certain pollution type, such CO_2 or—as in this study—aquatic micropollutants (Jänicke and Weidner 1995; Scruggs 2003).[7]

Pape (2009, p. 7) distinguishes between two types of literature studying effec-tiveness: Qualitative approaches that carry out comparative case studies in order to explain policy outputs; and quantitative approaches that seek to explain environ-mental performance (for example Scruggs 2003, or research on environmental indices such as the Environmental Sustainability Index, State of the Nation's Ecosystems, Ecological Indicators for the Nation). Qualitative studies focus on effective policymaking or implementation, and simply assume that a well-designed policy (Andersen 2001) or a well-implemented policy (Hill and Hupe 2002) will also deliver the desired policy impact on problem solving. However, these studies neglect the impact of policy design or implementation on problem solving, since well-designed policies can run into the danger of not being implemented—and implemented policies may not necessarily alleviate an underlying problem as intended when the policy was adopted. To date, qualitative studies have been unable to clearly define and quantify a policy design's prospective ability to effectively reduce an underlying problem, or to increase its probability of being effectively implemented.

Quantitative studies such as environmental performance indices, on the contrary, help evaluating the state of the environment in terms of pollution reduction. However, they neglect to examine environmental concerns from a policy per-spective. One cannot simply assume that an introduced policy is causally related to the improvement of the state of the environment, because recognizing the improvement of environmental conditions might come long after a policy is introduced. At this point, many variables have already intervened and affected the state of the environment aside from the introduced policy. Thus, it is too strong to assume that one may attribute the improved environmental conditions solely to the introduced policy (Koontz and Thomas 2006, p. 114). The EU Emissions Trading Scheme introduced 2005 is often stated as an example of CO_2 reductions resulting from the economic crises that started in 2008, and less so from the introduced policy (Laing et al. 2013).[8] Hence, it is argued here that an *ex-ante* evaluation that enables the assessment of a policy design's problem-solving abilities is valuable, but still missing in the literature.

Critics might argue that defining an optimal policy design is strongly subjective. Clearly, the definition of goals, such as improving water quality or ensuring gender

---

[7]More technically, one can define the reduction of pollutants in waters, for example, as the total emission reductions in relation to economic growth (Jänicke and Weidner 1995, p. 14).

[8]See also the Website of the European Commission: http://ec.europa.eu/clima/policies/ets/index_en.htm (last access 6.8.2015).

quality, is a political decision. Nevertheless, once this political goal is defined, it is possible to evaluate policy design's prospective ability to successfully achieve the defined goal. For example, the EU decided that it was a political goal of theirs to combat climate change. To this end, the EU adopted its *20–20–20 policy* that involves a 20% reduction in greenhouse gas emissions, a 20% increase in renewable energy sources, and a 20% increase in energy efficiency by 2020.[9] One could discuss whether climate change mitigation should really be a political goal. However, the question about broader political goals differs from the one regarding the means of how to achieve defined goals. This study does not include the discussion of broad political goals, but rather concentrates on the evaluation of the means that may or may not enable goal achievement. More generally, this work adopts a policy definition where public policymaking concerns finding solutions for those societal problems of which the public sectors are responsible. If one accepts that the aim of policymaking is to find solutions for societal issues, then the concept of comprehensive policy designs is considered a reflection of the ultimate aim of policymaking.

A further gap in research concerns the focus of existing studies on single policy instruments, rather than the study of entire policy designs. Implementation research in particular argues that the literature has failed so far in identifying the constellations in which certain instruments prove to be more effective than others (Knill 2006). In their study, Knill and Lenschow (2003) demonstrate that command-and-control as well as incentive-based instruments both perform well depending on a set of criteria. Diverse instrument choice scholars have also argued that the types of policy instruments do not matter for problem solving as much as the broader policy design does (Linder and Peters 1989, p. 45; Bressers and Huitema 2000). While it remains valuable to understand which categories of policy instruments exist to address public problems, there already is a solid base of knowledge regarding diverse aspects of policy instruments (Lowi 1972; Wilson 1986; Linder and Peters 1984; Dahl and Lindblom 1953; Howlett 2011; Bressers and O'Toole 1998; Crawford and Ostrom 1995). Nevertheless, if analysts wish to evaluate the prospective impact of policies on reducing societal issues, it is necessary to move away from instrument categorization and toward the study of entire policy designs. As such, there exists a gap in research as most studies rely on policy instrument categories or their mixes (Howlett 2005; Salamon 2002; Lascoumes and Le Gales 2007; Knill and Lenschow 2003; Metz and Ingold 2014), but rarely consider the details of policy design from a problem-solving perspective. For example, Pape argues that it 'might be impossible altogether [...] to make a clear connection between instrument choice and ideal regulation [...] since it depends on the design of the instrument [...] to assess their effect' (Pape 2009, p. 29). Even though Pape stresses the importance of policy design, her analysis (2009, p. 62) continues to use instrument categories by distinguishing standards from charges in

---

[9]See Website of the European Commission: http://ec.europa.eu/europe2020/europe-2020-in-a-nutshell/targets/index_en.htm (last access 6.8.2015).

water protection policy. The present study seeks to take seriously the criticism concerning instrument categories and therefore explores a new approach that considers the entire design of a policy in order to ex-ante evaluate its expected impact on reducing a public policy problem.

Moreover, policy scholars who have studied policy content largely rely on typologies (for example Knill and Lenschow 2003; Howlett 2005; Bressers and O'Toole 1998). By definition, typologies separate a given set of dimensions and through cross-tabulation consider all possible combinations of dimensions. The dimensions that researchers highlighted, such as coerciveness, proportionality, or inclusiveness, are valuable for understanding the nature of policy designs. The reliance on typologies as an analytical tool, however, is problematic as they are of purely descriptive nature and largely fail at enabling the formulation of hypotheses (Smith 2002). Therefore, the present work argues that rather than relying on typologies, an index approach is needed. A policy index has the potential to enable (a) evaluation of policy designs' problem-solving performances and (b) formulation of hypotheses.

To explore this path, Sect. 2.1.2 proposes an index of policy comprehensiveness for quantifying the prospective performance of policy designs in alleviating an underlying policy issue, e.g., reducing pollutants in waters. It is evident that the evaluation of a policy design's prospective impact on problem solving is a complex task. While the present work does not claim to be an end point in research, it constitutes a first step in exploring a new research path. This first step involves a deductive research approach, which deduces composite indicators of the policy comprehensiveness index from established knowledge of previous thinkers on characteristics of policy instruments. In the absence of other information or empirical evidence about valid composite indicators, this deductive approach is considered most appropriate here.

From a theoretical perspective, it is relevant not only to evaluate a policy design's prospective impact on problem solving, but also to uncover the social mechanisms behind policymaking. One must consider, in which social setting is it possible to achieve a comprehensive policy design? While there are many valid explanations for policy design, policy networks as one possible explanation among several others have largely been neglected in the literature. Although there exists a broad literature on policy networks, scholars mainly studied policy networks as the dependent variable, but did not link networks to policy design questions (Menahem 1998; Luzi et al. 2008). Among others, the emergence of coalitions, power structures, or (belief) homophily in policy networks has been investigated (Henry et al. 2010; Henry 2011). In this study, on the contrary, I focus on networks as the independent variable in order to explain whether and how network structures may affect policy design.

The authors of the very few studies that establish a link between networks and policy design have often been forced to simplify their definition of policy networks to the mere presence or absence of policy networks, or to one (or two) structural characteristics due to data or methodological constraints (Bressers and O'Toole 1998; Marin and Mayntz 1991; Marsh 1998; Daugbjerg and Marsh 1998). For example, Bressers et al. (1995) undertook a comparative case study on water policy networks in Europe. However, it was not the aim of this study to engage in a formal social network analysis, nor to compare specific structural network properties quantitatively. In another study, Bressers and O'Toole (1998) conceptualize policy networks as a relation between state and society, with the state on one side and society on the other. Subsequent policy network studies portray relations as more complex than a state-society opposition (Ingold 2008). Governmental actors responsible for the protection of the environment can form a coalition with environmental NGOs, while governmental actors responsible for economic growth can ally with industrial associations. Aside from a simple environment–economy opposition, many more competing interests can be involved in policymaking processes, such as health concerns or workplace safety, and they can form a complex network of interrelated actors. In order to take the network approach seriously, it is necessary to move beyond a simple state–society or environment–industry dichotomy with hypotheses stating the participation of environmental actors in decision-making leads to stricter environmental standards, while involvement of economic actors leads to laxer environmental standards (Pape 2009, p. 63). While such hypotheses attribute a high importance to actors' interests, they neglect the relational aspect of the network approach. In general, microlevel concepts explain policy design by the choices individual actors make based on their preferences (Sandström and Carlsson 2008). Certainly, actors' preferences matter as one among other valid explanations to policy design. However, Granovetter (1985, 1992) could demonstrate that individual choices and actions also have a strong social foundation, meaning that interactions among network members socially influence actors' attitudes, perceptions, behavior, and policy preferences (Richey and Ikeda 2006; Knoke 1990; Zuckerman 2005). Hence, not only actor types and their preferences need to be understood in order to explain policy choices, but also actors' ties and their embeddedness in the overall network.

Compared to purely microlevel explanations, the advantage of the network approach is that it goes beyond the mere aggregation of policy actors' attributes by taking into consideration actors' interdependencies (Sandström and Carlsson 2008). Network scholars conciliate elements of individualism with a structural approach and demonstrate that networks create their own governing structures (Lubell et al. 2012; Granovetter 1985, 1992). Hence, policy networks and their structural features may provide a promising explanation for policy design that is worth further studying. The arising question is: Does the logic of interaction between network members also affect the comprehensiveness of policy designs?

Bressers and O'Toole (1998, 2005) inspiring articles are among the very few studies that establish a link between network structure and policy outputs, without analyzing policy change, as, for instance, Howlett (2002) does. However, their ideas have not been rigorously tested or approved by applying a formal social network analysis to date. The present study seeks to close this research gap by

(a) operationalizing policy network concepts through a quantitative social network analysis, and (b) testing the link to comprehensive policy design.

While past research repeatedly postulated the need to take the network approach seriously (O'Toole 1997; Robinson 2006), to date, no hypotheses have been put forward that systematically link the structure of a policy network with comprehensive policy designs (Börzel 1998, p. 258). This gap may be due to methodological issues associated with the reliance on typologies that hamper systematic hypotheses testing. For example, in their article, Bressers and O'Toole (1998) define six characteristics of policy instruments where each instrument characteristic can adopt two values at least (e.g., high or low) resulting in 64 (i.e., 2^6) combinations of instrument characteristics and values. Moreover, Bressers and O'Tooles' network typology relies on the degree of belief cohesion and interconnectedness (high or low) of state-society relations. Linking policy network and instrument characteristics in the form of hypotheses leads to more combinations than are testable with conventional research on policy networks involving only a small number of cases. In order to operationalize Bressers and O'Tooles' research idea of linking policy networks to policy design, it is necessary to adapt methodologies by moving away from policy typologies and, instead, adopting a policy design perspective. This shift is necessary because it is difficult to argue that certain network structures promote the emergence of certain instrument categories, as the same instrument types can have very different consequences depending on the exact design of the policy, i.e., the definition of the target group, the degree of sanctions, and alike.

The literature has failed so far in establishing a systematic link between networks and policy design, among others, because of a lack of conceptualizing the performance of policy designs, be it their effectiveness (Newig and Fritsch 2009; Sandström and Carlsson 2008) or problem-solving capacity (Klijn et al. 2010). For example, Newig and Fritsch's study includes formulations in their hypotheses, such as 'participation of non-state actors leads to more ecologically rational decisions' or to 'better outcomes and impact in ecological terms' (Newig and Fritsch 2009, p. 200), but does not operationalize what these 'ecologically rational decisions' mean, or how to evaluate them.

From a theoretical perspective, the present study seeks to close existing research gaps by exploring whether a link can be established between structural network characteristics and comprehensive policy designs. To do so, valuable structural network concepts, such as coalition structure, interconnectedness, and belief similarity, are borrowed from policy change research (see Chap. 3). Moreover, a new policy index is introduced (see Chap. 2) that aims at quantifying policy design's problem-solving performance. As a result, independent and dependent variables are continuous, for example, more or less interconnectedness (independent variable) and more or less comprehensive policy design (dependent variable). In this way, the empirical relationship between the variables can be determined in a bivariate analysis (rather than in typologies), and hypotheses of association are testable. For example, it should be possible to test whether more network interconnectedness is linked to more comprehensive policy designs.

1.3   Structure of This Book

Each chapter of this book is structured such that theory-based evidence introduces the concepts employed. These concepts are then operationalized in a methodological section and discussed with regard to empirical results.

Chapter 2 is about policy design in theory and practice. It applies theoretical considerations on policy design to the case of water protection policies by comparing micropollutants policies across the Rhine River riparian countries Switzerland, Germany, France, and the Netherlands. The definition of public policies adopted in Sect. 2.1 implies that public policies represent some sort of solution for societal issues. Many policy theories conceptualize policy content in the form of policy change, which involves policy alteration but not necessarily problem solving. The present research, on the contrary, conceptualizes policies as more or less comprehensive solutions to an underlying policy problem. To render policy designs measurable in terms of their prospective ability to solve the underlying policy issue, I introduce an index of policy comprehensiveness. The index builds on the work of previous thinkers on policy content as exposed in Sect. 2.1.1, which distinguishes the first-, second-, and third-generation policy scholars. Based on insights into strength and weaknesses of previous research, I propose a new path for analyzing policies in Sect. 2.1.2, which considers the entire design of the policy. Rather than sorting single policy instruments into a typology, this approach highlights that policies are designed as an instrument mix with defined conditions under which the instruments apply. By taking these specific conditions into consideration, the index enables one to evaluate a policy design's prospective impact on problem solving. In order to render complex, multi-dimensional aspects of the empirical reality, such as policy design and measurable, Sect. 2.2 explains how indexing is used as a method for quantifying policy designs by means of qualitative data. Sect. 2.3 applies the policy index to the case of water protection policy to evaluate the performance of the studied policy designs in reducing aquatic micropollutants.

Chapter 3 deals with policy networks in theory and water policy networks in practice. Section 3.1 argues that policy network structures constitute a promising explanation to policy design, but remain understudied. Toward this end, it is necessary to understand the structural aspects of policy networks. Those network structures are conceptualized here on the basis of a literature review. To operationalize network structure, social network analysis is employed as a quantitative method of data analysis as exposed in Sect. 3.2. Section 3.2.1 explains in depth how the network data was gathered by the author, and Sect. 3.2.2 then provides a short introduction to social network analysis as a method for the visualization and mathematical analysis of relational data. Sect. 3.2.3 shows in detail the different types of network statistics applied to the gathered network data in order to operationalize variables capturing different structural network properties. By relying on network statistics, Sect. 3.3 illustrates the structural network characteristics, including belief cohesion, interconnectedness, coalition structures, brokerage, and entrepreneurship, for each country separately; it then adopts a comparative perspective by working out distinctive structural characteristics of the Swiss, German, French, and Dutch water policy networks in the field of micropollutants.

Chapter 4 explores the theoretical link between network structures and the choice of policy design by formulating hypotheses on their relationships. Today's literature remains unclear whether specific types of network structures tend to produce certain kinds of policy designs. To close this gap, hypotheses are established in a deductive manner by borrowing insights from policy process research about important structural network properties. Compared to the policy process research focusing on policy change, the network approach is employed here in order to explore the validity of a new link—namely the one between structural network properties and the comprehensiveness of policy design in solving an underlying policy issue. Section 4.2 explains how the method of structured focused comparison is employed to establish the link between network structure and policy design. Section 4.3 systematically evaluates empirical findings against the hypothesized linkages between policy network structures and policy design. The empirical results of the four studied cases are treated here as pretests of the formulated hypotheses. Similar to exploratory research, empirical findings are thought to provide first indications of a potential relationship between structural network properties and policy designs. While hypotheses were established deductively from existing literature, Sect. 4.4 examines the relevance of the network approach for future research in a more inductive manner and explores the explanatory strength of the network approach to policy design.

The conclusion summarizes main findings and implications of this research in Sect. 5.1. Moreover, Sect. 5.2 points to empirical and methodological limitations of this study, outlines the need and paths for future research, and also highlights alternative explanations for policy designs next to the present analytical focus on network structures. Finally, Sect. 5.3 examines the contribution of this research to science, as well as its relevance to policy practitioners.

References

Altmann, D., Schaar, H., Bartel, C., Schorkopf, D. L., Miller, I., Kreuzinger, N., et al. (2012). Impact of ozonation on ecotoxicity and endocrine activity of tertiary treated wastewater effluent. *Water Research, 46*(11), 3693–3702.

Andersen, M. S. (2001). *Economic instruments and clean water: Why institutions and policy design matter.* Paris: OECD.

Auberson-Huang, L. (2002). The dialogue between precaution and risk. *Nature Biotechnology, 20* (11), 1076–1078. doi:10.1038/nbt1102-1076

Bendz, D., Paxéus, N., Ginn, T., & Loge, F. (2005). Occurrence and fate of pharmaceutically active compounds in the environment, a case study: Höje River in Sweden. *Journal of Hazardous Materials, 122*(3), 195–204.

Bercu, J., Parke, N., Fiori, J., & Meyerhoff, R. (2008). Human health risk assessments for three neuropharmaceutical compounds in surface waters. *Regulatory Toxicology and Pharmacology, 50*(3), 420–427.

Bressers, H., & Huitema, D. (2000). What the doctor should know: Politicians are special patients. The impact of the policy-making process on the design of economic instruments. In M. S. Andersen & R.-U. Sprenger (Eds.), *Market-based instruments for environmental management* (pp. 67–88). Cheltenham: Edward Elgar.

Bressers, H., & O'Toole, L. (1998). The selection of policy instruments: A network-based perspective. *Journal of Public Policy, 18*(3), 213–239.

Bressers, H., & O'Toole, L. (2005). Instrument selection and implementation in a networked context. In P. Eliadis, M. Hill, & M. Howlett (Eds.), *Designing government: From instruments to governance* (pp. 132–153). Montreal, Kingston: McGill-Queen's University Press.

Bressers, H., O'Toole, L., & Richardson, J. (Eds.). (1995). *Networks for water policy: A comparative perspective*. London: Frank Cass.

Brodin, T., Fick, J., Jonsson, M., & Klaminder, J. (2013). Dilute concentrations of a psychiatric drug alter behavior of fish from natural populations. *Science, 339*(6121), 814–815.

Buser, H.-R., Balmer, M., Schmid, P., & Kohler, M. (2006). Occurrence of UV filters 4-methylbenzylidene camphor and octocrylene in fish from various Swiss Rivers with inputs from wastewater treatment plants. *Environmental Science & Technology Online News, 40*(5), 1427–1431.

Carter, N. (2007). *The politics of the environment: Ideas, activism, policy*. Cambridge: Cambridge Univiversity Press.

Chapman, P. M., & Wang, F. (2001). Assessing sediment contamination in estuaries. *Environmental Toxicology and Chemistry, 20*(1), 3–22. doi:10.1002/etc.5620200102.

Cleven, C. D., Howard, A. S., Little, J. L., & Yu, K. (2013). Identifying "Known unknowns" in commercial products by mass spectrometry. *LCGC Chromatography, 26*(3).

Crawford, S., & Ostrom, E. (1995). A grammar of institutions. *The American Political Science Review, 89*(3), 582–600.

Cunningham, V., Binks, S., & Olson, M. (2009). Human health risk assessment from the presence of human pharmaceuticals in the aquatic environment. *Regulatory Toxicology and Pharmacology, 53*(1), 39–45.

Dahl, R., & Lindblom, C. (1953). *Politics, economics and welfare*. Chicago: The University of Chicago Press.

Daugbjerg, C., & Marsh, D. (1998). Explaining policy outcomes: Integrating the policy network approach with macro-level and micro-level analysis. In D. Marsh (Ed.), *Comparing policy networks* (pp. 53–71). Philadelphia: Open University Press.

Daughton, C. G. (2004). Non-regulated water contaminants: Emerging research. *Environmental Impact Assessment Review, 24*(7–8), 711–732. doi:10.1016/j.eiar.2004.06.003.

Doern, B., & Niosi, J. (1996). Flexible innovation: Technological alliances in Canadian industry. *Canadian Public Policy, 22*(4), 407.

Doern, B., & Phidd, R. (1983). *Canadian public policy: Ideas, structure, process* (2nd ed.). Michigan: University of Michigan.

Fischer, M. (2013). *Policy network structures, institutional context, and policy change*. Paper presented at the COMPASSS working paper 73.

Fischer, M. (2014). Coalition structures and policy change in a consensus democracy. *Policy Studies Journal, 42*(3), 344–366.

Förstner, U., Heise, S., Schwartz, R., Westrich, B., & Ahlf, W. (2004). Historical contaminated sediments and soils at the river basin scale. *Journal of Soils and Sediments, 4*(4), 247. doi:10.1007/bf02991121.

Götz, C., Kase, R., & Hollender, J. (2010). *Mikroverunreinigungen - Beurteilungskonzept für organische Spurenstoffe aus kommunalem Abwasser. Studie im Autrag des BAFU*. Dübendorf: Eawag.

Grabosky, P. (1995). Counterproductive regulation. *International Journal of the Sociology of Law, 23*, 347–369.

Granovetter, M. (1985). Economic action and social structure: The problem of embeddedness. *American Journal of Sociology, 91*, 481–510.

Granovetter, M. (1992). Economic institutions as social construction: A framework of analysis. *Acta Sociologica, 35,* 3–11.

Gunningham, N., Grabosky, P., & Sinclair, D. (1998). *Smart regulation: Designing environmental policy.* Oxford: Clarendon Press.

Gunningham, N., & Sinclair, D. (1991). Regulatory pluralism: Designing policy mixes for environmental protection. *Law and Policy, 21*(1), 49–76.

Gunningham, N., & Young, M. (1997). Toward optimal environmental policy: The case of biodiversity conservation. *Ecology Law Quarterly, 24,* 243–298.

Hahn, R. (1989). *A primer on environmental policy design.* Chur: Harwood Academic Publications.

Hardin, G. (1968). The tragedy of the commons. *Science* (162), 1243–1248.

Henry, A. D. (2011). Ideology, power, and the structure of policy networks. *Policy Studies Journal, 39*(3), 361–383.

Henry, A. D., Lubell, M., & McCoy, M. (2010). Belief systems and social capital as drivers of policy network structure: The case of California regional planning. *Journal of Public Administration Research and Theory, 21*(3), 419–444.

Hill, M., & Hupe, P. (2002). *Implementing public policy. Governance in theory and in practice.* London: Sage Publishing.

Hogenboom, A., van Leerdam, J. A., & de Voogt, P. (2009). Accurate mass screening and identification of emerging contaminants in environmental samples by liquid chromatography-hybrid linear ion trap Orbitrap mass spectrometry. *Journal of Chromatography A, 1216*(3), 510–519. doi:10.1016/j.chroma.2008.08.053.

Hollender, J., Singer, H., & McArdell, C. (2008). Polar organic micropollutants in the water cycle. In P. Hlavinek, O. Bonacci, J. Marsalek, & I. Mahrikova (Eds.), *Dangerous pollutants (Xenobiotics) in urban water cycle* (pp. 103–116). Dordrecht: Springer.

Howlett, M. (2002). Do networks matter? Linking policy network structure to policy outcomes: Evidence from four Canadian policy sectors 1990–2000. *Canadian Journal of Political Science, 35*(2), 235–267.

Howlett, M. (2004). Beyond good and evil in policy implementation: Instrument mixes, implementation styles, and second generation theories of policy instrument choice. *Policy and Society, 23*(2), 1–17.

Howlett, M. (2005). What is a policy instrument? Tool, mixes, and implementation styles. In P. Eliadis, M. Hill, & M. Howlett (Eds.), *Designing government. From instruments to governance* (pp. 31–49). Montreal, Kingston: McGill-Queen's University Press.

Howlett, M. (2011). *Designing public policies: Principles and instruments.* New York: Routledge.

Howlett, M., & Rayner, J. (2007). Design principles for policy mixes: Cohesion and coherence in 'New governance arrangements'. *Policy and Society, 26*(4), 1–18.

ICPR. (2010). *Evaluation report radiocontrast agents* (Vol. Report Number 187e). Koblenz: International Commission for the Protection of the Rhine.

ICPR. (2012). *Evaluation report on industrial chemicals* (Vol. Report Number 202e). Koblenz: International Commission for the Protection of the Rhine.

Ingold, K. (2008). *Analyse des mécanismes de décision: Le cas de la politique climatique suisse.* Zürich and Chur: Rüeggger Verlag.

Ingram, H., & Fraser, L. (2006). Path dependency and adroit innovation: The case of California water. In R. Repetto (Ed.), *Punctuated equilibrium and the dynamics of U.S. environmental policy.* New Haven: Yale University Press.

Jänicke, M., & Weidner, H. (Eds.). (1995). *Successful environmental policy. A critical evaluation of 24 cases.* Berlin: Edition sigma.

Johnson, A., Jürgens, M., Williams, R., Kümmerer, K., Kortenkamp, A., & Sumpter, J. (2008). Do cytotoxic chemotherapy drugs discharged into rivers pose a risk to the environment and human health? An overview and UK case study. *Journal of Hydrology, 348*(1–2), 167–175.

Jordan, A., Wurzel, R., & Zito, A. (2013). Still the century of 'new' environmental policy instruments? Exploring patterns of innovation and continuity. *Environmental Politics, 22*(1), 155–173.

Kidd, K., Blanchfield, P., Mills, K., Palace, V., Evans, R., Lazorchak, J., et al. (2007). Collapse of a fish population after exposure to a synthetic estrogen. *Proceedings of the National Academy of Sciences, 104*(21), 8897–8901.

Kingdon, J. (1984). *Agendas, alternatives, and public policies*. Boston: Little, Brown.

Klijn, E.-H., Steijn, B., & Edelenbos, J. (2010). The impact of network management stratergies on the outcomes in governance networks. *Public Administration, 88*(4), 1063–1082.

Knill, C. (2006). Theoretical perspectives on the implementation of European policies. In J. Richardson (Ed.), *European union: Power and policy-making* (3rd ed.). New York: Routledge.

Knill, C., & Lenschow, A. (2003). Modes of regulation in the governance of the european union: Towards a comprehensive evaluation. *European Integration Online Papers, 7*(1), 4–15.

Knoke, D. (1990). *Political networks. The structural perspective*. New York: Cambridge University Press.

Koontz, T., & Thomas, C. (2006). What do we know and need to know about the environmental outcomes of collaborative management? *Public Administration Review, 66,* 111–121.

Kortenkamp, A., Faust, M., Scholze, M., & Backhaus, T. (2007). Low-level exposure to multiple chemicals: Reason for human health concerns? *Environmental Health Perspectives, 115*(S-1), 106–114.

Krauss, M., Singer, H., & Hollender, J. (2010). LC-high resolution MS in environmental analysis: From target screening to the identification of unknowns. *Analytical and Bioanalytical Chemistry, 397*(3), 943–951. doi:10.1007/s00216-010-3608-9.

Laing, T., Sato, M., Grubb, M., & Comberti, C. (2013). *Assessing the effectiveness of the EU emissions trading system*. Working paper number 126 and 106 (C. f. C. C. E. a. P. a. G. R. I. o. C. C. a. t. Environment, Ed.). London: London School of Economics and Political Science.

Lascoumes, P., & Le Gales, P. (2007). Introduction: Understanding public policy through its instruments—From the nature of instruments to the sociology of public policy instrumentation. *Governance-An International Journal of Policy and Administration, 20*(1), 1–21.

Leusch, F. L., Jager, C., Levi, Y., Lim, R., Puijker, L., & Sacher, F. (2010). Comparison of five in vitro bioassays to measure estrogenic activity in environmental waters. *Environmental Science & Science Technol, 44.* doi:10.1021/es903899d

Linder, S., & Peters, G. (1984). From social theory to policy design. *Journal of Public Policy, 4*(3), 237–259.

Linder, S., & Peters, G. (1989). Instruments of government: Perceptions and contexts. *Journal of Public Policy, 9*(1), 35–58.

Lowi, T. (1964). American business, public policy, case-studies, and political theory. *World Politics, 16*(04), 677–715.

Lowi, T. (1972). Four systems of policy, politics and choice. *Public Administration Review, 32*(4), 298–310.

Lubell, M., Scholz, J., Berardo, R., & Robins, G. (2012). Testing policy theory with statistical models of networks. *Policy Studies Journal, 40*(3), 351–374.

Luzi, S., Abdelmoghny Hamouda, M., Sigrist, F., & Tauchnitz, E. (2008). Water policy networks in Egypt and Ethiopia. *The Journal of Environment & Development, 17*(3), 238–268.

Marin, B., & Mayntz, R. (1991). *Policy network: Empirical evidence and theoretical considerations*. Frankfurt am Main: Campus Verlag.

Marsh, D. (1998). *Comparing policy networks*. Philadelphia: Open University Press.

Menahem, G. (1998). Policy paradigms, policy networks and water policy in Israel. *Journal of Public Policy, 18*(3), 283–310.

Metz, F., & Ingold, K. (2014). Sustainable wastewater management: Is it possible to regulate micropollution in the future by learning from the past? A policy analysis. *Sustainability, 6*(4), 1992–2012.

Miao, X.-S., Yang, J.-J., & Metcalfe, C. (2005). Carbamazepine and Its metabolites in wastewater and in biosolids in a municipal wastewater treatment plant. *Environmental Science and Technology, 39*(19), 7469–7475.

Mintrom, M., & Norman, P. (2009). Policy entrepreneurship and policy change. *Policy Studies Journal, 37*(4), 649–667.

Mintrom, M., & Vergari, S. (1996). Advocacy coalitions, policy entrepreneurs, and policy change. *Policy Studies Journal, 24*(3), 420–434.

Mintrom, M., & Vergari, S. (1998). Policy networks and innovation diffusion: The case of state education reforms. *The Journal of Politics, 60*(1), 126–148.

Mostafa, F., & Helling, C. (2002). Impact of four pesticides on the growth and metabolic activities of two photosynthetic algae. *Journal of Environmental Science and Health, Part B, 37*(5), 417–444.

Müller, M. S. (2011). *Polar organic micro-pollutants in the River Rhine: Multi-compound screening and mass flux studies of selected substances.* Eawag, Technische Universität Berlin Dübendorf: Berlin.

Newig, J., & Fritsch, O. (2009). Environmental governance: Participatory, multi-level—And effective? *Environmental Policy and Governance, 19*(3), 197–214.

O'Toole, L. (1997). Treating networks seriously: Practical and research-based agendas in public administration. *Public Administration Review, 57*(1), 45–52.

Pape, J. (2009). *Domestic driving factors of environmental performance: The role of regulatory styles in the case of water protection policy in France and the Netherlands.* Konstanz: Universität Konstanz.

Peel, J. (2005). *The precautionary principle in practice: Environmental decision-making and scientific uncertainty.* Sydney: The Federation Press.

Richardson, S., & Ternes, T. (2011). Water analysis: Emerging contaminants and current issues. *Analytical Chemistry, 83*(12), 4614–4648.

Richey, S., & Ikeda, K. I. (2006). The influence of political discussion on policy preference: A comparison of the United States and Japan. *Japanese Journal of Political Science, 7*(3), 273–288.

Robinson, S. (2006). A decade of treating networks seriously. *Policy Studies Journal, 34*(4), 589–598.

Rowney, N., Johnson, A., & Williams, R. (2009). Cytotoxic drugs in drinking water: A prediction and risk assessment exercise for the Thames catchment in the United Kingdom. *Environmental Toxicology and Chemistry, 28*(12), 2733–2743.

Sabatier, P. (2007). *Theories of the policy process.* Boulder: Westview Press.

Sabatier, P., & Jenkins-Smith, H. (1993). *Policy change and learning: An advocacy coalition approach.* Boulder: Westview Press.

Salamon, L. (2002). *The tools of government: A guide to the new governance.* Oxford, New York: Oxford University Press.

Sanchez, W., & Porcher, J. M. (2009). Fish biomarkers for environmental monitoring within the water framework directive. *Trends in Analytical Chemistry, 28.* doi:10.1016/j.trac.2008.10.012

Sandström, A., & Carlsson, L. (2008). The performance of policy networks: The relation between network structure and network performance. *The Policy Studies Journal, 36*(4), 497–524.

Schwaiger, J., Ferling, H., Mallow, U., Wintermayr, H., & Negele, D. (2004). Toxic effects of the non-steroidal anti-inflammatory drug diclofenac: Part I: Histopathological alterations and bioaccumulation in rainbow trout. *Aquatic Toxicology, 68*(2), 141–150.

Schwarzenbach, R., Escher, B., Fenner, K., Hofstetter, T., Johnson, A., Von Gunten, U., et al. (2006). The challenge of micropollutants in aquatic systems. *Science, 313*(5790), 1072–1077.

Scruggs, L. (2003). *Sustaining abundance: Environmental performance in industrial democracies.* Cambridge: Cambridge University Press.

Sedlak, D., Gray, J., & Pinkston, K. (2000). Peer reviewed: Understanding microcontaminants in recycled water. *Environmental Science and Technology, 34*(23), 508A–515A.

Stavins, R. (1989). Clean profits: Using economic incentives to protect the environment. *Policy Review, 48,* 58–63.

Thalmann, P. (2004). The public acceptance of green taxes: 2 million voters express their opinion. *Public Choice, 119*(1–2), 179–217.

Touraud, E., Roig, B., Sumpter, J., & Coetsier, C. (2011). Drug residues and endocrine disruptors in drinking water: Risk for humans? *International Journal of Hygiene and Environmental Health, 214*(6), 437–441.

Von der Ohe, P. C., Dulio, V., Slobodnik, J., Deckere, E. D., Kühne, R., Ebert, R.-U., et al. (2011). A new risk assessment approach for the prioritization of 500 classical and emerging organic microcontaminants as potential river basin specific pollutants under the European Water Framework Directive. *Science of the Total Environment, 409*(11), 2064–2077.

Wernersson, A.-S., Carere, M., Maggi, C., Tusil, P., Soldan, P., James, A., et al. (2015). The European technical report on aquatic effect-based monitoring tools under the water framework directive. *Environmental Sciences Europe, 27*(1), 1–11. doi:10.1186/s12302-015-0039-4.

Wilson, J. (1974). *Political organizations*. Princeton: Princeton University Press.

Wilson, J. (1986). *American government: Institutions and policies* (3rd ed.). Lexington, MA: D.C. Heath.

Wittmer, I., Bader, H.-P., Scheidegger, R., Singer, H., Lück, A., Hanke, I., et al. (2010). Significance of urban and agricultural land use for biocide and pesticide dynamics in surface waters. *Water Research, 44*(9), 2850–2862.

Zuckerman, A. (2005). *The social logic of politics: Personal networks as contexts for political behavior*. Philadelphia: Temple University Press.

Chapter 2
Comparing Policy Designs in Water Protection: Micropollutants Policies in the Rhine River Riparian States

2.1 Policy Design

Recurring debates in policy analysis revolve around the question, 'Which political solutions exist to solve a societal issue?' There still exist situations in which policy makers understand what options exist for reducing an environmental problem, but refrain from adopting any of these options. Here, policy analysts go one step further by asking: 'Why is little done politically to solve a specific societal problem when appropriate solutions exist and are known?' In order to find answers to these crucial questions, policy analysis unravels the mechanisms behind policymaking with the ultimate aim of contributing to better policies. According to the classical definition by Thomas Dye, '*[p]olicy analysis is what governments do, why they do it, and what difference it makes*' (Dye 1976). This definition demonstrates that policy analysis deals with three core topics:

1. Description and analysis of policy content and variation: Here, scholars analyze what kinds of policy instruments, programs, or goals exist to solve a societal problem and look at policy variations across sectors and countries (Eliadis et al. 2005; Salamon 2002; Hood 1986).
2. Finding explanations for the choice of policy content: In this field of research, analysts are interested in uncovering the reasons behind one solution being selected and adopted to solve a problem over another, or why no solution is adopted at all (Knill and Tosun 2012, p. 2; Ingold 2008; Howlett 1991; Bressers and O'Toole 1998; Varone 1998). Related to the question of policy selection is the question of policy change, where scholars aim to reveal the conditions under which long periods of stability (during which policies do not change) are interrupted by short phases of change (Sabatier 2007; Baumgartner and Jones 1991; Kingdon and Thurber 2011).
3. Description and analysis of policy effects: A central aim of policy analysis is to understand the effects of policy action, also called policy outcomes (Bressers 2004; Pressman and Wildavsky 1984; Knill 2006; Pollitt et al. 2006). Largely

© Springer International Publishing AG 2017
F. Metz, *From Network Structure to Policy Design in Water Protection*,
Springer Water, DOI 10.1007/978-3-319-55693-2_2

routed in implementation research, the following questions are addressed: Does the adopted policy actually reduce the problem in the intended way? Which factors contribute to the deviation from (or consistency with) the intended outcome?

Scholars of policy analysis have largely focused on public policies, i.e., actions that address societal problems and structural aspects of the public sphere. The present chapter focuses on public policies in the field of water protection by including the description and analysis of policy content. It seeks to evaluate whether policy content is well-designed in the sense that it has the potential to achieve a defined policy goal and alleviate an underlying water protection problem. This chapter also looks at policy variation across countries by comparing water protection policies of Switzerland, Germany, France, and the Netherlands. The effects of an adopted policy in the form of pollution reduction, on the contrary, are not within the realm of this study.

2.1.1 What Is Policy Design?

Many studies in policy analysis aim at explaining why certain policy content was chosen over another. To discern this content dimension of policy, it is helpful to understand the common distinction between policy, polity, and politics (Sciarini et al. 2004; Knill and Tosun 2012). The term *polity* refers to *rules, which define political structures* and characterize a political system. *Politics* points to the *procedural elements of policymaking*, i.e., power and conflict configurations. *Policy* concerns the solutions to societal problems and refers to the content dimension. Evidently, in policy analysis, the analytical focus is on the study of the policy content aspect. Nevertheless, the polity and politics dimensions are central explanatory factors for policy scholars. For example, if analysts aim at explaining a specific policy output such as the adoption of tax cuts, they might ask why those tax cuts were enforceable in one political system, but not in another (studying the influence of polity on policy). Scholars might also want to understand why those actors favoring tax cuts were able to impose their policy preferences in the policymaking process (studying the influence of politics on policies). Some academics put forward that policy analysis includes the study of polity and politics and is therefore broader than solely analyzing political structures, political parties, or interest groups (Knill and Tosun 2012, p.1).

The literature defines *a public policy as a collective course of action (or non-action) enacted by a set of actors, typically a government, a legislature, or an equivalent authority, to address a particular societal issue* (Howlett et al. 2009a, p. 4 ff.; Knill and Tosun 2012, p. 4). Implied in the definition is that public policies represent some sort of solution for a societal issue with the aim of improving and structuring life in a society. From this perspective, the goal of public policies is problem solving.

Many definitions place an emphasis on state actors, since they have the legal authority to adopt public policies (e.g., Knill and Tosun 2012, p. 4). Some researchers, especially in the governance literature, have questioned the predominance of the state (Howlett et al. 2009b; Hysing 2009; Arellano-Gault and Vera-Cortés 2005; Doern and Wilks 1998; Edelenbos et al. 2010; Esmark 2009; Foster and Plowden 1996). In fact, it is widely acknowledged in the policy literature that a broad range of actor types—governmental as well as non-governmental—are involved in policymaking in Western democracies and have an impact on how societal concerns are solved through public policies (Fischer 2012; Christopoulos and Ingold 2015; Ingold 2007; Weible 2007; Howlett and Ramesh 2003). Nevertheless, these studies also demonstrate that state actors are still highly influential today due to their legal authority and veto powers in decision-making.

Some authors distinguish between a broad and a narrow definition of policy. The broad definition includes the policymaking process (the process of finding solutions) and its result in the form of policy content (the solution itself). A narrow definition solely refers to policy content (Varone 1998) and suggests that a policy is the result of a policymaking process, but not the process itself (Jann and Wegrich 2014; Howlett and Ramesh 2003; Howlett and Giest 2012). To account for this consecutiveness, the word *policy output* is often used in the literature.

Moreover, the term *policy* is used to describe different levels of generalization of a course of action. On a very general level, the word *policy* refers to all those measures taken in a certain *policy field* (also termed *domain* or *sector*), such as economic policy, social policy, or environmental policy (Knill and Tosun 2012). A little less general, the term *policy* denotes public activities in *policy subfields*. In environmental protection policy, for example, subfields include water policy, air policy, or landscape policy. On a third level of generalization, the word *policy* stands for measures taken to address even more specific *policy issues* within the just-described policy subfields. In water policy, for instance, groundwater issues can be differentiated from floods or surface water quality issues. The least general use of the term *policy* refers to *single instruments,* also called *policy tools. Public policy instruments are single means through which collective courses of actions (behaviors or procedures) are structured to address a societal issue and achieve defined policy goals* (Salamon 2002; Lasswell 1958). Examples of such instruments in water protection policy include pollution charges or a bans, which restrict pollution to waters by placing a price on it or prohibiting it. Such policy instruments represent a way of impacting behavior in order to improve water quality. In the empirical reality, several policy instruments are usually bundled into policy programs (Howlett 2005). Therefore, scholars use the term *instrument mix, which denotes a bundle of several policy instruments.*

With regard to the distinction between *output, outcome,* and *impact*, the present study concentrates on policy *outputs*, termed *policy design*, and seeks to estimate policy outputs' prospective ability to produce the intended *outcome* and *impact*. Moreover, the present work adopts a narrow definition of policy, solely referring to policy content, which is considered an output of the policymaking process.[1] When referring to policy content, the word *policy design* is employed throughout the work. This term is similar but more specific than the word *policy instrument (mix)*. Like instrument mix, the term *policy design* also takes into consideration that policies are designed as a bundle of interrelated instruments. Additionally, the term emphasizes the specific provisions about where, how long, and to whom policies apply (Linder and Peters 1984; Schneider and Ingram 1988; Howlett 2009, 2014). As defined above, the essence of public policies is to find solutions for societal issues. Policy instruments are crucial elements of those solutions as they are the tools, which enable the pursuit of politically defined goals. However, policy instruments are integrated into wider policy programs, which define a number of conditions, and therefore, it is not enough to list or name single instrument categories from which the policy is composed (Bressers and Huitema 2000; Pape 2009, p. 2). To fully understand a policy's prospective ability to solve a societal problem, it is necessary to highlight the precise conditions under which an instrument or instrument mixes apply. For example, the word *policy instrument mix* could point to the combination of a subsidy and a best environmental practice. However, this information is not enough to capture its ability to solve the underlying issue because success depends on the precise design of the policy instrument mix. The term *policy design*, on the contrary, would further specify if the policy targets those groups causing the problem, if a positive behavior is incentivized, if negative behavior is constrained or even sanctioned, who is responsible for implementation, and alike. This way, the word *policy design* accounts for the fact that the very same instrument may have drastically different implications for problem-solving, depending on the conditions under which it applies. A regulatory instrument like a ban, for example, can have minor implications for problem-solving if it applies only to a minority of the groups causing a problem or if the enforcement authority lacks capacities. Only if we understand all the provisions of a specific policy design, can we then evaluate its prospective ability to perform best under a given situation in order to reach its defined goal.

Moreover, this study employs the term *policy design* by adopting a specific view on policy content: First, policy instruments have often been defined along their degree of state intervention. Here, on the contrary, policy design refers to intervention with regard to a specific issue. Hence, policies are understood in this study as a force for reducing a societal problem rather than a force by the state on a target group. From this perspective, the essence of public policies consists of alleviating

---

[1]Despite its focus on the policy content dimension, the present work also considers both the polity and the politics dimensions to explain policy variations across four countries: The polity dimension is crucial as the four countries under study display different political systems, which provide for different formal rules about decision-making processes and power relations, and thereby structure policy networks. Policymaking processes, i.e., the politics or procedural dimension of policy, are incorporated into the present work through the study of policy network structures, which reflect the aggregated result of multi-actor interactions in the policymaking process over time.

public issues, e.g., improving air or water quality, enabling gender equality on the job market, promoting economic growth, and alike. This definition may be a reduction of a more complex reality where problem-solving represents just one property of public policies among many others (e.g., re-election calculus, compromise-seeking). Nevertheless, problem solving represents an important part of public policy and offers one possible research perspective among many other valid viewpoints. Second, policy design is defined here as an impact on behavior and not as a constraint, because policies are designed not only such that they constrain undesired behavior, but also so that they promote desired behavior. Third, policies may be targeted at behavior, but they may also alter structures or a sequence of steps in order to create positive conditions for reducing a policy problem. Taking these three specificities into account, a *public policy design is defined here as a policy instrument (mix) and its specific conditions of application through which behaviors (or procedures) are impacted to address a societal issue.*

With its focus on policy design, this study does not seek to explain policy stability and change. Policy change and policy design are two related, but different concepts. While *change* refers to the alteration of policy between two points in time, *design* concerns the content of policy. This work evaluates policy designs with regard to their prospective ability to comprehensively solve a policy problem. While some policies are designed to comprehensively reduce a societal issue, others address a problem without being able to reduce or solve it. The concept of change, on the contrary, does not necessarily imply that a societal issue has been alleviated. Change simply concerns policy alteration between two points in time, which is not in all cases identical to problem solving. Some changes, however, may lead to problem solving, which therefore can be considered one type of policy change.

2.1.2   Lessons from Previous Research and Moving Toward a New Approach

The early, path-breaking work on policy design dates back to the 1950s (Lasswell 1956, 1958). Since then, a large body of literature developed that highlights different aspects of policy design.

An early generation of policy scholars studied entire public policy fields until Theodore Lowi claimed that public policy analysis should focus on the study of single policy techniques (Lowi 1964, 1972). Since then, a broad literature emerged, which terms these techniques *policy instruments*. Three broad goals of policy instrument studies can be distinguished (Linder and Peters 1989):

(a) Categorizing instrument types and characterizing policy instruments: The authors' common aim is to establish distinct *categories* of policies. While some scholars categorize single instruments (Vedung 1998), others focus on entire policy programs, i.e., a mix of instruments (Lowi 1972), and still others on national policy styles (Howlett 1991). Another group of researchers characterized policy instruments by a set of *attributes,* rather than by listing distinct categories of instruments (Linder and Peters 1989).

(b) Assessing complementarities and conflicts within bundles of policy instruments: In this stream of literature, scholars pay attention to the mix of different policy instruments and their fruitful, as well as unsuccessful, combinations (Howlett and Rayner 2007; Gunningham and Sinclair 1991).

(c) Characterizing the instruments' impact: A third body of literature focused on the *impact* of policy instruments. Some scientists studied the goals that policy instruments pursue, such as monitoring behavioral change or altering the behavior of target groups (Hood 1986), while others characterized efficiency or effectiveness of policy outcomes (Salamon 2002). The latter adopts an ex-ante approach, with policy instruments being evaluated with regard to their prospective ability to solve a policy problem. A different stream of literature, i.e., implementation and evaluation research, focuses on ex-post evaluation of policy instruments to discern whether a specific policy achieves the intended effects (Pressman and Wildavsky 1984; Hupe 2011; Falkner et al. 2005; Hill and Hupe 2009).

The following paragraphs provide a brief overview of the work of policy scholars on all three of the aforementioned aims (excluding ex-post evaluation research). The literature review is subdivided into first-, second- and third-generation policy design scholars. While first-generation literature started with categorizing and characterizing policy instrument types (see point a), second-generation scholars went on to assess policy instrument mixes and their cohesive or conflictual combinations (point b). Research on policy instruments' impact (point c) was initiated by the first-generation policy scholars and further expanded by the third generation.[2]

2.1.2.1 First-Generation Policy Design Scholars

The first generation of public policy scholars mainly addressed two basic questions: What kind of policy instruments do decision-makers have at their disposal for addressing public problems and for pursuing political goals? And into which basic categories can these instruments be grouped? In order to answer these questions, a large body of literature on policy instrument was developed that originated from the USA, Europe, and Canada.

---

[2]The differentiation between first-,second-, and third-generation policy scholars is ideal-typical since all three streams of literature built on each other, and therefore, second-generation scholars also do research about what I labeled first-generation topics; third-generation literature includes what I labeled first- and second-generation topics. Even though these literatures overlap, their distinction is nevertheless helpful here to illustrate how the research discipline evolved over time in a simplified way.

US American Schools

Theodore Lowi's seminal work argues that four broad types of policies can be identified, namely distributive, redistributive, regulatory, and constituent policies that produce particular patterns of political conflict (Lowi 1964, 1972). *Distributive policies* stand for all policies, which distribute resources from the government to a relatively wide group of beneficiaries. *Redistributive policies*, in contrast, reassign resources from one group or social class to another. *Regulatory policies* constrain behavior in order to protect specific groups of people (such as consumers) from other groups or sectors. Finally, constituent policies create or modify the state's institutions (Knill and Tosun 2012, p. 16). According to Lowi, redistributive and regulatory policies produce winners and losers, and therefore lend a higher potential to conflict than do distributive and constituent policies. Hence, Lowi deduced his famous statement that 'policies determine politics.' Critics of Lowi's typology argue that it is 'difficult to assign policies to just one category' (Birkland 2010). Most significantly, however, Lowi was the first to draw attention to the coerciveness of policy intervention, and more specifically to two dimensions, he termed *likelihood of coercion* and *applicability of coercion* (Lowi 1964, 1972). *Coercion (or likelihood of coercion) describes the degree to which an instrument constrains individual or group behavior* (Salamon 2002). *Applicability of coercion* refers to whether the policy identifies specific target groups or whether it applies to society in general. Ever since the genesis of Lowi's term, coercion has been considered 'the most common basis for classifying instruments in the literature' (Salamon 2000).

As a response, Wilson (1974, 1986) established a typology based the degree to which *costs and benefits* are distributed or concentrated across targets (see Table 2.1). Extending Lowi's idea of the impact of policies on politics, Wilson characterized the policymaking process (politics) rather than policy content (policies). Hence, Wilson distinguished four types of politics, i.e., majoritarian, entrepreneurial, interest group, and clientelistic (Wilson 1986, 1974; Schneider and Ingram 1993; Knill and Tosun 2012).

Another innovation to policy design literature came with *Robert Alan Dahl and Charles Edward Lindblom's* approach, which places policy instruments on a *continuum* rather than into discrete categories (Dahl and Lindblom 1953). Dahl and Lindblom characterized policies on five continua, emphasizing different aspects of governments' capacity to exert coercion. Among others, they point to the *nature of government influence*, which can range from persuasion (low coercive capacity) on one side of the continuum to compulsion (high coercive capacity) on the other extreme of the continuum. The fine-grained analysis of different aspects of coercion

Table 2.1 Wilson's cost-benefit typology

		Costs	
		Concentrated	Distributed
Benefits	Concentrated	Interest group politics	Clientelistic politics
	Distributed	Entrepreneurial politics	Majoritarian politics

represents a key contribution from their work. The authors look into *instrument ownership*, which denotes whether private enterprises or public agencies are responsible for the implementation of adopted policies. They draw attention to indirect versus direct government control and to voluntary versus compulsory *instrument membership* as well as to *instrument autonomy*, which ranges from autonomous agencies to bureaucratic ones.

Stephen Linder and Guy Peters synthesized the different aims of previous researchers (Howlett 1991). They identified seven general categories of policy instruments (direct provision, subsidy, tax, contract, authority, regulation, exhortation) and identified *four general instrument attributes* (Linder and Peters 1989): (1) *resource-intensiveness*, which measures the degree to which a policy instrument involves administrative costs and is simple (or complex) to operate; (2) *targeting*, which takes into account the precision of a policy instrument and its selectivity toward target populations; (3) *political risk*, which refers to the chances of failure of a policy, and also to an instrument's visibility to the public; and 4) *constraint,* meaning an instrument's coerciveness.

Linder and Peters' attempt to synthesize comes at the expense of analytical clarity: Some of the aforementioned dimensions are attributes of policy instruments (coerciveness, targeting), while others deliver explanations for instrument choices (political risk, resource-intensiveness).

Salamon (2002) offers another approach to characterizing policy instruments, by focusing on *four attributes*. First, Salamon's conception of *coerciveness* captures the degree to which an instrument restricts behavior. At the higher end of the coercion scale are instruments that limit or prohibit undesired activities; at the lower end are instruments that rely on voluntary cooperation of target groups. Second, Salamon emphasizes the *directness* of a policy instrument, which measures the extent to which the entity deciding upon the course of collective action is also involved in carrying it out (Salamon 2002). For instance, a direct instrument is one where the decision upon a course of action, funding, and implementation is all carried out by the same entity. On the contrary, a policy instrument is considered indirect when it is publicly financed but privately delivered—or financed nationally, but operated on the local or regional level. *Automaticity* represents the third-key dimension that Salamon uses to characterize policy instruments. Automaticity 'measures the extent to which a tool utilizes an existing administrative structure for its operation rather than creating its own special administrative apparatus' (Salamon 2000). The fourth dimension Salamon identifies is the *visibility* of a policy instrument. He writes '[i]nvisible tools are […] the easiest to pass.' This reasoning suggests that *visibility* is not an instrument characteristic, but rather an explanation for why policy makers choose certain instruments over others.

Crawford and Ostrom (1995) took another approach to the question of policy design by establishing a *syntax of a grammar of institutions*. The idea here is to identify the main components that characterize a policy by answering five basic questions: Who is allowed/obliged/forbidden to do what, under which condition, in order to fulfill which aim, and what sanctions are involved? The authors make the analogy to grammar, where each sentence always contains certain components, which deliver information about the subject, the object, the verb, and alike.

European Schools

The British scholar *Christopher Hood* is the author of a classic piece of instrument categorization (Hood 1986). Hood argues that governments have four resources at their disposal—*nodality/information, treasure/money, authority, and organization/ delivering services*—and can use them to fulfill two purposes: either monitoring (detectors) or altering the behavior (effectors) of target populations (see Table 2.2).

The typology has often been criticized for not being mutually exclusive. Hybrid policy designs are difficult to disentangle in order to fit into the typology (Hood 2007).

Despite these critiques, Vedung (1998) established a similar policy typology, which developed into the best-known and perhaps most influential one in the field (Hood 2007). Vedung identifies three classes of policy instruments—*carrots, sticks, and sermons*—based on the extent of coercion, defined as the degree of state intervention, that each instrument involves. With *carrots*, Vedung refers to a family of incentive-based financial policy instruments, such as charges or trading schemes. The term *sticks* is an analogy for regulative, also called command-and-control instruments, which rely on the use of authority by the state, such as prohibitions or authorization restrictions. *Sermons* refer to persuasive, information-based instruments, such as public campaigns or best environmental practices.

Many European scholars, notably Mayntz and Scharpf (1995), Varone (1998), and Ingold (2008), have thought about policy instruments and instrument selection. Among the European stream of policy research, the approach of two Dutch scholars, Bressers and O'Toole (1998, 2005), is particularly relevant to this book's research goal. Their work conceptualizes policy instruments as a set of rules, which specifies relations in a social setting. Bressers and O'Toole put forward six instrument attributes, which capture the degree to which the behavior of certain target groups is limited or expanded in relation to other societal groups. The first instrument attribute that the authors propose is *normative appeal*, which seizes the ideological constraints of a policy instrument toward targets. The more an instrument condemns a defined behavior as 'good' or 'bad', the more it is considered normative. A second attribute concerns *providing or withdrawing resources* to or from the target groups and establishes a resource-related (financial, personnel, authority, etc.) relationship between different policy addressees. A third characteristic involves the target group's *freedom to opt for or against the application of policy measures*. Some policy instruments leave target actors the choice of whether or not the instrument applies to them. In the case of subsidies for organic

Table 2.2 Hood's eight basic types of government tools (Hood 1986)

	Nodality	Treasures	Authority	Organization
Effectors	Advice	Grants, loans	Laws	Service delivery
Detectors	Surveys	Consultants	Registration	Statistics

agriculture, for example, farmers benefit from subsidies when they adopt organic farming practices, but farmers are also free to choose not to change their practices and to abstain from the subsidy (Bressers and O'Toole 1998: 224). A fourth dimension refers to the *proportionality* between a target group's behavior and the policy response. A fifth attribute is labeled the *role of decision-makers in implementation*. Policy designs often name the organization responsible for implementation. Bressers and O'Toole distinguish situations in which decision-makers assign themselves or closely affiliated organizations the task of implementation from situations in which lower levels of government, private agencies, or government corporations carry out implementation. A sixth instrument feature is called *bilateral or multilateral arrangements* and captures the degree to which policies create a direct (or indirect) relationship between the government and the target group.

Canadian Schools

There is a rich body of literature on policy instruments from Canadian scholars (Doern and Phidd 1983). Howlett (2000) has added a new element to the instrument typology schools. He argues that the literature has so far concentrated on *substantive* instruments and neglected *procedural* ones (Howlett 2000). Substantive instruments can be understood as those techniques that aim at reducing a policy problem. Instruments, which alleviate pollution, for instance, can be considered substantive. Procedural instruments, on the contrary, are not intended to reduce a policy problem directly, but they create structures or procedures that establish positive conditions for reducing a policy problem. Examples of procedural policy instruments include reporting or monitoring by following a defined timetable over a period of several years, as well as the formation or reform of administrative structures in order to act more efficiently in addressing a policy problem.

Summary of First-Generation Policy Design Scholars

A large number of scholars, more than could be presented here, have expressed valuable ideas concerning types of policy instruments or instrument characteristics. The literature overview brings to light that most policy analysts developed *typologies* of policy instruments. Those typologies allow for a classification of virtually unlimited policy tools into a limited number of general categories using a common language (Howlett 2005; Lowi 1972). Generic classification schemes help with identifying long-term patterns of public policymaking through comparisons across time, country, or policy field. Categorization also promotes lesson-drawing from past experiences with the performance of specific instruments under given circumstances (Linder and Peters 1984; Howlett 2005; Salamon 2002; Hood 2007). Howlett (2005) argues that policy instrument classifications do not only aim at better descriptions but also aim at providing better prescriptions or recommendations to decision-makers on how to best address a policy problem.

Most of the aforementioned prominent literature on policy instruments has its origins in the 1980s and 1990s (Howlett 2011b). After the early 1990s, scholars moved away from a single instrument focus of earlier works and drew attention to instrument mixes (Howlett 2005).

2.1.2.2 Second-Generation Policy Design Scholars

The first-generation policy scholars already criticized a pure focus on single instrument types and moved on to assess characteristics of policy instruments [see, e.g., Bressers and O'Toole (1998) and Salamon (2002) as exposed above]. Hence, a second generation of policy design scholars emerged, who assessed in more detail policy instrument mixes, their complementarities, and conflicts. Salamon (2002), for example, convincingly demonstrates that there is a growing number of instruments at the disposal of governments, which come bundled in programs and only very rarely appear in a pure form (Salamon 2002, p. 21). Most of today's policies are composed simultaneously of regulatory, incentive, and information elements. Hence, Vedung's (1998) distinction of sticks, carrots, and sermons should not be regarded as a policy typology with mutually exclusive categories. For example, pollution reduction policies are typically composed of the following: (a) emission limits setting a defined pollution cap (sticks); (b) financial incentives for polluters to invest in technologies or practices that will allow them to reduce their emissions and thus comply with the emission limit (carrots); (c) information about how to operate green technologies or practices (sermons); and finally, d) new governmental agencies being created in order to control, fund, or support citizens or businesses in their efforts to reduce pollution (organization). Hence, research has evolved to ask the question: Which mix of instruments (rather than which instruments) do we have at our disposal to solve societal problems?

One new insight this literature gained compared to the first-generation research is that policy instruments can undermine each other's effects. Instrument mixes fail to deliver desired policy goals when, for example, new instruments are simply added to existing ones (called *layering* in the literature) or when new policy goals are formulated without abandoning previously adopted policy instruments (called *drift*) (Howlett and Rayner 2007). As a result, policy goals and means, i.e., the policy instruments by means of which policy goals are to be achieved, do not match. To avoid such incoherence, a literature on *integrated strategies* emerged (Howlett and Rayner 2007; Gunningham and Sinclair 1991; Gunningham and Young 1997; Gunningham et al. 1998; Grabosky 1995). Scholars in this discipline analyze complementarities and conflicts within bundles of instruments. More concretely, they study how policy instruments can be combined such that they support one another in the pursuit of a common goal. Underlying research questions include the following: a) Which policy instruments can be mixed? and b) Which factors allow us to evaluate whether a policy mix is coherent? Scholars refer to carefully arranged instrument mixes as *New Governance Arrangements* (NGAs) (Howlett and Rayner 2006). NGAs are ideal types of policy designs which combine policy instruments into a cohesive and holistic strategy to optimally reach a defined policy goal (Howlett and Rayner 2007). Hence, NGAs are conceptual reference points of policy design rather than empirically observable policies. Some authors, such as Michael Howlett in his book titled 'Designing Public Policies' (2011a), base their definition of policy design on ideal-typical arrangements. From that perspective, policy designs are concepts emerging from an intellectual exercise;

they are underlying but not representing real-world policies. Accordingly, the study of policy designs is viewed as antecedent to the study of policy content and policymaking. The present work, on the contrary, conceives of policy design as a real-world, empirically observable arrangement of policy instruments.

2.1.2.3 Third-Generation Policy Scholars

There is a broad consensus among first- and second-generation policy scholars that the level of coercion or state intervention used to alter, limit, or create behavior or processes is a chief criterion for categorizing policy instruments (Linder and Peters 1989; Salamon 2002; Vedung 1998; Howlett and Ramesh 1995). Implicit to this definition of coercion, as with the degree of state intervention, is the perspective that the state restricts society in a top-down fashion. In reaction, a new third generation of policy research emerged which drew attention away from the state in favor of (a) 'more global' and (b) 'more local' policies (Howlett 2011a). The former stream of research highlights the effects of globalization in undermining state capacities. The latter stream of research 'decentered' policy studies away from analyzing policies based on central state authority toward studying policies that emerge locally, or from a bottom-up approach. A large body of literature developed that studied local governance, decentralization, and collaborative governance (Arellano-Gault and Vera-Cortés 2005; Doern and Wilks 1998; Edelenbos et al. 2010; Esmark 2009; Foster and Plowden 1996; Gibbs et al. 2002; Ingold 2014; Lemos and Agrawal 2006). This research demonstrates that various actors are involved in policy decisions today (Ingold 2008, 2011) and that contemporary governments are unable to move unilaterally without incorporating other social forces (Bressers and O'Toole 1998; Sabatier and Jenkins-Smith 1993). Hence, the view of a managerialist state was replaced by a more deliberative model, where multiple state and non-state actors participate in policymaking. Some scholars interpreted the involvement of non-state actors, along with deregulation and privatization trends, as a further sign for reduced government capacity to govern, aside from globalization (Howlett et al. 2009b; Hysing 2009; Provan and Kenis 2008; Eliadis et al. 2005; Jordan et al. 2005). Most famously, the move away from *government to governance* was proclaimed. This statement went along with a new perspective on policy instruments, which are no longer viewed as instruments of governments alone. Policymaking is rather conceived of as a participative process out of which policies *emerge bottom-up* rather than being designed top-down (Howlett 2011a). Consequently, parts of the literature employ the terms *formal rules* or *institutions*, rather than the expression *policy design* (Ostrom 1990, 2009). Since policies are considered emerging bottom-up, the question of which instruments do governments have at their disposal was replaced by the question of which instruments do *we* have at *our* disposal to best solve societal issues (Metz and Ingold 2014). Researchers then examined if the government-to-governance statement goes along with a turn from command-and-control instruments to more participative policy designs. In this regard, empirical studies demonstrate that in many countries and in multiple realms, this shift did not occur (see, e.g., Sager 2009; Jordan et al. 2003, 2013). The domestic arena is still important for policy design, along with (and not despite of) trends of globalizing and localizing policymaking (Howlett 2011a). Governance research demonstrated that policymaking is more complex today and involves multiple actors and levels of governance, i.e., from global to local. In light of such complexities, it remains relevant today to increase and deepen our understanding of policy design.

2.1.2.4 Lessons from Previous Research and Moving Toward a New Approach

Howlett and Rayner had already postulated in 2007 that a new generation of policy design theory was necessary, which conciliated the thinking of first- and second-generation policy scholars with insights from the third generation (Howlett and Rayner 2007). In order to reconcile those literatures, the governance perspective must incorporate the insight that governance does not replace government, but rather governments are 'part of' governance. This inclusive view on governance does not imply that policy research should recenter on solely studying state authorities, but rather that research should reconsider state actors as one of many actors shaping policy designs.

Conciliating a governance perspective with a focus on policy design is possible, for example, when conceiving of policy design as a result of network interactions, where state and non-state actors interact. The present work considers both and disentangles their relationship by conceiving of complex decision-making processes as policy networks (independent variable) out of which policy designs emerge (dependent variable).

Moreover, current research has to move beyond the search for instrument categories. Understanding which types of policy instruments exist evidently remains relevant today. However, past research already provides a solid base of knowledge concerning instrument categories, characteristics, mixes, and more complex processes of deregulation, privatization, and new governance modes. Instead of continuing to distinguish different categories of policy instruments, research could evaluate the expected impact of policy designs on reducing societal issues. If we define policies in terms of comprehensive problem solving (and not coercion), then it is a valid research aim to be able to say *how much* or *how little*—in other words how *comprehensively*—a policy design contributes to addressing a collective problem (as opposed to measuring the degree of state intervention). In order to understand a policy's prospective ability to comprehensively solve a societal problem, it is crucial, but not enough, to list or name single instrument categories out of which the policy is composed. Necessary is also to highlight the precise conditions under which instrument mixes apply.

The present work aims to contribute to the advancement of policy design research in two ways: First, instead of sorting policy instruments into boxes, this research aims at examining the ability of policy designs to comprehensively address societal issues. This research goal involves a change in perspective on policy designs, which are not viewed as channels of governmental coercion, but as means to solve societal issues. Underlying this research is a definition of policy design, which differs from the conventional coercion approach, with its problem-solving perspective rather than having a state-power perspective.

Secondly, to evaluate policy designs' comprehensiveness, the present work argues that typologies are not suitable and a different analytical tool is needed. Although typologies have commonly been used in political science, past research has made apparent that inherent limitations exist concerning typologies (Smith 2002). By definition, a typology separates a given set of dimensions and considers all possible combinations of dimensions through cross-tabulation. The resulting categories are ideal-typical, and therefore, they cannot necessarily be found in the empirical reality. Policy instruments do not neatly fit into conventional boxes, which can be seen in cases where command-and-control instruments are adopted by a government lacking the necessary enforcement capacities (Bressers and O'Toole 2005, p. 142). Here, command-and-control instruments may be much less compulsive than the label indicates (Bressers and O'Toole 2005: 142–143). Voluntary instruments or incentives, on the other hand, can be very compulsive when there is a high pressure to conform to a certain norm. With high societal or cultural pressure to adopt an environmentally friendly behavior, for example, one might feel more compelled to adapt the own conduct than by regulatory instruments. Another such example is subsidies, which are conventionally classified as economic instruments. Subsidies do not only involve resource exchanges, but also involve the provision of information on how to change behavior in order to be eligible to the subsidy. Thus, a subsidy does provide target actors not only with financial resources, but also with information. As such, a subsidy could also be classified as a persuasive instrument. These examples illustrate that existing instrument typologies have not proven adequate in portraying the complexity of policy content. Real-world policies are multi-dimensional and therefore do not neatly fit into typologies. To cope with this restriction, scholars produced more fine-grained typologies (see, e.g., Linder and Peters 1989; Bressers and O'Toole 1998). Nevertheless, many categories impair conceptual clarity and analytical parsimony, and complicate the task of classifying instruments into categories. Reducing the number of dimensions to broad categories leads to typologies that show as much variation within cases than between them (Linder and Peters 1989). When categories are not mutually exclusive, scholars can classify the same instrument in different ways (Smith 2002). Hence, simpler typologies might provide misleading information and obstruct comparison across countries or time (Linder and Peters 1989, p. 43). Additionally, these reduced categories bear the risk of not being exhaustive enough. A carrot–sticks–sermon typology does not even consider services provided by the state, such as wastewater treatment, street lighting, or education (Hood 2007). Their purely descriptive nature, along with their lack of predictive or explanatory power, represents another difficulty scholars encountered with typologies. Scientists have expressed difficulties to infer hypotheses from instrument typologies. Finally, most policy typologies insufficiently take into account instrument mixes since they rely on individual instruments as the unit of analysis, rather than on packages of instruments (exceptions include, e.g., Bressers and O'Toole 1998; Salamon 2002).

In summary, identifying categories of instruments by means of a typology is not a straightforward endeavor (Linder and Peters 1989). If policy instruments cannot be clearly classified using one common language, any typology loses its claim in supporting the search for a general understanding of policy instruments (Smith 2002).

Some alternatives to the typology approach have been proposed. Smith (2002), for instance, argues that taxonomies are preferable to typologies. Taxonomies classify items on the basis of empirically observable and measurable characteristics, similar to a cluster analysis. Since empirically derived categories are less ideal-typical, it could be more straightforward to classify instruments, thereby placing taxonomies at an advantage to typologies. A shortcoming of taxonomies, however, is that they are case- and context-specific. Hence, taxonomies do not necessarily lead to the establishment of a common language, which allows for comparisons across countries and time. Moreover, categorizing instruments into single categories never provides satisfactorily answers, not even by means of taxonomies, because policy instruments are inherently multi-dimensional.

We have seen so far, that, on the one hand, all the previously exposed schools of thought bring us closer to obtaining a detailed picture of policy design. On the other hand, the reliance on typologies is a great source of criticism within the existing literature. Those criticisms have shown that any new attempt to characterize policy design has to combine the following features:

1. In order to recognize the multi-dimensionality of policy designs, a new approach should not rely on ideal-typical categories.
2. A new approach should be exhaustive and, simultaneously, provide conceptual clarity.
3. It should overcome pure description. Rather than categorizing instrument types, a new approach should be able to evaluate a policy design's prospective ability to solve an underlying societal issue.
4. The new attempt should not purely focus on single policy instruments as the unit of analysis, but should also be applicable to an instrument mix. Additionally, it should consider all the arrangements specifying how long, to whom, and on which level the policy applies.
5. Policy typologies have been criticized for not allowing causal analysis and hypotheses testing. One obstacle in this regard is the fact that typologies are multi-dimensional, which makes relating several independent variables to one dependent variable difficult, as is the case with hypothesis testing. To fill this void, capturing multi-dimensionality in the form of categorical or continuous variables, rather than as mutually exclusive categories (such as typologies), would prove useful. Moreover, a new attempt to conceptualize policy designs should rely on numerical values and not distinct categories to allow for statistical analyses. Such a new approach should give clear guidance on how to objectively quantify policy.

2.1.3 A New Approach to Policy Design: The Policy Comprehensiveness Index

Building on the lessons of previous research, this book explores a new approach to policy design, which incorporates insights on policy instruments from primarily *first-* and *second-generation* scholars. Along with these discernments, the new approach examines the ability of policy designs to comprehensively address public problems.

Evaluating the performance of an instrument design for solving an underlying issue requires a new analytical tool. Such a tool should overcome shortcomings of typologies and comply with the five premises formulated previously in Sect. 2.1.2.4. The following items seek to highlight explanations for why an index approach is a powerful tool for arriving at each of the five premises:

1. An index approach does not predefine ideal-typical mutually exclusive categories, but rather highlights different facets of a policy in a composite way. Because indices summarize a multitude of indicators (Hajkowicz 2006), there is no artificial reduction into one single category, as is the case with typologies.
2. Indices capture complexity by looking at multiple dimensions all the while producing one synthetic, representative result. In this way, policies may parallel on another along some facets and differ from each other along others, in multiple combinations, and without creating a new category for each combination.
3. An index also complies with the present study's aim to moving beyond instrument categorization and toward a measure of performance. By means of indices, a policy design's performance can be evaluated such that it becomes apparent how much or how little a policy design contributes to addressing a public policy problem. Likewise, policy designs may be ranked with regard to their prospective ability to solve an underlying issue.[3]
4. Moreover, an index can be applied to a single policy instrument, as well as to policy mixes. Through its multiple composite indicators, an index can take into account the specific provisions that come along with policy instrument mixes.
5. Finally, an index may assign qualitative values (low–medium–high) or numerical ones (e.g., 0–0.5–1, or from 0 to 10). Creating one numerical index value enables one to rank policies' performance, as well as to compare policy designs over time. Moreover, hypotheses are testable in quantitative ways, e.g., in statistical analyses.

In conclusion, the ultimate aim of a policy index is not only to establish an instrument typology in the form of discrete categories, but also to create a single synthetic measure of policy performance. More specifically, here-proposed index

---

[3]When two policy designs perform equally, this can be due same scores for different composite indicators. If the aim was to categorize policy design, this would clearly be a problem. But, the aim here is to evaluate performance. Two equal scores reflect that policies can be designed very differently and still address the policy problem equally well.

seeks to evaluate a policy design's prospective performance in comprehensively addressing an underlying policy issue.

As opposed to environmental performance, the index captures *policy performance, which involves an ex-ante estimation about the chances of a policy design to (a) cope with an underlying problem and (b) to be implemented.* Both are necessary in order to problem-solve: First, a policy design has to be well adapted to the underlying policy problem. However, even such a well-adapted policy may not contribute to problem solving if it is not implemented. Hence, to achieve problem–solving, a policy also has to be designed so that it increases its chances for implementation.

An *ex-ante* evaluation is followed here to account for the analytical difference between policy *output, outcome,* and *impact* (Schneider and Janning 2006, p. 15). *Policy output* refers to an adopted policy and its content as a result of a policy-making process. *Policy outcome* reflects the policy results after implementation, and policy impact concerns the societal or environmental changes induced by a policy. Take the example of a crossroad where accidents happen regularly, and therefore, actors wish to adopt measures to reduce accidents. The decision to install stop signs, for example, is equivalent to a *policy output.* The installation of stop signs on the crossroad corresponds to a *policy outcome,* and the reduction of accidents represents the *policy impact.* The index proposed here exclusively evaluates *policy outputs,* defined as policy designs. As such, the index seeks to estimate whether the arrangements of policy designs are able to achieve the desired *policy outcome* and *impact.* Since the index is not tailored to evaluating actual policy outcomes or impact, it is considered an ex-ante estimation here.

In order to address a policy problem in a *comprehensive* way, policy designs need to fulfill three criteria: First, policy designs have to be *effective*; second, *efficient;* and third, *compelling* (see Sect. 1.2 for a definition of the terms *effective* and *efficient*). *Effectiveness* suggests that, ideally, a policy addresses the entirety of the sources contributing to a problem in order to comprehensively solve an underlying policy problem. At the same time, however, more regulation does not necessarily denote *efficient* problem solving. When only a subgroup of society causes a problem, it is not efficient if a policy targets society in general. Efficient policies, on the contrary, may be less costly and also receive a higher level of acceptance. A *compelling* policy is designed such that it increases its prospective chances in being applied in order to address an issue (without simply existing on paper). Hence, *policy comprehensiveness is defined here as the degree to which a policy design addresses an underlying policy problem in an effective, efficient, and compelling way.*

Since indices represent a summary of multiple indicators, it is fundamental to include the 'right' indicators. The literature advances two main requirements for choosing indicators: (a) the avoidance of redundancy and (b) exhaustiveness (Hajkowicz 2006). Redundancy occurs when two indicators measure the same phenomenon, which leads to double counting and a biased outcome. Exhaustiveness denotes that indicators should grasp all composite aspects of the overall phenomenon that the index aims to summarize. To comply with these requirements, index construction can be guided by either empirical data or a solid theoretical framework (Niemeijer 2002). For the present work, a theory-driven approach is adopted. The indicators that feed into the policy comprehensiveness index are derived from the rich literature on policy instrument characteristics developed by previous scholars (see Sects. 2.1.2.1 and 2.1.2.2).[4] As such, the proposal of indicators should be considered a starting point for future research that reflects the current knowledge of policy design.

Drawing on previous research about policy design characteristics, the following indicators are proposed to approximate the comprehensiveness of policy design, i.e., policy output (not policy outcome, impact, or change):[5]

- **Pressure on target group(s)**: The first indicator that feeds into the index can be found in above-described work of Linder and Peter under the term *constraint* (1989), in Dahl and Lindbom's (1953) idea about the *nature of government influence*, or in Salamon's (2002) work characterizing policies' *coerciveness*. Bressers and O'Toole (1998) formulate this aspect of policy design in a more positive way as *the freedom of target groups to opt for or against the application of policy measures*. Here, the term *pressure* is used, because policies do not only constrain undesired behavior, but also promote desired behavior. Both constraining and promoting involve a certain degree of pressure for adopting a desired behavior.
  Target groups are the addressees of policies and as such are supposed to adapt their behavior or procedures accordingly. Depending on the exact design of a policy, there can be more or less pressure on target groups to adapt their behavior (or procedures). In other words, individuals or groups may be granted more or less freedom for choosing whether or not to take action (Bressers and O'Toole 1998). Persuasive instruments, such as information campaigns or subsidies, typically do not force the target group to adopt a particular behavior. Likewise, product charges leave some degree of choice between continuing to use the product (and pay the charge) and ceasing to use it.

---

[4]An alternative approach would be to look into the indicators proposed by other policy indices. To my best present knowledge, there is currently no other index measuring the general ability of policy designs to comprehensively address a societal issue. Existing indices are specifically designed for measuring defined problems such as corruption (Corruption Perceptions Index), economic uncertainty (Economic Uncertainty Index), migrant integration (Migrant Integration Policy Index), or environmental sustainability (Environmental Sustainability Index, State of the Nation's Ecosystems, Ecological Indicators for the Nation, Environmental Policy Stringency, Index of Environmental Regulations, Climate Laws, Institutions and Measures Index, Environmental Regulatory Regime Index). In absence of an exemplary index, the theory-guided approach to select indicators, chosen here, seems to be the most appropriate one to establish a first index proposal.

[5]As mentioned above, *comprehensive* emphasizes that a policy design has to be effective, efficient, and compelling. These criteria apply to the overall index, but not each indicator. The indicators composing the index can fulfill one or two out of the three criteria, because through the addition of the composite indicators the final index fulfills all three criteria.

Based on the assumption that policies aim at alleviating policy problems, the more a target group has the choice about whether or not to respond to the policy, the less pressure there is to comprehensively reduce a societal problem. And vice versa, the more a policy design places pressure on target groups to adopt desired or quit undesired behavior, the more likely the underlying problem is to be alleviated. The amount of existing pressure is a first indicator that a policy design has the potential to comprehensively address a societal issue by compelling the target group to change behavior or processes.

- **Sanctions** are a decisive element of policy designs and can be found, for example, in the work of Crawford and Ostrom (1995). In theory, target groups could deliberately choose not to comply with a policy. To prevent such situations, sanctions impose a financial or social cost on defection. Consequently, *sanctions* limit the freedom to choose non-compliance. The higher the costs of violating a policy compared to the gains of non-compliance, the more likely a policy is to not only exist on paper, but also be applied in reality. Policy designs that include sanctions are able to compel target groups to take a desired course of action, and this, in turn, can promote comprehensive problem solving.
  The threat of sanctions alone does not yet guarantee policy implementation. Enforcement requires the existence of an agency responsible for the control of compliance, as well as the imposition and pursuit of sanctions. Hence, when a policy provides for an enforcement agency, its ability to comprehensively reduce a societal issue is considered larger than without such provisions.
- **Inclusiveness**: Lowi (1964, 1972), Linder and Peters (1989, pp. 43, 46), as well as Bressers and O'Toole (1998) theorized about how precisely policy designs target the root causes of a problem (see also Metz and Ingold 2014). Policy problems can be caused by the behavior of (or processes practiced by) a few individuals, a specific or several target groups, or society in general (Linder and Peters 1989; Schneider and Ingram 1993; Howlett and Ramesh 2003; Lowi 1964). Precise policies target those units, which cause the underlying policy problem, while imprecise policies apply to other units (more or less units) than just the ones causing the problem. Ideally, policies are effective in that they address the entirety of target groups contributing to a problem. In reality, distortions might occur if an instrument addresses symptoms rather than causes of a problem, or if directly addressing the root cause is politically unfeasible. Hence, some policies are more adept than others at including liable target groups. The idea here is that the broader the target group relative to those units contributing to a problem, the more comprehensive the policy design; the more partial the target groups, the less comprehensive the policy. Thus, establishing policy *inclusiveness* denotes achieving a point of maximum policy efficiency. More concretely, policy inclusiveness concerns *horizontal* policy efficiency as it refers to addressing different societal groups or sectors (such as agriculture and industry) that cause a policy problem.
- **Proportionality**: While inclusiveness takes a horizontal perspective, proportionality takes a vertical one. This indicator is derived from Bressers and

O'Toole's (1998) work who refer to the correspondence between the scope of a policy issue and the scope of the regulative answer to it. The underlying approach takes up this idea of proportionality between problem and solution and adds the aspect of governance level. Accordingly, proportional policies apply to the same jurisdictional level (local, regional, national, international) as the elements contributing to the regulated aspect of the problem. Precisely, proportionality deals with the level on which the policy has to be implemented, and not the level upon which the policy is decided. For instance, a law adopted by a national parliament pertaining to the promotion of economic growth in exclusively one region in the country would be considered a regional-level policy for this indicator.

A further difference between inclusiveness and proportionality should be pointed out: Inclusiveness examines all potential causes of a policy problem and evaluates the ratio of liable target groups considered in a policy design. The indicator *proportionality*, on the contrary, restricts its perspective to the selected target groups of a policy and disregards all elements not targeted by the policy, even if they contribute to a policy problem. For example, if a defined climate change policy targets the transportation sector and disregards the industrial or energy sector, only the transportation sector is taken into consideration in order to evaluate the proportionality between the scope of the transportation policy and the scope of the transportation problem.

There are two possible disproportionalities between the policy level and the problem scale. First, the policy deals only partly with the problem when the policy applies to a smaller scale compared to the extent of the problem (see also Metz and Ingold 2014). In that case, a policy design is not effective enough. Second, disproportionalities exist when a policy applies to a larger scale than the extent of the problem. In such a case, a policy design is too extensive and, therefore, not efficient. This phenomenon may occur when a national-level policy deals with a geographically concentrated problem.

In short, the indicator proportionality is the difference between the extent of the problem and the jurisdictional level of its solution. The smaller the difference, the higher the score for proportionality. Take the example of a municipality that adopts measures to solve its own local issue. Despite the fact that the policy applies only locally, the policy would still receive a high score for proportionality because problem scale and solution level are identical. An example of disproportionality would be an environmental quality norm that applies and has to be monitored in an entire country, whereas the emitters of that pollution are regionally concentrated in one jurisdiction.[6] In that case, a regional

---

[6]Note that the problem scale of the pollution considered in the policy design is contrasted to its solution level. What is not considered here are the types of pollution that may occur, but that are not taken into consideration in the policy design.

environmental quality norm would be more proportional. Hence, establishing proportionality means working toward a point of maximum vertical policy efficiency. The more proportional a policy design, the more comprehensively (meaning efficiently) it addresses a policy problem. Likewise the less proportional, the less likely a comprehensive policy design is to transpire.

- **Directness**: This indicator is derived from Dahl and Lindblom's (1953), Salamon's (2002), and also Bressers and O'Toole's (1998) work, who themselves build on Pressman and Wildavsky's research (1984). Thanks to the latter, it is well known that a huge number of organizations participate in implementation and adjust, alter, or even obstruct policy design. Implementing a policy in the intended way depends largely on the design of the implementation duties. A policy design characterized by resource shortages for implementation (financial, time, personnel, intellectual) or competing loyalties can impair interactions among implementers and hence can lead to gaps in the implementation process. Pressman and Wildavsky conceived of implementation as a problem stemming from too many *clearance points*, defined as the number of individual decision points that must be agreed to before any policy decision can be translated into action (Peters 2013, p. 139). When numerous actors participate in the implementation process, there are many clearance points and opportunities for altering policy design.

Building on this line of thought, Bressers and O'Toole (1998) wrote about the *degree of state involvement in implementation*. Similarly, Salamon popularized the term *directness*, which 'measures the extent to which the entity authorizing, financing, or inaugurating a collective activity is involved in carrying it out' (Salamon 2000, p. 1654). When a policy is designed such that the same institution is involved in all the three tasks (i.e., deciding, financing, and implementing), there are fewer opportunities for clearance points. A policy is classified here as *direct* when the same authority *carries out* all the three tasks, or when the authority that adopted the policy *supervises* implementation, while other levels of government or private actors actually implement and finance the policy. Generally, a policy design is considered indirect when the decision-making entity plays no or little role in supervising implementation, whereas other levels of government or private actors are assigned key responsibilities in controlling or carrying out implementation. Insights from implementation studies and principal-agent theories demonstrate that indirect policy designs bare the risk that goals of those deciding (principals) and those implementing (agents) diverge. With principal-agent difficulties, the number of clearance points increase, as well as the risk of goal displacement; goal achievement, on the contrary, becomes more difficult to achieve. Hence, an indirect instrument design is considered less compelling because the diffusion of tasks increases the chances of principal-agent difficulties. With dispersed implementation duties, there is some leverage to adapt, change, or soften policy design so that its impact on behavior or processes differs from its intended outcome. On the contrary, when a policy is designed such that the same authority is involved in all three tasks (deciding, funding, and implementing), exceptions, adaptations, or dilution of it in the implementation process are less likely. Thus, the more direct a policy is designed, the higher the chances of comprehensive problem solving.[7]

- **Bindingness**: Western democracies generally assign a hierarchy to their norms (Shelton 2006; Rüthers et al. 2011). The constitution is placed on the highest rung followed by laws, and finally ordinances or decrees. Oral agreements are unwritten laws and ranked below written ones. Either of these types of legal acts or informal arrangements may deal with policy problems.
  Policymaking processes are distinct among the different types of legal acts. In general, policymaking processes that adopt or amend a law are much more formal; i.e., they follow a more precise constitutionally given procedure involving the legislature, than the adoption of decrees, for example. Therefore, decrees (and even more so unwritten law) can be changed more easily and in a shorter timeframe than laws or constitutions. Constitutions and laws, on the other hand, are generally characterized by more stability. Because of their higher rank and structure, constitutions and legal acts enjoy higher authority and can be more compelling. When a policy problem is dealt with on the level of a legal act (rather than in a decree or an action plan), it signals policy commitment on a longer-term horizon. The hierarchy of norms can be considered an indicator of how compelling a policy is. The indicator builds on the relationship that the higher the level of bindingness, the more comprehensively a policy problem is addressed.

In total, six indicators can be deduced from existing policy design theories to establish an index of policy comprehensiveness.[8] From here on, this index will be referred to as *policy comprehensiveness index*, which is a tool for evaluating and comparing policy designs' prospective ability to comprehensively address an underlying issue. The index serves to evaluate policy outputs, which passed the

---

[7]Some streams of literature highlight self-regulation (Howlett 2004, p. 6) or adjustment flexibility, defined as the possibility to adapt regulations to local circumstances or new developments (Knill and Lenschow 2003, p. 8). Note that these concepts do not necessarily contradict a direct policy design. For example, a community that self-regulates its matters may adopt *direct*—as defined here—policy designs when it decides, finances, and self-implements its own policies.

Also note that directness is not in opposition to privatization. A policy design can be *direct*—as defined here—even if a private actor is entrusted with the implementation of a public policy, under the condition that correct implementation is supervised by the entity that adopted the policy.

[8]Some policy design features put forward by previous scholars, such as visibility or political risk (Linder and Peters 1989; Salamon 2002), have deliberately not been included in the index. Rather than providing information about the comprehensiveness of a policy design, these factors help to explain why one instrument is chosen over another.

policymaking process and have been adopted, but not yet implemented. As such, the quality of the policy design is evaluated rather than the quality of its implementation. However, the policy comprehensiveness index is not tailored to evaluate policy outcomes, policy impact, or policy change.

The indicators capture the degree to which policy designs address underlying policy problems in an effective, efficient, and compelling way. The first two indicators, i.e., *pressure on target groups* and *sanctions*, measure the degree to which policy designs compel target groups to respond to a policy as intended. The indicator *directness* points to the degree to which policy designs are compelling for implementers. *Bindingness* highlights the degree to which policy designs compel the state to enforce its own laws to addresses an underlying policy problem. The indicators *inclusiveness* and *proportionality* both capture whether a policy is designed to address the underlying problem in an effective and efficient way. Effectiveness and efficiency are necessary for highlighting the fact that comprehensive policy design establishes congruence between the scope of the problem and its solution. The difference between the two indicators is that *inclusiveness* concerns achieving a point of maximum horizontal policy effectiveness and efficiency; proportionality, on the other hand, deals with achieving a point of maximum vertical policy effectiveness and efficiency. Taken together, the composite indicators result in an index that matches the definition of comprehensiveness according to whether a policy design should address an underlying policy problem in an effective, efficient, and compelling way.

In summary, the index proposes six indicators evaluating a policy design's prospective performance in comprehensively addressing a societal problem. Accordingly, comprehensive problem solving depends on the degree of pressure on target groups, and on deterrence from non-compliance through sanctions enforced by agencies staffed with enough resources to control compliance and impose sanctions. Additionally, comprehensive problem solving is evaluated by looking at a policy design's inclusiveness and proportionality with regard to addressing the entirety of sectors and levels (not more and not less) contributing to the policy problem. Another indicator for a policy design's prospective performance is its directness, i.e., the degree to which the authority deciding is also involved in supervising implementation. The indicator *bindingness* captures the degree to which a legally binding formal law addresses a policy problem or an informal policy document.

2.2 Policy Comprehensiveness Index: Data, Method, Operationalization

The previous paragraph introduced the indicators that feed into the policy comprehensiveness index. Subsequently, methodological questions of index construction are addressed.

2.2.1 Data Collection

The assessment of policy designs according to the policy comprehensiveness index relies on qualitative data gathered through content analysis of policy documents and through interviews conducted by the author. More concretely, information about the policies that each country adopted (or is in the process of adopting) toward micropollutants was gathered. To ensure that the gathered, qualitative data is comparable across cases, the exact same information about different aspects of micropollutants policies across the four countries was collected. To establish a uniform and in-depth qualitative and descriptive analysis of a policy design, it is proposed here to establish a *policy design profile*. It then serves as underlying 'raw data' for the policy comprehensiveness index. A policy design profile lists the most important aspects of a policy design, including the instrument types or instrument mix adopted (e.g., ban, trading scheme, and norms) and other specifications about target groups, sanctions, competent enforcement agency, and alike. The literature presented under Sects. 2.1.2.1 and 2.1.2.2 drew attention to the important characteristics of policy designs. Accordingly, profiling may start with naming single instrument types out of which the policy design is composed (Q 1, Table 2.3). There is sufficient theoretical knowledge available to gather information on the latter by using a universal or common language. For example, one could distinguish regulative, economic, persuasive, and organizational instruments like in Hood's (1986) or Vedung's typology (1998) (see Sect. 2.1.2.1 for alternatives). Moreover, the questions asked by Crawford and Ostrom (1995) offer sufficient guidance on the aspects that need to be understood regarding policy design, aside from naming policy instrument types: If or what sanctions are involved (Q 2, Table 2.3)? Who enforces sanctions in cases of infringement (Q 3, Table 2.3)? Who is the target group (Q 4, Table 2.3)? What are the criteria defining target groups and do they enable exceptions (Q 5, Table 2.3)? To profile a policy design, it is additionally useful to answer the following questions: Which jurisdictional level is responsible for implementation (Q 6, Table 2.3)? Who adopted the underlying policy (termed decision-making entity in Q 7, Table 2.3)? Who bears the costs of the policy (Q 8, Table 2.3? Which entity implements or controls implementation (Q 9, Table 2.3)? What is the type of policy document cementing a policy design (Q 10, Table 2.3)?

For the policy design profile, two further questions are deduced from the literature on characteristics of policy problems, as they enable one to compare the cause of the problem with the adopted policy solution (Metz and Ingold 2014; Peters and Hoornbeek 2005): Which are the sectors or target groups contributing to the occurrence of the underlying policy problem (Q 11, Table 2.3)? On which scale (local, regional, national, international) does the policy problem occur (Q 12, Table 2.3)? In total, analysts must answer twelve questions for each policy design under investigation, in order to establish uniform profiles and to acquire sufficient information for ranking policy designs according to the policy comprehensiveness index.

To gather information about policy designs and answer question 1 through 10 of Table 2.3, an in-depth content analysis of policy documents was conducted. Policy documents studied here include laws, ordinances or policy plans, legal drafts

Table 2.3 Policy design profile

Profiling a policy design
Q1: Policy instrument(s)
Q2: Sanctions for non-compliance
Q3: Competent enforcement agency
Q4: Target group(s)
Q5: Criteria defining target groups/exceptions
Q6: Competent jurisdictional level for implementation
Q7: Decision-making entity
Q8: Distribution of costs
Q9: Competent agency for implementation/supervision of implementation
Q10: Type (and name) of policy document
Q11: Target groups/sectors contributing to the underlying policy problem
Q12: Problem scale

Q Question

reflecting all stages of the process, meeting protocols of parliamentary committees or working groups, letters such as invitations to public hearings or exchanges between the legislature and the executive, and reports, i.e., evaluation or implementation reports written by the bureaucracy or external advisors. Access to these kinds of policy documents was offered by the Web sites of the environmental ministries in the four countries and the European Commission, databases provided by parliamentary or the executive services, and also Google searches. In cases where a database was not public, access through e-mails, phone calls, or personal meetings was requested and usually enabled by the respective organization.

While policy documents were used to characterize policy designs in terms of instruments, sanctions, target groups, and implementation duties, (Q1–Q10 of the policy design profile, Table 2.3), other sources of information are necessary to describe the environmental problem itself (Q11–Q12, Table 2.3). To characterize micropollutants as a policy problem, and to identify emitters and the scale of the problem for waters, scientific sources of information, such as scientific articles, were consulted (Hollender et al. 2008; Schwarzenbach et al. 2006; Cunningham et al. 2010; Touraud et al. 2011; Sacher et al. 2008; Bach and Frede 2012; Müller 2011; Altmann et al. 2012; Reungoat et al. 2011; Clarke and Smith 2011; Valiente Moro et al. 2012; Sedlak et al. 2000; Rowney et al. 2009; Johnson et al. 2008; Bercu et al. 2008; Richardson and Ternes 2011; Kortenkamp et al. 2007; Mostafa and Helling 2002; McGonigle et al. 2012). Additional sources of information include analytical policy reports (ICPR 2003, 2010a, b, c, d, e, f, 2011a, b, 2012a, b, c; WHO 2008, 2012; BAFU 2012; Abegglen and Siegrist 2012; Götz et al. 2010b; EEA 2011; IAWR 2007), or results of statistical offices (Annual reports of the International Rhine Monitoring Station in Weil am Rhein 2010).

In a second step, the information from content analysis was validated and complemented with detailed insights from a total of 41 semi-structured in-person interviews with high-ranking state officials (see Annex 1 for a full list of interviews).

2.2.2 Index Construction and Operationalization
of Indicators

Indices are useful tools for summarizing multiple indicators by isolating key aspects
from an otherwise overwhelming set of information (Niemeijer 2002). The literature describes index construction as a series of five steps (Keeney and Raiffa 1993;
Hajkowicz 2006; Von Neumann and Morgenstern 1944):

1. The analyst must identify relevant indicators and obtain data for each indicator.
2. Data units (e.g., money, time, distance) have to be transformed into a uniform
   unit.
3. The relative importance of each indicator can be weighted.
4. The analyst must determine whether an additive, multiplicative, or hybrid utility
   function is appropriate for the aggregation of the underlying indicators.
5. A sensitivity analysis should be conducted.

Each of those five steps is subsequently explained in more detail and applied to
construct the policy comprehensiveness index.

2.2.2.1 Identification of Relevant Indicators

The identification of relevant indicators can be guided either by empirical data or by
a solid theoretical framework (Niemeijer 2002). In both cases, indicators should
exhaustively grasp all composite aspects of the overall phenomenon that the index
aims to summarize, without being redundant (Hajkowicz 2006). For the present
study, indicators were deduced from theory. The theory-driven approach was
chosen because existing literature on policy designs (see Sect. 2.2.1) provides
reliable information about the relevant indicators necessary to construct an
exhaustive index (Niemeijer 2002). Section 2.1.3 explained in detail how relevant
indicators were deduced from theory in order to establish the policy comprehensiveness index. In total, six indicators were identified:

1. pressure on target group,
2. sanctions,
3. inclusiveness,
4. proportionality,
5. directness, and
6. bindingness.[9]

The next paragraphs explain the operationalization of each of the six components in more detail.

---

[9]Since the present study relies on only four cases, it was not possible to test for correlation between
indicators (Hajkowicz 2006). Researchers working with the index should test for redundancy and
feel free to drop certain indicators.

2.2.2.2 Transformation into Uniform Units

The second step in index construction consists of transforming diverse data units into a uniform unit. Since composite indicators of indices usually rely on different units (e.g., hours, miles, hectares, and liters), a transformation function is necessary, which places the scores of each policy scenario at some point on a scale between zero (lowest utility) and one (highest utility) (Hajkowicz 2006). The simplest transformation technique is a linear one, which adjusts lowest and highest indicator values on a linear scale. An alternative to the linear scale is the logarithmic one, which takes into consideration diminishing marginal utility. A linear transformation function $v_i(x_i)$, where composite indicators of the index are denoted by i and indicator raw scores by x_i, can be written as follows (Hajkowicz 2006):

$$v_i(x_i) = \frac{x_i - \min x_i}{\max x_i - \min x_i}$$

where min x_i and max x_i denote minimum and maximum indicator scores across all observed scenarios.

For the present study, the underlying raw data is qualitative, and hence, the 'units' are not conventional ones, such as hours, euros, kilograms, or kilometers. In order to maintain assigned scores to indicators, it is particularly important to clearly define what the qualitative 'units' and what the highest or lowest level of performance corresponds to for each indicator of the policy comprehensiveness index. The next step explains (a) which piece of information from the policy design profile (see Sect. 2.2.1) feeds into building each indicator (see Tables 2.16, 2.18, 2.20, 2.22, 2.24, 2.26) and (b) how to assign scores to each of the six indicators.

Tables 2.17, 2.19, 2.21, 2.23, 2.25, and 2.27 highlight, for each indicator, the corresponding qualitative type of 'unit.' Moreover, these tables portray how the qualitative units are transformed into quantitative, measureable units by translating minimum and maximum performance levels into a linear scale between 0 and 1, where no performance corresponds to a score of 0 and complete performance corresponds to a score of 1: low to 0.25, medium to 0.5, and high to 0.75.[10]

Pressure on Target Groups

The first indicator of the policy comprehensiveness index consists of evaluating the degree to which a policy places pressure on target groups to abandon undesired behavior or to adopt desired behavior. In order to carry out such an evaluation, it is helpful to resort to information from the policy design profile concerning single *policy instruments (and instrument mixes)* contained in policy designs (Table 2.4).

The present work broadly distinguishes between command-and-control, economic, and persuasive instruments (Vedung 1998). The degree of pressure on target

---

[10]The performance levels of all six indicators are already scaled between 0 and 1. Therefore, it was not necessary to transform units by means of the transformation function.

Table 2.4 Information from the policy design profile for the indicator 'pressure on target group(s)'

Indicator of the policy comprehensiveness index	Information from the policy design profile
Pressure on target groups	Q1: Policy instrument(s)

Table 2.5 Pressure on target group(s): units, levels of performance, and scores

Indicator (qualitative 'unit')	Pressure on target groups (likelihood that target groups take action)
No performance (score 0)	No policy instruments adopted
Low performance (score 0.25)	Policy instrument mix relying on persuasion at most [*sermons* according to Vedung (1998), or *nodality/information* according to Hood's (1986) instrument typology]
Medium performance (score 0.5)	Policy instrument mix relying on economic incentives at most [*carrots* according to Vedung (1998), or *treasure* according to Hood (1986)]
High performance (score 0.75)	Policy instrument mix combining economic incentives with some sort of control/state authority
Complete performance (score 1)	Policy instrument mix mostly relying on state authority [*sticks* according to Vedung (1998), or *authority* according to Hood (1986)]

groups is considered gradually decreasing from the first to the last instrument category. In water protection policy, command-and-control instruments include, for example, substance bans, authorization restrictions for marketing substances, environmental quality norms, emission limits, mandatory best environmental techniques, or other forms of compulsory technical standards (see Metz and Ingold 2014 for an encompassing overview). Examples of economic instruments in water protection include product or substances charges, subsidies for behavioral changes, or improved wastewater treatment and effluent charges. Persuasive instruments may include voluntary best environmental practices, information campaigns or consulting, research, or suggestions for the correct disposal of waste or private–public partnerships.

The next question is then how to translate the degree of pressure that a policy design puts on target groups into quantifiable performance levels. Table 2.5 highlights the underlying qualitative unit of the indicator, namely the types of policy instruments that make up an instrument mix. Additionally, the table links instrument mixes to different performance levels and scores. Performance levels range from no performance (score of 0), where a policy design does not fix any policy instruments, to complete performance (score of 1), where a policy design consists of a policy instrument mix largely relying on command-and-control types of policy instruments. Instrument mixes that are composed of persuasive instruments at most, but do not contain policy instruments putting more pressure on target groups than persuasion, obtain a low performance level (score of 0.25). Instrument mixes that rely on economic incentives at most, i.e., there are no policy instruments placing a higher level of pressure on target groups than economic incentives, attain a medium-level performance (score of 0.5). Instrument mixes composed of economic incentives paired with some sort of control attain a high performance level.

An example of a policy design that places a high degree of pressure on target groups could be an emissions trading system, which sets a politically defined cap to yearly CO_2 emissions and, as such, controls the maximal amount of emissions. Nevertheless, individual emitters can choose to continue emitting higher amounts of CO_2 by paying for emission allowances. Taking the same example of CO_2 reduction policy, a score of 1 could be attributed to a policy design, which legally capped the yearly amount of allowed CO_2 emissions for individual emitters. In that scenario, emitters would not be able to continue emitting CO_2 at discretion, but would be highly compelled to change their production processes in order to comply with the defined cap. A CO_2 policy design could attain a medium level of performance if it relied on a fuel tax to reduce emissions from transportation. Although fuel taxes set a price signal and, therefore, encourage people to switch from cars to alternative means of transportation, these incentives cannot ensure such a change in behavior.[11] If people choose to pay the fuel tax, CO_2 emissions do not diminish as a consequence of the introduced policy. Policy design that relies on voluntary CO_2 reduction measures, or on public campaigns, would attain a low level of performance here, since it is not unlikely that target groups choose to abstain from taking action by ignoring a public campaign or a voluntary engagement.

Sanctions

The second indicator of the policy comprehensiveness index concerns sanction. To evaluate the effectiveness of sanctions, it is necessary to know *if sanctions exist* at all and, if so, whether there is a *competent agency* with enough resources to control compliance and enforce sanctions where necessary (Table 2.6). Without sanctions and enforcement agencies, there is the danger that a policy is not implemented and exists solely on paper, which would considerably reduce the comprehensiveness of a policy in addressing a societal problem.

Sanctions include all types of measures that can deter target groups from adopting or continue adopting an undesired behavior. If we establish a continuum measuring, the stringency of sanctions, imprisonment, or severe monetary fines would lie on the high side, while appeals to moral conscience would lie on the low side of the continuum. To determine performance, the absolute stringency of sanctions is less decisive than its relative impact on target groups' behavior. For example, a policy design defining a one million dollar fine might be a very effective way to deter

---

[11]Note that the assignment of policy instrument types to different performance levels is meant as a guideline for analysts. However, the overarching logic, which should finally guide analysts' decision on how to assign scores, is the *likelihood that target groups take action*. A policy instrument mix that relies on economic incentives at most could theoretically attain a complete level of performance if economic incentives were high enough to oblige target groups to take action. A hypothetical example would be if fuel taxes were constantly increased, so eventually no one would be able to use cars as a means of transportation anymore.

Table 2.6 Information from the policy design profile for the indicator 'sanctions'

Indicator of the policy comprehensiveness index	Information from the policy design profile
Sanctions	Q 2: Sanctions for non-compliance Q 3: Competent enforcement agency

Table 2.7 Sanctions: units, levels of performance, and scores

Indicator (qualitative 'unit')	Sanctions (portion of deterred target groups)
No performance (score 0)	No existing sanctions
Low performance (score 0.25)	Existing sanctions deter small parts (less than half) of the target group
Medium performance (score 0.5)	Existing sanctions deter half of the target group
High performance (score 0.75)	Existing sanctions are high enough to deter important parts (more than half) of the target group
Complete performance (score 1)	Existing sanctions are high enough to deter the entire target group

individuals, small farms, and enterprises. However, companies with very large budgets may purposefully take hefty fines into account in order to continue with a politically undesired, but economically profitable behavior. To take into consideration that the impact of sanctions on target groups' behavior is sensitive to its context of application, the 'unit' of the indicator is defined here as a relative value, i.e., the proportion of target groups that sanctions achieve to deter and not as an absolute value (e.g., amount of a fine, years of imprisonment). Table 2.7 provides an overview of how to translate different levels of sanctions into performance levels and scores. Accordingly, complete performance (score of 1) for the indicator exists when sanctions are high enough to deter the entire target group of the policy. Low performance, on the contrary, implies that sanctions deter only small parts of the target group (score 0.25). One can assign a medium-level performance if a policy is designed such that existing sanctions deter half of the target group (score 0.5). If sanctions deter important parts of the target group, a policy design is characterized by high performance on this indicator (score 0.75). No performance is equivalent to an absence of sanctions in the policy design (score of 0).

Inclusiveness

The third indicator of the policy comprehensiveness index is called *inclusiveness* and addresses the question of how precisely policy designs target the root causes of a problem. To gauge inclusiveness, two steps are necessary: First, analysts must identify the *sectors contributing to a policy problem*; second, they must assess the *target group(s) addressed by a policy* (see Table 2.8). Inclusiveness then builds on the idea of comparing the sectors causing a problem with the groups targeted by a policy design. A policy design is considered inclusive when it targets the entirety of groups contributing to a problem. Who the target groups of a policy really are

Table 2.8   Information from the policy design profile for the indicator 'inclusiveness'

Indicator of the policy comprehensiveness index	Information from the policy design profile
Inclusiveness	Q 4: Target group(s) Q 5: Criteria defining target groups/exceptions Q 11: Target groups/sectors contributing to the underlying policy problem

Table 2.9   Inclusiveness: units, levels of performance, and scores

Indicator (qualitative 'unit')	Inclusiveness (ratio of target groups to causes)
No performance (score 0)	No defined target groups
Low performance (score 0.25)	Small parts (less than half) of the groups/individuals causing the problem are targeted by the policy
Medium performance (score 0.5)	Half of the groups/individuals causing the problem are targeted by the policy
High performance (score 0.75)	Important parts (more than half) of the groups/individuals causing the problem are targeted
Complete performance (score 1)	The entirety of groups/individuals causing the problem are targeted

highly depends on the *criteria defining target groups*, which refers to exceptions or restrictions of the applicability of the policy design. For example, a policy might apply only to particularly large operators or certain sectors, rather than to society as a whole, which would restrict the inclusiveness of a policy design in cases where the very vast majority of society contributes to causing a policy problem.

The underlying 'unit' of inclusiveness is defined here as the ratio of groups targeted by a policy design to the groups causing a policy problem (Table 2.9). Accordingly, a policy design attains complete performance for this indicator (score 1), when the entirety of groups causing a problem is defined as target groups by a policy design. A policy that is designed such that only small parts of the groups causing a problem are targeted attains a low performance level (score 0.25). Medium-level performance (score 0.5) indicates that half of the groups causing the problem are targeted by the policy. High performance (score 0.75) is attributed to policy designs which target important parts of the groups causing an underlying policy problem. No performance (score 0) signifies that a policy design does not define any target groups.

Let us take the example of a policy design, which aims at the reduction of gender inequalities on the job market and introduces a 50% women quota for the boards of directors of the 30 biggest stock companies. Since gender inequalities largely exist on the job market outside of the 30 biggest stock companies and outside of boards of directors, policy design would attain a low score for the level of inclusiveness.

Proportionality

The fourth indicator of the policy comprehensiveness index is labeled *proportionality*. A policy is considered proportional when it must be implemented at the same jurisdictional level (local, regional, national, international) as the elements contributing to the problem. A two-step approach is necessary for evaluating proportionality: First, one needs information on the *scale of the problem;* secondly, on the *competent jurisdictional level for implementation* (see Table 2.10). In many countries, different levels of government have constitutionally or legally given competencies, such as education, police, water, or waste management. In accordance with a country's division of competencies, policies are implemented at a defined jurisdictional level, such as the municipal, state, federal, or European level, termed here *competent jurisdictional level for implementation* (in short, *solution level*). The proportionality of a policy design is measured as the difference between the extent of the problem and the jurisdictional level of the solution. The smaller the difference between solution level and problem scale, the higher the score for *proportionality*.

Building on this idea, the underlying 'unit' for the indicator proportionality is the ratio of the solution level to the problem scale Table 2.11). Accordingly, a policy design performs best (score 1) when the defined solution level corresponds exactly to the problem scale; it performs least well when the defined solution level is much larger or much smaller than the problem scale (score 0.25). A medium-level performance for that indicator exists when policies are designed such that the defined

Table 2.10 Information from the policy design profile for the indicator 'proportionality'

Indicator of the policy comprehensiveness index	Information from the policy design profile
Proportionality	Q 6: Competent jurisdictional level for implementation Q 12: Problem scale

Table 2.11 Proportionality: units, levels of performance, and scores

Indicator (qualitative 'unit')	Proportionality (ratio of solution level to problem scale)
No performance (score 0)	No defined solution level
Low performance (score 0.25)	The defined solution level is much smaller/larger than the problem scale (e.g., less than half the size or more than 1.5 times the size)
Medium performance (score 0.5)	The defined solution level is somewhat smaller/larger than the problem scale (e.g., less than quarter the size or more than 1.25 times the size)
High performance (score 0.75)	The defined solution level is only slightly smaller/larger than the problem scale (e.g., less 0.125 times the size or more than 1.125 times the size)
Complete performance (score 1)	The defined solution level corresponds exactly to the problem scale

solution level is somewhat larger or smaller than the problem scale (0.5); high performance exists in cases where solution level is only slightly larger or smaller than the problem scale (0.75). No performance at all signifies that a policy design does not define any solution level at which the policy has to be implemented. To define how 'larger' or 'smaller' play out in the empirical reality is likely to be context-sensitive and needs to be assessed by future research. For now, Table 2.11 provides preliminary guidelines, which should be applied with flexibility until further research approves their validity. Take the example of a policy that aims at securing a country's endangered rain forest from being cut or burned. If a policy design achieved securing half of a country's total rain forest acreages, one could argue that the defined solution level is much smaller than the problem scale (i.e., half the size). However, this evaluation of proportionality might not result in equal accuracy when considering the context of the policy. If we consider securing half of the Amazonian rain forest, this acreage might be large enough to achieve a high level of performance on the indicator *proportionality*. It is not possible to define, in absolute terms, how a smaller or larger solution level in comparison with the problem scale is realized. Rather, analysts should evaluate the ratio between problem scale and solution level relative to the underlying context.

Directness

The fifth indicator of the policy comprehensiveness index is *directness*, which measures the extent to which the entity deciding upon and financing a policy design is also involved in (the supervision of) its implementation. To assess directness, the analyst requires information regarding who *authorized or decided upon a policy design*, as well as *who pays for* and *implements a policy* or *supervises correct implementation* (Table 2.12).

The decision-making entity in Western-style democracies is usually the legislature, the executive, or the bureaucracy. The supervision of implementation is usually the responsibility of state bodies, which oversee lower levels of government, or the private sector tasked with implementation. A hypothetical example of a direct policy would be municipalities levying their own fee in order to finance garbage collection carried out by municipal workers, who are directly supervised by the municipality itself. The policy design would be considered less direct if a private company carried out garbage collection without any supervision by the municipality deciding upon and financing the policy. With dispersed implementation duties, there are more clearance points that provide some leverage to adapt,

Table 2.12 Information from the policy design profile for the indicator 'directness'

Indicator of the policy comprehensiveness index	Information from the policy design profile
Directness	Q 7: Decision-making entity Q 8: Distribution of costs Q 9: Competent agency for implementation/supervision of implementation

Table 2.13 Directness: units, levels of performance, and scores

Indicator (qualitative 'unit')	Directness (number of clearance points through dispersion of implementation duties)
No performance (score 0)	No defined implementation duties
Low performance (score 0.25)	Implementation and funding are passed on to private sector or to other levels of government without control by authority deciding
Medium performance (score 0.5)	Implementation is controlled by authority deciding, and implementation and funding are passed on to private sector or other levels of government
High performance (score 0.75)	Implementation is controlled and funded by authority deciding, and implementation is passed on to private sector or other levels of government
Complete performance (score 1)	Authority deciding is also funding and implementing

change, or soften the policy design. Hence, directness is quantified here via the number of clearance points accumulating through the dispersion of implementation duties (Table 2.13). Accordingly, a policy is thought to perform strongest on that indicator if it is designed such that the authority deciding is also funding and implementing a policy (score 1). A policy is thought to perform weakly on directness when implementation and funding are passed on to the private sector or to other levels of government without any control by the authority deciding (score 0.25). A policy design attains a medium-level performance when implementation is controlled by the deciding authority, but implementation and funding are passed on to the private sector or other levels of government (score 0.5). High performance indicates that implementation is controlled and funded by deciding authority, and implementation is passed on to the private sector or other levels of government (score 0.75). No performance for directness signifies that implementation duties are not clearly defined in the policy design (score 0).

Bindingness

The last indicator of the policy comprehensiveness index carries the label *bindingness*. The information contained in the policy design profile on the *type of policy document* provides important insights into the evaluation of a policy's *bindingness* (Table 2.14).

Laws or ordinances, which have formally been adopted by the legislature, demand more commitment to a policy by the state than informal policy plans or unwritten arrangements. A policy design attains best performance scores for bindingness (score 1) when it is long-term binding through the adoption or amendment of constitutional law (Table 2.15). Low performance scores are

Table 2.14 Information from the policy design profile for the indicator 'bindingness'

Indicator of the policy comprehensiveness index	Information from the policy design profile
Bindingness	Q 10: Type of policy document

Table 2.15 Bindingness: units, levels of performance, and scores

Indicator (qualitative 'unit')	Bindingness (type of policy document)
No performance (score 0)	No existing legal document
Low performance (score 0.25)	Existing document, but non-binding
Medium performance (score 0.5)	Short-term binding (e.g., ordinance and degree)
High performance (score 0.75)	Medium-term binding (e.g., law)
Complete performance (score 1)	Long-term binding (e.g., constitution)

attributed to policies anchored in a non-binding documents (score 0.25). Medium-level performance exists in cases where policies are adopted through ordinances or decrees, which can be altered more easily than laws or constitutions and, therefore, can be considered short-term binding (score 0.25). Policies ratified through laws point to bindingness on a longer time horizon and to high performance (score 0.75). No performance is equivalent to a situation in which there is no legal document codifying the design of a policy (score 0).

In summary, policy analysts may rely on information gathered through the policy design profile (Q 1–12, Table 2.3) in order to evaluate performance levels of the six indicators composing the policy comprehensiveness index. In this study, the assessment of policy designs' performance on each of the six indicators builds on careful estimation. The scores attributed to the indicators of the policy comprehensiveness index were cross-validated and approved by a top-ranked bureaucrat in a personal interview. Future research could alternatively rely on surveys, in which policy actors themselves evaluate policy designs according to the six indicators of the policy comprehensiveness index.

2.2.2.3   Weighing Indicators

The third step in index construction consists of attributing weights to indicators if the relative importance of indicators differs. It is possible that composite indicators are not equally important predictors of what the overall index captures. In such cases, indicators can be assigned different weights. Weights strongly impact the overall index score and therefore must be well justified through empirical verification (Hajkowicz 2006). To do so, test data should be used, which is different from the final data to which the index is applied (Lockwood 2004, p. 508). Weights can be identified, for example, by asking survey respondents to weigh the importance of different indicators.

Since the policy comprehensiveness index constitutes a novel approach, and this study represents its preliminary test, a lack of evidence exists to justify the greater importance of some indicators over others. To minimize the risk of falsely overestimating the importance of certain indicators, this work refrains from making a priori assumptions about the indicators' relative importance. Instead, equal weights are attributed to the six index components. With more data, future research could close the gap in empirical evidence and assess the relative importance of each indicator for the overall index of policy comprehensiveness, using, for example, a factor analysis. It is also possible that 'weighing needs to be based on the character of the policy problem to be solved' (Knill and Lenschow 2003, p. 15).

2.2.2.4 Generating an Indexed Measure of Utility

The fourth step in index construction involves choosing an adequate utility function for the aggregation of the composite indicators. To generate a one-dimensional summary of the underlying multi-dimensional data, the analyst has the choice between an additive, multiplicative, and hybrid utility function. Where indicators interdepend on each other; i.e., a high score for one indicator depends on a high score of another, multiplicative, or hybrid utility functions can be used (Hajkowicz 2006). The consequence is that poor performance on any one indicator makes scoring high in the overall index difficult. Most commonly, an additive utility function is applied, which implies that poor performance for one indicator can be compensated by high performance for others. Von Neumann and Morgenstern's (1944) famous utility theorem, along with their multi-attribute utility theory, constitute the theoretical underpinnings of a simple and commonly used form of an additive utility function, which is written as follows (Hajkowicz 2006):

$$u_j = \sum_{i=1}^{N} w_i v_i(x_i)$$

Here, u_j stands for the utility of policy scenario j. A policy scenario represents a particular set of indicator values describing the empirical reality (Hajkowicz 2006). Different composite indicators of the index are denoted by i, indicator weights by w_i, indicator raw scores by x_i, and a transformation function by $v_i(x_i)$ (explained above). Consequently, one can determine the utility u_j of a policy scenario j from indicator weights (w_i) and a transformation function $v_i(x_i)$ of indicator raw score x_i (Hajkowicz 2006, p. 124). The resulting indexed measure of utility allows for the evaluation of one policy scenario as 'better' or 'worse' over another scenario and by what quantum (but not whether it is 'good' or 'bad' in an absolute meaning). Hence, the indexed measure of utility allows for relative statements of utility, but not for absolute ones (Hajkowicz 2006, p. 124).

The present study adopts the idea of an additive utility function, because poor performance at any one indicator does not necessarily imply zero overall policy comprehensiveness. Low performance at the indicator *pressure on target group* indicates that a policy design poorly compels target groups to adopt a desired behavior by its choice of policy instruments. For example, a country's production industry could be left free to adopt voluntary CO_2 reduction measures in a time-frame of 5 years. In this example, the regulator abstains from adopting policy instruments that define how industry would have to change their production processes, and therefore, the indicator *pressure on target groups* would attain a low score. However, the policy design could compensate for this lack of compellingness by defining very high sanctions (e.g., in the form of a tax on production goods) after the 5-year period, if the producing industry does not achieve any CO_2 reductions. In the awareness of being sanctioned after 5 years in the case of non-action, sanctions are likely to compel the industry to take action for the reduction of CO_2 emissions. Another example would be if a policy design mostly relying on persuasion, and at the same time non-compliance with a desired behavior, is highly discouraged by constant reminders, pleas, or alike. For instance, smoking might not be prohibited, but constant warnings on packets of cigarettes, posters in public spaces, TV commercials, educational school programs, and alike might very effectively compel people to stop smoking. Hence, low performance at the indicator *pressure on target groups* can be compensated by high performance at *sanctions*. Nevertheless, the possibility of interdependencies between indicators should be kept in mind, and if proven by future research, a hybrid form of utility function could be considered. The present study applies the aforementioned utility function to the six indicators ($N = 6$) so that $w_i = 1$, because as explained previously in this chapter, no weights are attributed to the composite indicators so far, and $v_i(x_i) = x_i$, as the units of x_i are already scaled between 0 and 1.

2.2.2.5   Sensitivity Analysis

The last step in index construction comprises some form of sensitivity analysis for identifying the most influential indicators (Hajkowicz 2006). Sensitivity analyses serve to identify those indicators that most influence changes in overall index results. The impact of individual indicators can be assessed by the percentage of change in overall results, caused by a percentage change in an individual indicator, holding all other indicators constant. Indicators causing higher changes in overall results are most crucial for the overall index.

As the present study relies on a non-weighted additive utility function, each indicator should have equal importance in the overall index, and therefore, the sensitivity analysis, in a conventional sense, provides no additional benefit. Alternatively, a sensitivity analysis might serve to assess the degree to which eventual estimation errors of indicator performance levels impact overall index results. This assessment might be relevant because some indicators of the index rely on estimations made by the policy analyst, which might be subject to estimation biases. For example, the aim of the indicator *sanctions* is to evaluate whether existing sanctions are high enough to deter the entire target group causing the underlying problem. Some analysts might evaluate the portion of deterred target groups by sanctions slightly differently than others. Let us take the example of analyst A attributing a score of 0.25 to the indicator sanctions and analyst B a score of 0.5, while all other indicators attained a score of 1. In such a case, analyst A would estimate the comprehensiveness as high with a score of 0.91 and analyst B would also conclude a highly comprehensive policy design with a score of 0.87. The low difference of 0.04 units, or 4%, indicates that small estimation biases of one indicator have a very small influence on overall index results. Even if two analysts attributed a score of 0 versus a score of 1 to any one indicator while all others would be kept constant, overall results would vary by 0.1 units or 10%. The evaluation of overall policy comprehensiveness would most likely result in the same conclusion, e.g., a highly comprehensive policy design in both cases.

2.3   Micropollutants Policies Along the Rhine River

2.3.1   *Micropollutants and Rhine River*

With 200,000 km^2 and 58 million inhabitants, the Rhine catchment area is, together with the Wolga and Danube, one of Europe's three biggest river systems (ICPR 2010e). It unites four riparian states (Switzerland, France, Germany, and the Netherlands) with five basin states (Italy, Austria, Lichtenstein, the Belgian region of Wallonia, and Luxembourg).[12] Intensive economic and agricultural activities, as well as population density, pose great pollution threats to the Rhine waters. In fact, a large-scale inventory of the Rhine River basin mandated by the EU Water Framework Directive in 2004 concluded that the chemical status of the Rhine was not sound in 88% of the water body and that various micropollutants were widespread and exceeded threshold values throughout the whole basin.

The high concentration levels of micropollutants can be attributed, among others, to a high consumption of pharmaceuticals by an increasingly elderly population (Sacher et al. 2008). There are over 3000 different pharmaceuticals on the European market today, and general trends suggest an increased consumption in the future with more drugs being used by more people (Clarke and Smith 2011).

A further source of pharmaceutical micropollutants is the intensive livestock raising in the agricultural lands along the Rhine. When manure is applied to agricultural fields, estrogenic and antibiotic substances drain and contaminate both groundwater and surface water (UBA 2014). Even more significant is pollution coming from plant protection products, which have a particularly long half-life in waters and therefore considerably impact the Rhine's ecosystem (ICPR 2010e). In measurements taken between 1990 and 2006, only 7 out of 36 (19%) plant protection products showed statistically significant downward trends in load, which reveals the difficulties in reducing discharges from agriculture (Bach and Frede 2012).

Another compelling reason to examine the case of the Rhine is its great density of industrial plants along its shores. There are six main industrial centers distributed along the course of the Rhine from Switzerland to the Netherlands. Basel, Strasbourg, the Rhine-Neckar area, the Frankfurt-Rhine-Main, as well as the Rhine-Ruhr region, and Rotterdam-Europort are famous for their industrial production, starting from agro-chemicals, medicinal chemical manufacturing, and

---

[12]ICPR Web site: http://www.iksr.org/index.php?id=32&L=3 (last access 20.7.12).

nano- and biotechnology, to textiles, metals, construction materials, petrochemicals, refineries, and the food industry. For more detailed insights into the extent and magnitude of diverse pollution sources in the Rhine, see Metz and Ingold (2014).

When considering the entire Rhine basin, all potential sources of micropollutants emissions—residents, agriculture, and businesses—challenge water quality in the countries located along the Rhine. In conclusion, the Rhine River basin constitutes a suitable case study area, because micropollutants pose a considerable and similar threat to the aquatic environment as well as humans living on the shores of all riparian countries. As will be explained in the next paragraph, the Rhine provides a unique case study, not only because waters are affected by micropollutants, but also because the issue of micropollutants has already been noted on political agendas of the Rhine riparian countries.

The Rhine River basin distinguishes itself from other trans-boundary basins suffering from micropollutants, because the riparian states have made a transnational effort to achieve good water quality through their involvement in the International Commission for the Protection of the Rhine (ICPR). In 2008, the ICPR installed a project group on micropollutants that had the task of elaborating on a common river basin strategy as well as policy recommendations to the Rhine community for the reduction of micropollutants from diverse sources.[13] The existence of this common river basin strategy is pivotal for case selection as it indicates that the issue of micropollutants has been placed on political agendas of the ICPR member states.

Despite the common river basin strategy, the ICPR does not have the competence to adopt legally binding decisions. The ICPR can provide policy recommendations, but has no decision-making power. It is up to ICPR member states to decide upon and implement precise pollution reduction measures. Members of the ICPR include Switzerland, Germany, France, Luxembourg, the Netherlands, and the European Union (Convention on the Protection of the Rhine, Berne, April 12, 1999).

While the Rhine riparian countries, Switzerland, Germany, France, and the Netherlands, are considered for analysis here, the EU[14] and Luxembourg[15] have been excluded. Despite their ICPR membership, the issue of micropollutants does

---

[13]Mandate for the MIKRO project group of the ICPR, see: http://www.iksr.org/index.php?id=317&L=3 (last access 13.9.15).

[14]This research focuses on country-level policies, because the EU-level regulation is neither encompassing nor satisfactory in addressing Rhine-specific pollution problems. In fact, a comparison between the ICPR's so-called *Rhine 2020 list of substances* and the priority substances under the EU Water Framework Directive (WFD) shows that many Rhine-relevant substances are not subject to EU legislation (see also Sacher et al. 2008; Müller 2011). The WFD requires member states to define river basin-specific substances at the national level to complement the EU legislation (WFD Art. 2 (18), Annex VIII). Hence, the WFD constitutes a baseline that ensures water protection with regard to the most crucial common problems, while detailed provisions must be adopted by the EU member states.

[15]Luxembourg was excluded from analysis, because preliminary interviews with the director of the Luxembourg Administration for Water Management of the Ministry of Interior (November 19, 2011) and with two scientists from the Research Centre Henri Tudor (May 11, 2012) revealed that the problem of aquatic micropollutants was known in Luxembourg, but not yet on the political agenda.

not rank high enough on the political agendas of the EU and Luxembourg to justify analysis in the framework of this research. Nevertheless, an overview about the EU legislative framework on water protection is provided in Sect. 2.3.2.

In the four countries under investigation, micropollutants are treated as an issue predominantly belonging to water protection policy. Hence, the present investigation is limited to the policy field dealing with the protection of surface waters. A consequence of this restriction is that the present work does not provide an encompassing overview of the regulatory framework of micropollutants in all relevant policy fields. Instead, the goal here is to establish an in-depth understanding of the latest revision of water protection legislation concerning the reduction of micropollutants in surface waters for each of the four countries. The term *latest revision* is to be understood in relation to this study's data gathering period, between 2013 and 2014.

The study is further restricted to the analysis of national-level policymaking processes, because the adoption of legally binding water protection legislation is to date a responsibility of national governments rather than of the international level, represented by the ICPR or the EU.

In summary, the present cross-sectional analysis relies on the comparison of the same policy issue across Swiss, German, French, and Dutch cases. In order to rule out competing explanations to the key independent variable, cases are selected that display constant factors regarding the policy issue, as well as the field, timeframe, and governance level.

2.3.2   The EU Policy Framework on Water Protection

Policies on micropollutants should be understood in the context of the EU regulatory framework, because the EU is considered an important policy driver in environmental protection policy, at least for the EU members Germany, France, and the Netherlands (Switzerland is not a EU member). Although there is no EU directive that specifically targets the reduction of aquatic micropollutants, numerous EU documents deal with the issue in different policy fields. The IPCC Directive[16] (2008/1/EC), for instance, regulates production processes; REACH[17] (EC No 1907/2006) restricts hazardous chemical substances from entering the market, and Directive 2004/27/EC restricts pharmaceuticals from entering the market; the EU Directives 98/8/EG on Biocides and 91/414 EEC on the Certification of Pesticides regulate agricultural uses of chemicals; the Urban Waste Water Directive (91/271/EEC) sets technical standards for sewage treatment plants.

---

[16]EU Directive concerning the Integrated Pollution Prevention and Control.

[17]EU Directive concerning the Regulation on Registration, Evaluation, Authorization and Restriction of Chemicals.

With regard to water policy, the latest encompassing EU reform, the adoption of the Water Framework Directive (WFD 2000/60/EC), aims at reaching a good status of surface waters by 2015. One measure in attempting to achieve this goal is the reduction or phasing out of 33 so-called *priority or hazardous substances*, defined as toxic, persistent, and bioaccumulating, listed in Annex X of the Directive (WFD Art. 1c, Art. 16 and Art. 2(29)).[18] To complement the list of priority substances with concrete norms limiting concentration levels, Article 16(7) of the WFD requires that the EU sets environmental quality norms (EQNs) in a 'daughter directive,' i.e., Directive 2008/105/EC, also termed Environmental Quality Standards Directive. Member states are entrusted with the adoption of concrete policy measures for pollution reduction in order to meet the EQNs of the listed substances (WFD Art. 11 and 4).

During the 2012 revision of the priority substance list, twelve new compounds were added to the list (2013/39/EC) after long discussions about the inclusion of pharmaceuticals.[19] As pharmaceuticals were contested by some EU member states, the potential candidate substances, i.e., the anti-inflammatory agent, diclofenac, and two estrogens (17-beta-Östradiol, 17-alpha-Ethinylöstradiol), were finally not included in the priority substance list. Instead, a 'watchlist' of substances was established, which entails the mentioned pharmaceuticals. Member states are required to gather monitoring data for all substances on the watchlist in order to support future prioritization (EQSD 2013/39/EC). The inclusion of substances on the watchlist is a signal to EU member states that these substances may soon be regulated.

While the priority substances are of EU-wide concern, member states are required to select additional substances of national or local concern, so-called *river basin-specific substances*, and to define corresponding EQNs (WFD Annex VIII).[20] Member states achieve the good state of surface waters only if concentrations of both EU priority substances and river basin-specific substances do not exceed defined quality norms (WFD Art. 2 (18)).

All in all, the WFD requires member states to control the imissions of defined chemical substances in waters. If concentrations exceed EQNs, member states are free to choose how to achieve a decline of pollution levels. Hence, concrete micropollutants reduction policies are still mainly adopted on the member states rather than on the EU level. The following chapters therefore illustrate the national-level policies of the countries bordering the Rhine River, including the EU members Germany, France, and the Netherlands, and the non-EU member Switzerland.

---

[18]See 'Summaries of EU legislation' Web site of the European Commission: http://europa.eu/legislation_summaries/environment/water_protection_management/l28180_en.htm (last access May 2012).

[19]See Web site of the European Commission: http://ec.europa.eu/environment/water/water-dangersub/lib_pri_substances.htm#prop_2011_docs (last access May 2012).

[20]http://ec.europa.eu/environment/water/water-dangersub/ (last access 15.10.2014).

2.3.3 The Swiss Micropollutants Policy

2.3.3.1 Switzerland: The Amendment Process of the Waters Protection Act and Ordinance

In Switzerland, several research projects (Projekt Fischnetz[21] on declining fish populations; NFP 50 on endocrine disruptors[22]) proved that the presence of micropollutants in Swiss water bodies has a negative impact on the aquatic environment (Burkhardt-Holm et al. 2002). The results of the research projects revealed that municipal sewage treatment plants were a major source of micropollutants in Switzerland. Consequently, the issue entered the political agenda. The Department for Water of the Federal Office for the Environment (BAFU) mandated follow-up research to investigate potential policy options. In 2007, BAFU launched a project titled *Strategy Micropoll for the Optimization of sewage treatment* (see Table 2.17). Several working groups, composed of BAFU, cantons, operators of treatment plants, drinking water associations, science, and industry, were installed to discuss and develop further policy action. As of 2009, BAFU prepared a draft proposal for a revised Waters Protection Ordinance. The draft proposed a technical standard according to which wastewater treatment plants had to reach an 80% cleaning effect of organic micropollutants (draft WPO Annex 3.1, paragraph 2, number 8). This technical standard is held for sewage plants with a population equivalent (PE) of more than 100,000 or to plants between 10,000 and 100,000 PE if they drain into waters used for drinking water purposes or small streams with low dilution ratio (draft WPO Annex 3.1, paragraph 2, number 8).

Applying these selection criteria, about 100 of the 700 existing sewage plants in Switzerland would have needed to upgrade to a new filter in order to comply with the technical standard. The estimated total cost of about 1.2 Billion Swiss Francs was to be paid by the households connected to those 100 treatment plants that upgraded their technology. Between November 2009 and March 2011, the draft ordinance was submitted for a public hearing. A total of 94 actors were officially asked to provide a statement,[23] and others were invited as well to share their opinions concerning the micropollutants policy draft. The majority of the respondents were in favor of political action toward the reduction of aquatic micropollutants, but criticized the unequal distribution of costs.

While the Swiss Parliament had initiated several inquiries in the past, which required the Federal Government to provide information about the existence and negative effects of micropollutants, only in August 2010 did the Parliament's involvement in the policymaking process gain momentum. Both parliamentary chambers adopted a motion (Motion 10.3635) by March 2011 in which they charged the Federal Government with finding a financing solution that respects the

---

[21]www.fischnetz.ch (last access November 13, 2013).

[22]http://www.nrp50.ch (last access November 13, 2013).

[23]Source: Liste der Adressaten der Anhörung, GSchV; SR 814.201.

polluter pays principle. Since all Swiss households (not just those connected to the 100 selected plants) emit wastewater, BAFU proposed the introduction of a Swiss-wide wastewater charge, capped at a maximum of 9 CHF per year and inhabitant. The revenue raised from the charge was to be used to cover up to 75% of the costs that sewage plant operators pay for the upgrade of their technology. In order to adopt a new charge, not only the Waters Protection Ordinance, but also the Act had to be amended.

At the end of 2011 and beginning of 2012, BAFU worked in consultation with the other federal departments on a proposal to amend the Waters Protection Act. The proposal was discussed in the Swiss Federal Council on April 25, 2012, and was subsequently submitted for a public consultation (April–August 2012). A total of 90 actors were invited to provide a statement, and in the end, 158 actors shared their opinions.[24] The respondents still criticized the financing solution. Some actors put forward that the problem should be addressed at the source, and they supported the idea of a charge on those products that contain harmful substances (BAFU 2012). The cantons argued that some open questions remain with regard to the distribution of the costs, as well as the eligibility criteria for reimbursements. In order to integrate the diverse criticisms, BAFU worked on a revised legal draft of the Waters Protection Act. By the end of 2013, the Federal Council sent the finalized draft of the Waters Protection Act to the Parliament. Both parliamentary chambers adopted the legal revisions by mid-2014, and the changes were estimated to come into force by 2016. In the Waters Protection Ordinance that was under revision in 2014, technical standards, selection criteria for treatment plants, and indicator substances have been redefined.

The Swiss policy network, which will be analyzed in Sect. 3.3, represents an aggregated summary of actors' interactions in the aforementioned policymaking process. At the time of the present study, implementation of the adopted policy lay in the future, and hence, network interactions exclude the implementation phase. Also in the future, lay a shift in policy focus from point-source to diffuse pollution. Until 2014, the Swiss regulation of aquatic micropollutants focused on an end-of-pipe approach to reduce point-source pollution from wastewater treatment plants. BAFU's strategy, however, was to shift its focus from point-source micropollutants toward micropollutants from diffuse sources, such as from agriculture, urban areas, and roads, once the revision of the Waters Protection Act and Ordinance is completed.[25]

---

[24]Source: Liste der Adressaten der Vernehmlassung, GSchG; 814.20.

[25]See Web site of the Swiss Federal Office for the Environment: http://www.bafu.admin.ch/gewaesserschutz/03716/index.html?lang=de (last access May 2012).

2.3.3.2 Indexing the Swiss Waters Protection Act and Ordinance

The revised version of the Swiss Waters Protection Act entered into force in March 2014. The aim of the amendment was to guard sensitive aquatic plants, animals, and microorganisms from harmful human-emitted trace substances, i.e., micropollutants (draft WPO, Annex 2.1, paragraph, 11.1f). The Swiss policy design on micropollutants is evaluated below by means of the policy comprehensiveness index and its six composite indicators.

Pressure on Target Groups

The first composite indicator of the policy comprehensiveness index is *pressure on target groups*. One can evaluate the pressure that a policy design places on target groups by assessing the types of policy instruments composing the policy design, and therefore stimulate a reaction by target groups.

- **Policy instruments**: The Swiss policy design on micropollutants builds on a *technical standard* for wastewater treatment plants, a *charge* and *subsidy* to protect waters from negative impacts of point-source micropollutants emitted by sewage plants. More specifically, the technical standard takes the form of filtering requirements for selected plants treating municipal wastewater. According to the Waters Protection Ordinance (draft WPO, Annex 3.1, paragraph 2.8), selected wastewater treatment plants are required to filter 80% of trace substances from raw sewage, which is measured on the basis of indicator substances (Götz et al. 2010a), which will have to be determined.[26] Treatment plants are required to upgrade their treatment technology within 20 years, starting from 21.3.2014 (WPA Art. 61a, paragraph 2).
  The Waters Protection Act lays down that the investments needed for additional filtering technologies are funded through a charge levied by all Swiss wastewater treatment plants, and respectively, connected inhabitants, through 2040 (WPA Art. 60b). The Federal Government then redistributes the generated funds to cantons, which, in turn, subsidize 75% of the investments of treatment plants for new filters (WPA Art. 61a, paragraph 3).

The Swiss policy design places a high level of pressure on target groups (score of 0.75 in Table 2.16), because wastewater treatment plants are obliged to meet the technical standard as soon as they match the selection criteria (size of sewage plant, sensitivity of waters). There is not much freedom for operators to choose whether or not to comply with the standard once they meet the selection criteria. Subsidies further support target groups in adopting the desired technologies by reducing their investment costs, but the subsidies do not provide more freedom of choice.

---

[26]The following Web site states that a total of 12 indicator substances will have to be determined: https://www.micropoll.ch/faq/ (last access 16.8.15). The exact number of indicator substances to finally enter into the Waters Protection Ordinance remains, however, a matter of political debate.

Table 2.16 Evaluation of the Swiss policy design's pressure on target groups

Indicator	Level of performance	Score
Pressure on target group	High performance: policy instrument mix relying on *state authority* through a technical standard and relying on *economic incentives* through a wastewater charge and subsidies	0.75

Sanctions

For the evaluation of the second indicator of the policy comprehensiveness index, one needs information about sanctions as well as the competent enforcement agency of the Swiss policy design.

- **Sanctions**: Article 71 of the Swiss Waters Protection Act levies a fine of 20,000 CHF on infringements. A deliberate offence against the Act can lead to imprisonment for up to 3 years (WPA Art. 79, paragraph 1).
- **Competent enforcement agency**: The Swiss judiciary is responsible for imposing and pursuing sanctions in cases of infringements.

Existing sanctions are high enough to insure that the majority of Swiss wastewater operators will comply with the formulated technical standard for the reduction of micropollutants (score 0.75 in Table 2.17).

Inclusiveness of Target Groups

The third composite indicator assesses the inclusiveness of the Swiss approach with regard to targeting those groups that emit micropollutants, and relies on the following information:

- **Target groups**: The target groups of the studied policy design are *wastewater treatment plants* and their respective operators, such as municipalities, that have to change their purification procedures. Swiss inhabitants, who finance the technical upgrade, do not qualify as a target group, because they are not required to change their behavior as a consequence of the introduced policy.
- **Criteria defining target groups and exceptions**: The technical standard only applies to the largest sewage plants, measured in population equivalents, or to smaller plants if they drain into small rivers, with a low dilution ratio, or into waters used for drinking water purposes. The exact selection criteria are to be determined in a revised Waters Protection Ordinance as of 2015. Explicitly excluded from the policy are industrial sewage treatment plants, which must neither comply with the technical standard nor pay the 'micropollutants charge.' Moreover, the agricultural sector does not represent an explicit target group of the above analyzed policy design, because agricultural emissions are mostly diffuse. This exemption might change in the future, though, as the Swiss policy strategy was explicitly to focus, first, on point-source pollution from wastewater treatment plants and, second, to expand policy action to diffuse sources.
- **Target groups contributing to the policy problem**: Micropollutants are emitted by point sources of pollution including households, as well as by

Table 2.17 Overview of the Swiss policymaking process

Amendment of the Swiss Waters Protection Ordinance	
Trigger Before 2007	Research (NFP 50, Project *Fischnetz*)
	Administrative decision about putting micropollutants on the agenda
Concept phase 2007–2009	*Strategy Micropoll for the Optimization of sewage treatment* (2007–2012)
	BAFU report (Umweltwissen 17/09) and mandated studies on the reduction of micropollutants from sewage
Elaboration Nov. 2009	Draft of the amended Waters Protection Ordinance
Consultation Nov. 2009–July 2010	Public hearing on the draft ordinance (November 27, 2009–April 30, 2010)
Amendment of the Waters Protection Act	
Trigger August 2010– March 2011	Motion of the Council of State's Committee on the Environment, Spatial Planning and Energy about a funding solution for the elimination of trace substances in wastewater that respects the polluter pays principle, adoption by both parliamentary chambers (10.3635)
Concept phase April 2011–April 2012	BAFU reports (Umweltwissen 1214) and mandated studies on financing technical solutions for the reduction of micropollutants
Elaboration 2011–2012	Draft of the amended Waters Protection Act
Consultation Spring 2012– Spring 2013	Interdepartmental consultation on the draft Waters Protection Act
	Federal Council discusses the draft Waters Protection Act (April 25, 2012)
	Public consultation on the draft Waters Protection Act (April 25, 2012–August 31, 2012)
Finalization May 2013–Spring 2014	Revision of the draft Waters Protection Act
	Federal Council receives revised draft of Waters Protection Act (June 2013)
Parliamentary Phase End 2013–Spring 2014	Federal Council forwards draft and dispatch to the Parliament (BBl 2013 5549, June 26, 2013)
	Discussion in the Council of States (October 10, 2013) Discussion in the National Council (March 3, 2014)
	Adoption by both parliamentary chambers (March 21, 2014)
Implementation	Coming into force of amended Waters Protection Act March 31, 2014
	Implementation from 2016 (estimated) until December 31, 2040

agricultural and industrial producers connected to municipal and industrial wastewater treatment plants (for a more detailed description of emission sources, see Sect. 1.1). Moreover, micropollutants are emitted by diffuse sources of pollution including agriculture, settlements, or transportation.

A medium level of performance (score 0.5 in Table 2.18) is attributed to the indicator *inclusiveness* here, because scientific results suggest that the 50% load reduction of micropollutants can be achieved in surface waters through improved

Table 2.18 Evaluation of the Swiss policy design's inclusiveness

Indicator	Level of performance	Score
Inclusiveness	Medium performance: *Half* of the emission of micropollutants is targeted by the policy	0.5

wastewater filters (Abegglen and Siegrist 2012, p. 52). By targeting point-source pollution, emissions from all polluters connected to wastewater treatment plants, including households, industry, settlements, and agriculture, are reduced, which leads to a 50% load reduction. The other 50% of loads, which are not reduced by the Swiss end-of-pipe approach, stem from industrial wastewaters not connected to municipal treatment (so-called direct dischargers), and from diffuse sources of pollution.

In summary, a medium level of inclusiveness characterizes the Swiss policy (score of 0.5) because addressing point-source pollution leads to a 50% load reduction and includes household, urban, agricultural, and industrial sources of emissions, but excludes all diffuse emissions.

Proportionality: Solution Versus Problem Scale

The fourth indicator of the policy comprehensiveness index evaluates whether a policy design establishes a match between problem scale and its solution level and builds on the following information:

- **Competent jurisdictional level for implementation**: The Swiss Waters Protection Act is a federal law, which is applicable on the national jurisdictional level. Wastewater treatment plants that match the selection criteria must upgrade their filtering technology no matter where on the Swiss territory they are located.
- **Problem scale**: Wastewater treatment plants are scattered throughout Switzerland, and wherever they exist, they continuously emit micropollutants into waters (Gälli et al. 2009). Hence, emissions from sewage treatment plans cause problems on the national scale.

The Swiss policy applies to the national level, and likewise, emissions from micropollutants by wastewater treatment plants are of national-level concern. When restricting the focus to micropollutants emitted by wastewater treatment plants, the problem scale and solution level of the Swiss policy design correspond to one another (score of 1 in Table 2.19).

Table 2.19 Evaluation of the Swiss policy design's proportionality

Indicator	Level of performance	Score
Proportionality	Complete performance: *National solution* level corresponds to the *national problem scale*	1

Directness: Dispersion of Decision-Making, Financing, and Implementation

The fifth indicator, directness, concerns the dispersion of decision-making, financing, and implementation tasks. The less dispersed those tasks are, the less frequent are clearance points as well as a displacement goals between the original intentions of a policy design and its actual implementation. Conversely, a high score for that indicator signifies that one entity is involved in all three tasks, which should reduce the number of clearance points and, hence, increase the likelihood that a policy is implemented as intended when the policy was adopted. To evaluate directness in the Swiss case, one needs the following information about the entities of deciding, financing, and implementing:

- **Decision-making entities**: The legislature charged the Federal Government to propose a policy for the reduction of micropollutants in surface waters. The executive, i.e., the Swiss Federal Council, then accepted the draft proposal for an amended Waters Protection Act, prepared by the Federal Office for the Environment, and forwarded it to the legislature. Finally, both chambers of the legislature adopted the amendments for the reduction of micropollutants.

    While the executive and legislature decided on the Act, the bureaucracy, namely the Federal Office for the Environment, prepared and adopted the Waters Protection Ordinance, which entails more detailed and technical provisions.

- **Distribution of costs**: For the technical upgrade of wastewater treatment plants, an estimated sum of 1.2 Billion Swiss Francs must be invested in additional filtering technologies. The Waters Protection Act states down that the Federal Government levies a charge from all Swiss wastewater treatment plants until 2040 (WPA Art. 60b). The charge is passed on to the inhabitants connected to treatment plants. The severity of the charge is calculated based on the size in population equivalents of sewage plants and is legally capped to a maximum of 9 CHF per inhabitant per year. As soon as sewage plants adopt a filtering technology for the elimination of micropollutants, they (and their connected inhabitants) are exempted from the charge (WPA Art. 60 b, paragraph 2). The exemption also applies to those sewage plants that voluntarily invested in micropollutants filtering technology after January 1, 2012. In a second step, the Federal Government redistributes the generated funds to cantons, which, in turn, are able to compensate wastewater treatment operators for 75% of the investments for new filters (WPA Art. 61a, paragraph 3). To promote mergers, 75% of the costs for the construction of canalization, linking smaller treatment plants to larger and more efficient ones, may also be compensated.

    To summarize, Swiss inhabitants and selected sewage treatment plants bear the costs for investments. The Federal Government levies a charge from inhabitants and redistributes funds to cantons. Cantons then subsidize sewage treatment operators for 75% of their investment costs.

- **Competent agency for (supervision of) implementation**: According to Article 45 of the Swiss Waters Protection Act, the cantonal level is responsible for the implementation of sewage treatment plants' technical upgrade. Article 46 states that the Federal Government coordinates and maintains control of implementation. In the case of micropollutants, the Federal Government automatically controls implementation by levying the wastewater charge and redistributing it to the cantons (WPA Art. 60b, 61a).

From the aforementioned insights, one can conclude that the Swiss Federal Government is involved in all three tasks: deciding, financing, and implementing. The accumulation of tasks reduces the number of clearance points and the risk of goal displacement. As one authority is involved in all three tasks, a high score for directness is attributed to the Swiss policy design (score of 0.75 in Table 2.20) as one authority is involved in all three tasks (deciding, funding, and control of implementation).

Bindingness

To assess the degree to which a policy design is compelling, the policy comprehensiveness index builds on the indicator *bindingness*. In order to assess the level of bindingness of a policy design, the type of policy document is used as a proxy here.

- **Type of policy document**: The Swiss policy on micropollutants is a legally binding obligation laid down in the Waters Protection Act and Ordinance. Financial aspects (charge and subsidy) are dealt with in the Waters Protection Act, while technical requirements for wastewater treatment plants are fixed in the Waters Protection Ordinance.

Through its adoption on the level of a law, the Swiss policy on micropollutants signals a high level of commitment on a longer time horizon, and thus stability. Additionally, the amendment of the Swiss Waters Protection Act led to the involvement of the Swiss legislature in the decision-making process, which further indicates a high level of authority. In conclusion, the Swiss policy design signals both stability and authority and therefore attains a high level of bindingness (score 0.75 in Table 2.21).

Table 2.20 Evaluation of the Swiss policy design's directness

Indicator	Level of performance	Score
Directness	High performance: Implementation is *controlled and subsidized* by the *Federal Government* *Implementation* is passed on to the *Cantons*	0.75

Table 2.21 Evaluation of the Swiss policy design's bindingness

Indicator	Level of performance	Score
Bindingness	High performance: *medium-term* binding	0.75

Summary of Index Results for the Swiss Policy Design on Micropollutants

The index results summarized in Table 2.22 reveal that the Swiss policy design has the ability to comprehensively reduce the problem of micropollutants in surface waters (score of 0.75). The Swiss policy is designed as a compelling regulation for target groups (pressure 0.75, sanctions 0.75), implementers (directness 0.75), and the state (bindingness 0.75), which considerably increases its chances of being implemented in reality. Moreover, the Swiss policy design addresses aquatic micropollutants in an effective and efficient way by targeting a sufficient proportion of potential emitters (inclusiveness 0.5) and by having adopted a national-level solution for a national scale problem (proportionality 1.0). This design demonstrates that the Swiss end-of-pipe approach can be an effective way of designing a public policy, which has the potential to comprehensively reduce micropollutants in waters.

2.3.4 The German Micropollutants Policy

2.3.4.1 Germany: The Adoption Process of the Surface Water Ordinance

The German Surface Water Ordinance was mainly triggered by the need to respond to European legislation (see Table 2.23). Traditionally, the adoption of water legislation was controlled by the German Länder (German constituent states), and hence, every Land had its own water legislation.[27] In 2000, when the EU WFD (2000/60/EC) came into force, the transposition into Länder-level legislation lagged behind in several German Länder. Hence, the EU Commission sued Germany for insufficient implementation of the WFD in 2005. In response, Germany underwent a large federal reform in 2006 during which the competence to adopt water legislation was passed on to the national level.[28] In order to then transpose the WFD, and particularly its daughter directive, the Environmental Quality Standards Directive (EQSD 2008/105/EC), into uniform national regulation, a new federal ordinance was drafted as of 2008. Because of the many technical details required by the EU framework, the new regulation was formalized on the level of an ordinance and, thus, it complements the German Water Management Law (WHG). The so-called Surface Water Ordinance (OGewV) is relevant with regard to micropollutants, because the OGewV determines environmental quality norms for a list of

---

[27]The federal state had the right to enact framework directives (Rahmengesetzgebungskompetenz des Bundes), i.e., the Water Resource Act (WHG). Bound by this framework, the Länder adopted their own state-level water legislation.

[28]Water legislation today follows the 'competition principle' (konkurrierende Gesetzgebungskompetenz des Bundes) according to which the Länder adopt their own water laws until the federal state enacts a nationwide and uniform legislation overruling the state-level.

Table 2.22 Performance of the Swiss policy design after the policy comprehensiveness index

Indicators of the policy comprehensiveness index	Scores
Pressure on target groups	0.75
Sanctions	0.75
Inclusiveness	0.5
Proportionality	1.0
Directness	0.75
Bindingness	0.75
Overall index score	0.75 (high)

Table 2.23 Overview of the German policymaking process

Adoption of the German Surface Water Ordinance (OGewV)	
Trigger Before 2008	Implementation of the WFD 2000/60/EC and EQSD 2008/105/EC in Germany Reform of the federal system 2006
Concept phase 2008–2010	Working group on environmental quality norms (BLAK-UQN) (15./September 16, 2008)
	General meeting of the Working Group on Water (22./September 23, 2009; March 25./26, 2010)
	74th Conference of Environmental Ministers (July 11, 2010)
Elaboration 2010	Draft proposal prepared by the Environmental Ministry with support from the Environmental Protection Agency (March 29, 2010)
Consultation Nov. 2010	Interdepartmental consultation
	Public hearing (November 2010)
Finalization Spring 2011	Revised draft prepared for the Cabinet of Ministers
	The Cabinet adopts the Surface Water Ordinance (March 16, 2011) and forwards it to the Federal Council
Adoption Summer 2011	Debates in the committees of the Federal Council (March 17, 2011)
	The Federal Council's committees propose amendments (May 17, 2011, 153/11)
	Adoption of the Ordinance conditional upon 25 amendments by the Federal Council in its 883th session (May 27, 2011)
	Approval of the requested amendments by the Cabinet of Ministers (June 22, 2011)
Implementation As of Aug. 2011	Coming into force of the Surface Water Ordinance (July 27, 2011)

river basin-specific (micro)pollutants. As such, concentration limits are defined for a number of substances. If a substance exceeds its limit, further political measures have to be taken to reduce immissons.

In order to identify river basin-specific (micro)pollutants, Länder were asked to report those substances, which are particularly relevant to each Land's waters. In order to carry out these reports, the Länder established a working group on environmental quality norms (EQN) (*Bund-Länder-Arbeitskreis UQN*). In several meetings throughout 2009 and 2010 of the Working Group on Water (LAWA) and the Conference of Environmental Ministers of the Länder, the Länder discussed and decided upon a common position. The Länder had very divergent ideas about the strictness of EQN and the amount of substances to be regulated. Their common position, which posits a 1:1 implementation of European directives, can therefore be interpreted as the smallest common denominator (74. UMK, TOP 34). The reason for this goes back to a decision made in October 2006 during the 67th Environmental Ministers Conference of the Länder, according to which they opposed a general legally binding introduction of a fourth purification stage to eliminate micropollutants from sewage water.[29] For the same reason, the Länder opposed the listing of substances beyond EU legislation. EQNs for new types of pollutants could force the Länder to improve wastewater treatment technology.

To insure continuity, the Federal Environmental Ministry drafted the OGewV together with the Environmental Protection Agency on the basis of the water laws of the Länder (*LAWA Musterverordnung*, 2 July 2003). After a two-year long concept phase, the draft of the amended Surface Water Ordinance proposed EQNs for 13 new substances—new compared to the previously existing Länder-level legislation—because they were found in comparably higher concentration levels in waters (draft OGewV Annex 5). The Environmental Ministry submitted the proposal to an interdepartmental—as well as public—consultation in 2010. For the consultation, the Ministry invited 32 actors to provide a statement.[30] There is no information available about the number of statements actually given. After making some adjustments, the Environmental Ministry submitted the draft to the Federal Cabinet of Ministers. The Federal Government adopted the Surface Water Ordinance on March 16, 2011, and sent the draft to the Federal Council, the parliamentary representation of the Länder on the national level. Debates within the Federal Council's committees revealed that the Länder opposed the listing of some new substances, such as pharmaceuticals, because the costs of compliance were considered too high of a burden for households or other water users (153/11). Finally, the Federal Council adopted the OGewV, but requested a total of 25 changes on May 27, 2011. After the Federal Government adopted the changes, the amended Surface Water Ordinance came into force in July 2011. The final Ordinance excluded five out of the 13 newly proposed substances (carbamazepine, fenpropimorph, triphenyl phosphate, sulfamethoxasole, uranium) by referring to the ongoing revision of the EQSD on the European level. The German 2011 substance list is regularly updated according to the newest scientific insights, as well as to the revisions of EU legislation.

The German policymaking process on micropollutants was based on the claim that a broad societal debate would be necessary to define tolerable pollution levels in

---

[29]See Web site of the Environmental Minister Conference: http://www.umweltministerkonferenz. de/Dokumente-UMK-Dokumente.html (last access May 2012).

[30]Source: Schreiben an die Verbände from August 16, 2010.

waters prior to the need for political action. Moreover, the German Länder put forward that before measures on micropollutants would be taken by sewage plants, relevant substances, costs, and energy intensiveness had to be well understood. In order to comply with these requirements, a great amount of research on micropollutants was carried out by the German Federal Environmental Agency, the Environmental Agencies of the Länder, various universities, and water associations.[31]

The policy network emerging from the aggregated interactions among policy actors in the aforementioned policymaking process will be subsequently analyzed in Sect. 3.3. As with the case of Switzerland, the implementation phase is excluded from analysis as it lay in the future at the time of the present research project.

2.3.4.2 Indexing the German Surface Water Ordinance

In 2011, Germany adopted its new federal water ordinance. The aim of the German Surface Water Ordinance is the protection of surface waters in general (OGewV, Art. 1), which, among others, includes the protection of waters from micropollutants.

Pressure on Target Groups and Policy Instruments

The first indicator of the index captures the degree of pressure exerted by a policy design on target groups. In this regard, the types of adopted policy instruments provide a helpful indication for the level of pressure of a policy design.

- **Policy instruments**: The German Surface Water Ordinance sets *environmental quality norms* for a total of 162 river basin-specific substances against which the ecological status of water bodies must be evaluated (OGewV, Annex 5). The defined EQNs establish mandatory limits on acceptable concentrations of (groups of) substances in waters. EQNs being exceeded represent a signal to the federal government and Länder (constituent states) governments that further policy measures are needed (Metz and Ingold 2014).

  Although pollution reduction measures must be adopted by the German Länder if EQNs are violated, the Ordinance does not specify which measures are to be taken in such cases. In general, Länder deal with transcendences of emission limits on a case-by-case basis. If a responsible polluter can be identified, compliance with the discharge authorization and the best-available technology is verified. With no detected infringements, the responsible administration attempts to consult the respective polluter about voluntary retention measures (interviews with State of Rhineland-Palatinate (RLP), March 13, 2014; State of Baden-Württemberg (BAWÜ), April 4, 2012; State of Hesse (HES), March 13, 2014).

---

[31]Research funded by the Federal Ministry of Education and Research: http://www.riskwa.de/de/94.php; research funded by Federal Ministry of the Environment: http://www.bmu.de/forschung/ufoplan/doc/40881.php; research funded by NRW: http://spurenstoffe.net/index.php/de/projekte (last access May 2012).

Taking a closer look at the Länder located along the Rhine, one finds hetero-geneous approaches to the reduction of aquatic micropollutants. In North Rhine-Westphalia and Baden-Württemberg for instance, some operators of wastewater treatment plants *upgraded filtering technology on a voluntary* basis in order to eliminate micropollutants.[32] North Rhine-Westphalia launched a governmental *subsidy* program (*Ressourceneffiziente Abwasserreinigung*) in February 2012, which incentivizes sewage operators to invest in further filtering technology (interview with State of North Rhine-Westphalia (NRW), March 27, 2012). In addition, projects exist to introduce legally binding, stricter *technical standards* for wastewater treatment plants, which would require investments in new technology for the elimination of micropollutants. To date, upgrading treatment technology remains a voluntary action on behalf of operators. North Rhine-Westphalia, Baden-Württemberg, and Rhineland-Palatinate established a competence center (or *expert network*) on anthropogenic micropollutants in order to exchange experiences with the elimination of micropollutants in wastewater treatment plants (interviews with State of North Rhine-Westphalia (NRW), March 17, 2014; Württemberg (BAWÜ), April 4, 2012; Rhineland-Palatinate (RLP), March 13, 2014). Politically, however, Rhineland-Palatinate, Bavaria, and Hesse rejected the end-of-pipe solution to date. These Länder put forward that emissions from households were not the primary source of pollu-tion, but rather from agriculture or industry (interviews with State of Hesse (HES), March 13, 2014; State of Bavaria (BAY), April 26, 2012). Rhineland-Palatinate therefore pushed for a *restriction of pesticides and vet-erinary medicinal products* on the EU-level and postulated an EU-wide strategy to minimize pollution from antibiotics used in livestock farming (interview with Rhineland-Palatinate (RLP), March 13, 2014).

In summary, there is no common micropollutants policy on the Länder level, which is not surprising given the context of German federalism.

Although the policy design builds on state authority, the adoption of EQN represents only a 'soft' pressure on target groups to change their behavior. On the one hand, German Länder are under some pressure because they are required to monitor a list of substances and comply with specified EQN of the Surface Water Ordinance. On the other hand, they have ample freedom in defining concrete measures for the reduction of micropollutants that exceed defined concentration limits. Considering the type of policy instrument adopted, the German policy design places a medium degree of pressure on target groups to take action regarding micropollutants reduction measures (score of 0.5 in Table 2.24).

A general drawback of the EQN approach is its difficulty in compiling a com-prehensive list of all critical substances, which are present in undesirably high concentrations levels in waters. Pharmaceuticals, for example, were not yet listed in the 2011 version of the German Surface Water Ordinance, which also reduced

---

[32]See Web site of the Ministry of the Environment of NRW: http://www.umwelt.nrw.de/umwelt/wasser/abwasser/mikroschadstoffe/index.php (last access May 2012).

Table 2.24 Evaluation of the German policy design's pressure on target groups

Indicator	Level of performance	Score
Pressure on target group	Medium performance: policy instrument relying on '*soft*' *state* *authority* through environmental quality norms	0.5

pressure on the respective target groups to act. Wastewater treatment plants, for example, would be under much more pressure to act and invest in 'pharma filters' if pharmaceuticals had then been listed.

Sanctions and Competent Enforcement Agency

The Surface Water Ordinance does not specify how exceeding the defined EQN would be sanctioned. Hence, the level of performance on that indicator equals zero here (score of 0 in Table 2.25).

Inclusiveness of Target Groups

When contrasting all the groups contributing to the emissions of micropollutants to the target groups of the German policy design as below, the German approach cannot be considered particularly inclusive in addressing the entire sources of pollution (score of 0.25 in Table 2.26).

- **Target groups**: The primary target groups of the policy design comprise of all *German Länder*, which are required to comply with the EQNs. Although Länder represent the direct target group of the policy, they do not emit micropollutants. The 'real' emitters of micropollutants, however, are only indirectly targeted by the policy because they do not have to change their behavior and processes at first. Emitters only need to change their behaviors if concentration limits are exceeded. The *indirect target groups* of the policy design are defined by the substances listed in the Surface Water Ordinance and include point-source and diffuse emissions from agriculture and industry.
- **Criteria defining target groups and exceptions**: The EQNs apply to German surface waters in general, and therefore, all German Länder are required to monitor the listed substances. However, the costs and efforts of monitoring these substances might diverge for the different monitoring stations, because from the

Table 2.25 Evaluation of the German policy design's sanctions

Indicator	Level of performance	Score
Sanctions	No performance: *no* existing sanctions	0

Table 2.26 Evaluation of the German policy design's inclusiveness

Indicator	Level of performance	Score
Inclusiveness	Low performance: *Small parts* of the sources of emissions are targeted by the policy	0.25

162 listed substances, only those that are expected to exceed half of the formulated EQN in annual mean must be monitored (OGewV, Annex 5(2)). These specifications may lead to a situation where certain monitoring stations have to monitor fewer substances than others.

Implicit to the Surface Water Ordinance is another criterion defining exceptions of target groups. The list of 162 substances in the Surface Water Ordinance represents a selection of undesired substances in waters. Hence, all substance groups that are not listed also exclude the respective emitters as (indirect) target groups of the policy design.

- **Target groups contributing to the policy problem**: Micropollutants are emitted by point sources of pollution, namely municipal and industrial wastewater treatment plants, and all inhabitants, as well as agricultural, and industrial producers connected to the plants. Moreover, diffuse sources of pollution emit micropollutants including agriculture, settlements, or transportation.

The German Länder are responsible for complying with defined EQN, but the real emitters of micropollutants, i.e., the society or economy, are not directly required to adapt behavior or procedures as a consequence of the introduced policy. If Länder detect exceedances in waters, it remains a challenge to detect the responsible emitter. A further challenge then is to negotiate changes in procedures or behavior on behalf of the identified emitter. I therefore estimate that only a small fraction of the groups or individuals emitting micropollutants are actually targeted by the policy.

Proportionality: Solution Versus Problem Scale

When contrasting the jurisdictional solution level to the scale on which the listed micropollutants occur, one can evaluate the level of proportionality of the German policy design.

- **Competent jurisdictional level for implementation**: The German Surface Water Ordinance is a federal regulation that applies to Germany on the national jurisdictional level.
- **Problem scale**: The problem scale of those micropollutants that are listed in the Ordinance depends on the exact substance of interest. Although the substance list of the Surface Water Ordinance reflects only those selected substances that exhibited critical concentration levels on a larger scale in the past, the problem scale still varies depending on the exact substance (the selection of substances was explained in the previous Sect. 2.3.4.1). A precise chemical substance might cause problems on the local, regional, national, or even international scale.

In conclusion, a national solution level contrasts a mixed problem scale. Certain substances of the national list may be detected on the local or regional scale, but not on the national scale. For substances that are of regional or local concern, the solution level is larger than the problem scale, because in principle all listed substances have to be monitored. The German policy design attempts to circumvent such situations of disproportionality by requiring monitoring only when the concentrations are expected to exceed half of the formulated EQN in annual mean

Table 2.27 Evaluation of the German policy design's proportionality

Indicator	Level of performance	Score
Proportionality	High performance: The defined solution level can be larger than the problem scale for certain listed substances	0.75

(OGewV, Annex 5(2)). As such, the policy design enables a flexible adaptation of the regulation to local circumstances. Moreover, situations where the solution scale is larger than the problem scale are minimized. To summarize, the German policy design establishes a high level of proportionality between problem scale and its solution level (score of 0.75 in Table 2.27).

Directness: Dispersion of Decision-Making, Financing, and Implementation

One may evaluate the directness of the German policy design when analyzing the distribution of decision-making, financing, and implementation tasks.

- **Decision-making entities**: The draft proposal of the Surface Water Ordinance was prepared by the Federal Ministry for the Environment and then adopted by the Federal Government and the Council of Constituent States. The latter is the parliamentary representation of the German Länder on the federal level.
- **Distribution of costs**: German Länder are required to monitor the list of substances defined in the Ordinance and pay for the related costs. The Ordinance does not specify who bears the costs for pollution reduction measures in cases where concentration limits are exceeded.
- **Competent agency for (supervision of) implementation**: The German Länder are required to comply with defined EQN and are responsible for the implementation of micropollutants reduction policies if concentrations exceed the defined limits. Moreover, Länder are required to regularly report monitoring results to the Federal Ministry for the Environment that supervises the correct implementation of the Surface Water Ordinance.

In the German case, goal displacement is unlikely, because the Federal Ministry for the Environment controls the correct implementation of EQNs. Moreover, the policy design is highly direct, because the German Länder are involved in deciding, financing, and implementing the policy design. This way, the number of clearance points is reduced to a minimum. Since one authority is involved in all three tasks (deciding, funding, and implementation), the German policy design is evaluated as highly direct here (score of 0.75 in Table 2.28).

Table 2.28 Evaluation of the German policy design's directness

Indicator	Level of performance	Score
Directness	High performance: *Implementation* and *financing* are executed by the *German Länder*, and implementation is *controlled* by *Federal* Ministry for the Environment	0.75

Table 2.29 Evaluation of the German policy design's bindingness

Indicator	Level of performance	Score
Bindingness	Medium to high performance: *short-term to medium-term* binding	0.625

Bindingness and Type of Policy Document

The type of policy document adopted can be considered a proxy for the level of bindingness of the German policy design.

- **Type of policy document**: The German policy on micropollutants is a legally binding obligation laid down in the Surface Water Ordinance. Since ordinances can be amended much faster than constitutions or laws, they signal a lower degree of bindingness. In fact, the German substance list is regularly revised (approximately in intervals of four years) based on new scientific insights and amended EU priority substances.

On the one hand, the ordinance is binding because it was passed by the legislature and constitutes applicable law. On the other hand, ordinances can be amended more easily than constitutions or laws. Taking both into account, a level of performance between medium and high was attributed to the German policy design on the indicator *bindingness* here (score of 0.625 in Table 2.29).

Summary of Index Results for the German Policy Design on Micropollutants

Table 2.30 shows that the German policy design achieves a medium level of comprehensiveness in addressing the problem of micropollutants in waters (score of 0.48). The design's main weakness is its lack of inclusiveness in addressing liable target groups (inclusiveness 0.25). Rather than reducing emissions of micropollutants, the policy design only controls concentration levels in waters. Moreover, the policy design does not highly compel target groups (pressure 0.5, sanctions 0) to immediately take measures regarding the reduction of micropollutants. Nevertheless, the German policy is designed as a compelling regulation for implementers (directness 0.75), i.e., the Länder, and for the state (bindingness 0.625), which considerably increases the chances that the defined EQNs are respected. The policy design establishes vertical effectiveness and efficiency by enabling a flexible adaptation of monitoring requirements to local circumstances (proportionality 0.75).

Table 2.30 Performance of the German policy design after the policy comprehensiveness index

Indicators of the policy comprehensiveness index	Scores
Pressure on target groups	0.5
Sanctions	0
Inclusiveness	0.25
Proportionality	0.75
Directness	0.75
Bindingness	0.625
Overall index score	0.48 (medium)

2.3.5   The French Micropollutants Policy

2.3.5.1   France: The Adoption Process of the 'Plan Micropolluants'

In contrast to Germany, France holds a policy document that specifically targets micropollutants. The *Micropollutants Plan 2010–2013* explains the global national strategy toward aquatic micropollutants.[33] The Plan builds on previous actions of the French Government toward pollution of waters (see Table 2.31). Most importantly, the Micropollutants Plan complements and updates the National Action Plan against Pollution of Aquatic Environments from Dangerous Substances (PNAR, Décret n°2005-378, 20.4.2005; Arrêté 30.6.2005; Arrêtée March 21, 2007) from 2005, which transposes the EU Dangerous Substance Directive into national law (74/464/EEC). Crucial in this plan is the initiation of the National Research Action on Dangerous Substances in Water (RSDE) (Circulaire February 04, 2002; Circulaire January 05, 2009), according to which wastewater treatment and industrial plants are required to monitor their effluents and report the results to the French Government. The acquired knowledge of sources, entry paths, and concentrations of pollutants in waters revealed that micropollutants contaminate aquatic environments in France and that further action is needed. Hence, the Ministry of Ecology invited a large number of concerned actors, i.e., the national ministries of both health and agriculture, Water Agencies, and water providers, as well as scientific, societal, and economic actors, to meet and elaborate on measures for the reduction of micropollutants (meetings on July 9, 2009 and January 1, 2010).

Additionally, the National Agency for Water and Aquatic Environments (ONEMA) organized an event titled *Aquatic Micropollutants Days* (March 10, 2010–March 12, 2010) to gather expertise on the presence of micropollutants in waters from different scientific angles. Almost 150 experts from Water Agencies, research, laboratories, engineering, ministries, and governmental agencies presented their latest results and discussed potential policy measures to reduce micropollutants in waters. Based on these debates, the Ministry of Ecology drafted the Micropollutants Plan in March 2010 and submitted the draft for a public hearing in April. Annex 4 of the Micropollutants Plan indicates that 24 actors were consulted. The Department of Water and Biodiversity (DEB) of the French Ministry of Ecology reported in an interview, held in the framework of this research project on October 18, 2013, that only a few actors provided a statement. After revisions were made to the document between May and June 2010, it was sent to the National Water Committee (CNE). The latter is the French *water parliament*, which votes on all water-related policy documents of the French Government. On July 6, 2010, the CNE adopted the Micropollutants Plan, and on October 13, during the meeting of the Council of Ministers, the Government gave its consent. One year after the adoption of the Micropollutants Plan, a follow-up report was published on the

---

[33]See Web site of the French National Environmental Ministry: http://www.developpement-durable.gouv.fr/Les-micropolluants-dans-les.html (last access May 2012).

Table 2.31 Overview of the French policymaking process

Adoption of the French micropollutants plan	
Trigger Before 2009	Dangerous Substance Directive 76/464/EEC
	PNAR National Action Plan against Pollution of Aquatic Environments from Dangerous Substances
	RSDE National Research Action on Dangerous Substances in Water
Concept phase 2009–2010	Meetings to elaborate on the Micropollutants Plan (July 9, 2009 and January 1, 2010)
	Aquatic Micropollutants Days organized by ONEMA (March 10, 2010–March 12, 2010)
Elaboration March 2010	Draft proposal prepared by the Ministry of Ecology
Consultation April 2010	Public hearing (January 1, 2010–April 30, 2010)
Finalization May–June 2010	Revisions of the draft proposal
Adoption July–Nov. 2010	Approval by the National Water Committee (CNE) (July 6, 2010)
	Approval by the Council of Ministers (October 13, 2010)
Implementation Since 2011	Follow-up report on *Micropollutants in aquatic environments. Timeframe 2007–2009* (published on October 17, 2011)
	Conference on micropollutants *Micropollutants Plan 2010–2013: state of progress achieved in one year* (October 18, 2011)

presence of micropollutants in French waters (October 17, 2011) and a conference was organized to discuss the progress made in one year (October 18, 2011).

Overall, a comparably short policymaking process contrasts a rather broad and encompassing policy strategy toward micropollutants. Despite its non-bindingness, the Micropollutants Plan is a promising first step in preparing a broader toolbox of legally binding policies as of 2014. Section 3.3 presents an aggregated summary of actors' interactions during the aforementioned policymaking process in the form of policy networks. The implementation phase that took place after 2014 is not covered by the policy network, since it lay in the future of the present study.

2.3.5.2 Indexing the French 'Plan Micropollutants'

In France, a global national strategy toward aquatic micropollutants was adopted in the form of the *Micropollutants Plan* in 2010.[34] The Micropollutants Plan provides an overview of the already-existing legal instruments on both the EU- and national-level targeting micropollutants; it points to remaining regulatory as well as

---

[34]See Web site of the French National Environmental Ministry: http://www.developpement-durable.gouv.fr/Les-micropolluants-dans-les.html (last access May 2012).

knowledge gaps and establishes the future policy and research agenda. In the context of French water governance, such a strategic document provides guidance to the Water Agencies, which are otherwise free to adopt their own water protection policies. The Plan's purpose, therefore, is not only to propose measures for the reduction of micropollutants, but also most importantly to prioritize actions for Water Agencies.

Pressure on Target Groups and Policy Instruments

The Micropollutants Plan is structured into 4 axes, which reflect the priorities of the French Government: (1) reducing emissions at the source; (2) improving the knowledge of water quality and making data accessible; (3) improving scientific (technological) knowledge for monitoring waters; and (4) evaluating and communicating progress in pollution reduction. A total of 22 specific policy measures explain how each of these 4 axes is achieved in practice and provide an indication of the degree to which target groups feel pressured to take action. Hereafter, I outline a selection of the 22 proposed measures in order to provide a general understanding of the French policy design.

- **Policy instruments**: A first example of policy measures included in the Micropollutants Plan is action numbers 2 and 7 which deal with the introduction of national legally binding *EQNs* to reduce emissions from particularly relevant substances. Action number 3 concerns *banning* the most dangerous substances from being marketed on EU and national levels. The French Government mainly focuses on the ban of plant protection products and biocides, due to results from a large roundtable conference on the environment in 2009 (*Grenelle de l'environnement engagement n° 129*). The political aim is to prohibit a total of 40 pesticides from the market for which substitutes exist by 2010. For compounds lacking substitutes, the goal is to reduce their use by 50% over the next 10 years (compared to 2010). In order to achieve this goal, alternative agricultural practices are financially incentivized (*Plan ECOPHYTO 2008–2018*). *Subsidies* originate from the so-called *diffuse pollution charge* that farmers are required to pay on pesticides. The Micropollutants Plan also postulates the need to improve this financial incentives scheme for a more effective reduction of pesticides use (action number 4 and 11).

  In handling point-source pollutants, the Micropollutants Plan announces more detailed *monitoring* (action number 5) in order to better understand emissions into waters (action number 13). The French legislator already set a list of defined micropollutants that have to be monitored by specific industries and wastewater treatment plants (circulaire DGPR January 5, 2009 and circulaire DEB September 29, 2010). Only those industries, which represent a particular risk for the environment, so-called 'installations classées pour la protection de l'environnement (ICPE),' are required to self-monitor the defined compounds. To make monitoring information accessible for research and the government, a nationally integrated *database* on water was installed in 2010, the 'schéma

Table 2.32 Evaluation of the French policy design's pressure on target groups

Indicator	Level of performance	Score
Pressure on target group	Low performance: policy plan proposing an instrument mix and relying on persuasion	0.25

Table 2.33 Evaluation of the French policy design's sanctions

Indicator	Level of performance	Score
Sanctions	No performance: *no* existing sanctions	0

national des données sur l'eau (SNDE)' (action number 6), whereby industrials directly report their monitoring data into the national database on water.

While the Micropollutants Plan prioritizes source-related measures, it also includes some end-of-pipe ideas. For classified industrial plants, the Micropollutants Plan proposes the definition of *technical standards* to improve wastewater treatment (action number 7). Many more measures are outlined in the Plan; some target waste management, while others establish actor networks or private–public partnerships. The *Micropollutants Plan* also includes projects for future policy plans, for instance, on pharmaceuticals ("PNRM Plan National sur les Résidus de Médicaments dans les Eaux," May 30, 2011). In summary, the Micropollutants Plan represents a collection of proposed measures for the reduction of micropollutants. The adoption of the propositions into legally binding regulations remains to be seen.

The Micropollutants Plan fulfills an important signaling effect: It demonstrates that the topic of aquatic micropollutants is on the political agenda, and gives priority guidelines to the Water Agencies. So, even if Water Agencies are not constrained by the proposals of the Plan and are free to adopt their own water protection policies (interview with AGENCE, February 20, 2014), they may still be persuaded to take action. Therefore, a low score was attributed to the French policy design on that indicator (score of 0.25 in Table 2.32).

Sanctions and Competent Enforcement Agency

The Micropollutants Plan does not define sanctions. Therefore, the French policy design attains a score of zero for the indicator *sanctions* (score of 0 in Table 2.33).

Inclusiveness of Target Groups

The French policy design neither specifies target groups nor exceptions, but rather proposes a broad range of measures for the reduction of micropollutants from diverse sources. Hence, particularly positive to the French approach is its inclusiveness with regard to diverse sources of micropollutants, as well as the coverage of a wide range of potential policy measures targeting all parts of the society and the economy (score of 0.75 in Table 2.34).

Table 2.34 Evaluation of the French policy design's inclusiveness

Indicator	Level of performance	Score
Inclusiveness	High performance: *Important parts* of the sources of emissions may be targeted	0.75

Proportionality: Solution Versus Problem Scale

Proportionality can be evaluated when contrasting solution level with problem scale.

- **Competent jurisdictional level for implementation**: In the context of French water governance, policies are adopted on the level of water basins, which are delimited by Water Agencies. Although the Micropollutants Plan represents a national-level policy plan, it does not change the fact that policies are adopted and implemented on the basin level, because Water Agencies are not bound to the guidelines of the Plan.
- **Problem scale**: The French Micropollutants Plan places an emphasis on the complexity of the phenomenon of aquatic micropollutants and considers all types of substances, sources of pollution, and entry points into the aquatic environment. When considering micropollutants in its entirety, the scale of the problem depends on the exact chemical substance of interest and can range from local to international.

On the one hand, the French Micropollutants Plan leaves Water Agencies the flexibility to adjust their policies on micropollutants to the circumstances in water basins. In this regard, the solution level can be fine-tuned to the scale on which a problematic chemical substance causes problems. For substances detected in waters on the national scale, on the other hand, the basin-level solution might be smaller than the national problem scale and lead to policy designs, which are not encompassing enough to comprehensively reduce pollution. In conclusion, a medium performance level is attributed to the French policy design on that indicator (score of 0.5 in Table 2.35).

Directness: Dispersion of Decision-Making, Financing, and Implementation

To evaluate the directness of the underlying policy design, the architecture of the French water governance should be taken into account, which is highly decentralized compared to traditional French centralism.

Table 2.35 Evaluation of the French policy design's proportionality

Indicator	Level of performance	Score
Proportionality	Medium performance: Mostly, problem scale and solution level are likely to *match*, but for substances detected on the national scale, the defined solution level by water basin could be *smaller* than the problem scale	0.5

Table 2.36 Evaluation of the French policy design's directness

Indicator	Level of performance	Score
Directness	Low performance: *Implementation* and *financing* are executed by the French *Water Agencies*, and *implementation* is *not strictly controlled* by the national *Ministry* of Ecology	0.25

- **Decision-making entities**: The Micropollutants Plan was drafted by the Ministry of Ecology, approved by the cabinet of Ministers (government), and then adopted by the French water parliament called *Comité Nationale de l'Eau*. Nevertheless, it remains an enterprise driven by the Ministry of Ecology.
- **Distribution of costs**: Water Agencies, i.e., the water basin-level authorities, have budgetary independence from Paris as they levy their own tax, i.e., a pollution charge paid by classified industrial plants. It is their responsibility to finance pollution reduction measures.
- **Competent agency for (supervision of) implementation**: The implementation of water protection policies lies within the realm of Water Agencies. To date, it is not specified who would supervise the implementation of these policies if concrete reduction measures on micropollutants were adopted.

Theoretically, water protection policies are adopted, financed, and implemented (or supervised) by Water Agencies, which would strongly indicate for a direct policy. In the present case, however, the Micropollutants Plan is a national-level policy document, while implementation and financing is a task of Water Agencies. While the Plan was elaborated on the national level, the adoption of more concrete policies for the reduction of micropollutants in waters, as well as the funding of the measures, is left to the basin level. Since Water Agencies did not design the Micropollutants Plan, and additionally, implementation is not controlled by the national level, ample room is left for goal displacement. The underlying policy design is less direct because the tasks of deciding, funding, and implementing are dispersed between several authorities, and hence, a low level of performance is attributed to the French policy design on directness (score of 0.25 in Table 2.36).

Bindingness and Type of Policy Document

The score for bindingness is estimated by considering the type of policy document adopted.

- **Type of policy document**: The Micropollutants Plan is not a legally binding policy document, but rather an action plan, which suggests a collection of possible measures to Water Agencies.

Although the Plan is technically not legally binding, it still provides a signal to the Water Agencies, which are therefore more likely adopt micropollutants reduction measures. Therefore, a low score is attributed to the French policy design for bindingness (score of 0.25 in Table 2.37).

Table 2.37 Evaluation of the French policy design's bindingness

Indicator	Level of performance	Score
Bindingness	Low performance: existing document, but *non-binding*	0.25

Table 2.38 Performance of the French policy design after the policy comprehensiveness index

Indicators of the policy comprehensiveness index	Scores
Pressure on target groups	0.25
Sanctions	0
Inclusiveness	0.75
Proportionality	0.5
Directness	0.25
Bindingness	0.25
Overall index score	0.34 (medium/low)

Summary of Index Results for the French Policy Design on Micropollutants

The index results in Table 2.38 reveal that the French policy design achieves a medium to low level of comprehensiveness in addressing the problem of micropollutants in waters (score of 0.34). The policy is designed as a very flexible tool and therefore has the potential to address micropollutants in an effective and efficient way (inclusiveness 0.75, proportionality 0.5) depending on the exact substance of interest. The downside of the policy design is that it is not compelling to target groups (pressure 0.25, sanctions 0), nor to implementers (directness 0.25), or the state (bindingness 0.25).

2.3.6 The Dutch Micropollutants Policy

2.3.6.1 The Netherlands: Policy Project on Pharmaceutical Micropollutants

The Dutch policymaking process on micropollutants is different from the ones in Switzerland, Germany, and France because it has centered on a specific group of substances, namely pharmaceuticals. The other countries under investigation have framed the issue broadly, including plant protection products, as well as industrial and household chemicals. And yet, what renders the phenomenon of micropollutants new to policy agendas are pharmaceutical residues in water. The Dutch policy in particular has been addressing this new policy problem in a separate policy-making process, in addition to the already-existing policies on pesticides (Action Program Diffuse Sources September 30, 2009) or industrial pollution in waters (*Besluit kwaliteitseisen en monitoring water 2009*), since early 2000. The question as to whether pharmaceutical residues could cause a problem in the environment was taken up by the Dutch Parliament in 1997 (see Table 2.39). The Second

Table 2.39 Overview of the Dutch policymaking process

Dutch policymaking process on measures for the reduction of pharmaceutical micropollutants	
Trigger 1997–2004	1997–2001: Tweede Kamer inquiry to Minister of Environment concerning risks arising from endocrine substances in waters resulting in *Strategienota Omgaan Met Stoffen—SOMS*
	Report *Milieurisico's van geneesmiddelen*, Gezondheidsraad (no. 2001/17)
	RIZA reports *Vergeten stoffen in Nederlands oppervlaktewater* (no. 2001-020) and *Estrogens in the aquatic environment* (no. 2002-001)
Concept phase 2005–2013	Introduction of a working group on veterinary and human pharmaceuticals in water bodies (*Interdepartementale werkgroep (dier)geneesmiddelen en het watermilieu*, Kamerstuk no. 28808-35, April 26, 2005)
	Report concerning results of working group on February 21, 2007 (Kamerstuk no. 28808-39)
	Implementation of pilot measures, e.g., monitoring, wastewater treatment (elektronisch patiënten dossier, Kamerstuk no. 30535/27625-19, September 30, 2009)
	Position paper by VEWIN on diffuse water pollution (*Uitvoeringsprogramma Diffuse Bronnen Waterverontreiniging*, January 2010)
Elaboration 2007–2014	Environmental Ministry informs the Second Chamber (Tweede Kamer) on policy options and pilot measures (Kamerstuk no. 28808-39, February 21, 2007; no.30535, September 30, 2009; no. 27625-281, September 4, 2012; no. 27625-305, June 25, 2013)
Parliamentary Involvement Since 2013	Inquiry by Groenlinks on the state of improved wastewater treatment (no. 27625-281, March 25, 2010)
	Meeting of *Vaste Commissie voor Infrastructuur en Milieu* to discuss results of pilot measures (June 27, 2013)
	Motion 27625-299 and 27625-300 calling for regulation of pharmaceutical micropollutants in waters
	Parliamentary roundtable conference on pharmaceuticals and water quality (*Geneesmiddelen en waterkwaliteit*, January 30, 2014)
Further research Since 2013	Report on Screening micropollutants in Dutch surface water (*Evaluatie screening RWS 2011-2012. Rapportage screeningsonderzoek van microverontreinigingen in de Nederlandse oppervlaktewateren van Rijkswaterstaat*, September 20, 2013)
	BTO/KWR report pharmaceuticals in drinking water (*Vóórkomen en voorkómen van geneesmiddelen in bronnen van drinkwater*, Nov 2013)
	Position paper by VEWIN on pharmaceuticals (*Geneesmiddelen*, January 2014)

Chamber commissioned the Minister of the Environment to assess the potential risks arising from endocrine substances in the environment. As a result of the inquiry, hormone active substances were added to the Dutch strategy aiming at the correct and safe usage of chemical substances to protect humans and the environment in 1999 (*Strategienota Omgaang Met Stoffen—SOMS*).

When the Dutch Health Council, an independent scientific and highly influential advisory body in the Netherlands, released a report on the risks caused by pharmaceuticals in the environment in 2001, awareness of policy makers rose (*Gezondheidsraad Nederland, 2000: Milieurisico's van geneesmiddelen*). Hence, a number of research projects were launched in order to better understand the sources of the problem. Among others, the former[35] National Institute for Integrated Freshwater and Wastewater Management reported on estrogens in the aquatic environment (RIZA report no. 2002-001).

In response, an interdepartmental working group (*Interdepartementale Werkgroep (Dier)Geneesmiddelen in het Watermilieu*) was formed to prepare potential policy measures for the reduction of emissions from both human and veterinary pharmaceuticals in Dutch surface waters in 2005 (Kamerstuk nr. 28808-35). To integrate diverse fields of expertise, the working group consisted of the former Ministry of Housing, Spatial Planning and the Environment (VROM), the former Ministry of Transport and Water Management (V&W), the former Ministry of Agriculture, and Nature and Food Quality (LNV); moreover, public research institutes were part of the working group, i.e., the National Institute for Public Health and the Environment (RIVM), the former National Institute for Integrated Freshwater and Wastewater Management (RIZA), the Institute for Coast and Sea (RIKZ), and the Medicines Evaluation Board (CBG) which assesses the risks of pharmaceuticals (cf. document Tweede Kamer, 2005). The working group's results were reported to the Parliament in February of 2007 (Kamerstuk nr. 28808-39). Recommendations included a more purposeful usage and prescription of pharmaceuticals, and the consideration of environmental impacts by medical doctors; separate treatment of highly burdened urine or wastewater from hospitals and nursing homes; and investments in green pharmacy with higher biodegradability or absorption levels by the human body.

As a consequence, the Dutch Government made funds available to implement pilot measures in order to lay the ground for the formulation of policies (September 30, 2009 nr.30535/ 27625-19). A number of pilot measures were implemented. The Water Board Regge en Dinkel, for instance, launched a monitoring project in order to detect concentration levels of pharmaceuticals in surface waters. When the results suggested that a number of substances could be detected, some at comparably high levels (above 0.5 µg/l), the Water Board discussed potential measures with

---

[35]The Dutch administration was restructured in 2010, which lead to the dissolution and merger of several ministries and agencies. Hence, several Dutch actors, which participated in the policy process on pharmaceutical micropollutants, did not exist anymore at the time of data gathering. They obtained the prefix 'former' or 'ex' in their acronym.

pharmacists and medical doctors. As funds were lacking, no further policy action was undertaken (interview with Waterschap Vechstromen on March 11, 2014).

When similar experiences took place in the other parts of the country, the Association of Dutch Drinking Water Companies (VEWIN) published a position paper in 2010 calling for progress, and increased efforts in reducing micropollutants. Drinking water companies are confronted with rising costs as trace amounts of micropollutants in surface waters force drinking water plants to invest in more and more sophisticated (and expensive) filtering technology.

To meet the call for action of the drinking water sector, the Environmental Ministry published three consecutive letters, which inform (the Second Chamber) about the progress of pilot measures, and presented potential policy options (February 21, 2007 nr. 28808-39; September 30, 2009 nr.30535; September 4, 2012 no. 27625-281; June 25, 2013 no. 27625-305). The first letter pointed out three different strategies to reduce pharmaceuticals in waters: first, the *load approach* ('Vrachtenbenadering') which refers to introducing a fourth treatment step for the elimination of pharmaceuticals from wastewater; second, the *concentration approach* ('Concentratiebenadering') which implies the improvement of the dilution ratio of receiving water bodies; and third, separate treatment of highly pharma-contaminated urine through the separate collection (called urine separation), and treatment of hospital or nursing home wastewater (February 21, 2007 nr. 28808-39). When this agenda did not lead to a revision of legal texts, the political party Groenlinks addressed an official inquiry to the Government concerning the state of improved wastewater treatment in March 2010 (March 25, 2010 no. 27625-281).

The adoption of concrete policy measures was then delayed by the 2012 political turbulences in the Netherlands, which resulted in the resignation of the Dutch Government and new parliamentary elections. Only in 2013 did the policymaking process gain new momentum. On June 27, 2013, the parliamentary Committee for Infrastructure and Environment held a meeting to discuss the results of the pilot measures (September 30, 2009 nr.30535, Tweede Kamer, 2013b). As a result, the Dutch Parliament adopted a motion charging the Dutch Government to adopt EQNs to limit concentration levels of pharmaceuticals in surface waters (motion 27625-299 and 27625-300). To promote concerted action, the Dutch Parliament organized a roundtable conference—a stakeholder gathering—on pharmaceuticals and water quality on January 30, 2014. Politicians, governmental actors, researchers, medical experts, and members from the pharmaceutical industry were present and debated source-directed and end-of-pipe measures (document Tweede Kamer, 2014). During the discussions, pharmaceutical industry representatives expressed doubts concerning the existing scientific knowledge on the effects of pharmaceutical residues in waters, and thus, they subsequently called for further research. The Department for Water Management (RWS report September 20, 2013), the Watercycle Research Institute (BTO/KWR report *Vóórkomen en voorkómen van geneesmiddelen in bronnen van drinkwater*, Nov 2013) as well as numerous Dutch universities, continue to conduct research on pharmaceuticals in Dutch water bodies. At the same time, however, actors, such as the drinking water association VEWIN, maintain that measures should already be taken.

The 15-year-long policymaking process did not yet lead to the adoption of a concrete policy output. Rather, actors are still working on problem definition (e.g., research, pilot measures), which generally characterizes early stages of policy-making processes. Numerous actors have expressed their frustration regarding the lengthy process and the lack of concrete measures (interviews with Water Board March 11, 2014, UvW April 10, 2014, KWR April 11, 2014, RWS April 15, 2014, RIWA April 23, 2014, VEWIN April 28, 2014). Nevertheless, Sect. 3.3 analyzes the policy networks emerging from the described policymaking process. The continuation of the policymaking process after January 2014 lay in the future of the present study and is therefore not considered here.

2.3.6.2 Indexing the Dutch Policy Project on Pharmaceutical Micropollutants

Already in the early 2000s, the Netherlands put the topic of aquatic micropollutants on the political agenda and thus was the first Rhine riparian country that discussed the issue politically. Particular to the Dutch policy debate is its focus on pharmaceutical micropollutants, while other countries consider all types of pollutants in small concentrations. Agricultural or industrial pollution in waters is generally not a new phenomenon in European politics, while pharmaceutical residues represent a new challenge. Despite the progressive policy debate in the Netherlands, the 15-year-long policy process did not yet lead to the adoption of a concrete policy output, and actors are still working on problem definition at the time of the present study. Hence, there is not yet an adopted policy output that could be evaluated here, and therefore, the proposed policy ideas are classified below in a slightly abbreviated analysis.

Even if the Dutch policy process under investigation remained without concrete policy outputs, there are policies within other policy subsystems which might contribute to water protection from micropollutants, but that are not studied here. Among those are, for example, policies on pesticides that belong to the subsystem of agricultural policy (Action Program Diffuse Sources September 30, 2009) or policies on industrial pollution (*Besluit kwaliteitseisen en monitoring water 2009*).

Pressure on Target Groups and Policy Instruments

During the Dutch debate on pharmaceutical micropollutants in waters, a number of policy instruments were proposed.

- **Policy instruments in discussion**: The policy discussion evolved from a more source-directed to an end-of-pipe approach. An interdepartmental working group first proposed source-directed measures, such as green pharmacy or reduced prescriptions of pharmaceuticals in the year 2007, which was difficult to regulate. The Ministry for the Environment therefore proposed an end-of-pipe alternative, such as improving wastewater treatment and charged a consultancy to estimate the costs of upgrading. The engineering company Grontmij calculated the costs for technical improvement of all—not just selected—wastewater treatment plants, which resulted in very high cost estimates (interview with Union of Water Boards (UvW), April 10, 2014). As a result, the Government did not opt for this solution either. According to an interview with the Ministry of the Environment in 2014, the Dutch approach now considers the entire cycle, from the source to the end-of-the-pipe, and supports the society or the economy in their voluntary efforts to reduce pharmaceutical pollution in waters (Ministry for Infrastructure and Environment (IenM), April 10, 2014). The idea here is to take into account the changing role of government and not to construct policies from a top-down approach. Instead, the Ministry seeks to make use of the 'power of society,' by promoting a debate on potential societal solutions and connecting involved actors. Following the idea of an 'energetic society,' *voluntary measures* are preferred to legal state-imposed policies (interview with Ministry for Infrastructure and Environment (IenM), April 10, 2014).

As the proposed policy ideas are not constraining, a score of 0 was attributed to the indicator (score of 0 in Table 2.40).

Sanctions and Competent Enforcement Agency

Sanctions are not formulated as long as no policy is adopted (score of 0 in Table 2.41).

Inclusiveness of Target Groups

The Dutch perspective is to find a policy solution for human and veterinary pharmaceutical residues in water bodies. Since pharmaceuticals represent a specific subgroup of micropollutants, the Dutch policy is less inclusive compared to all chemical substances contributing to the problem. Nevertheless, a score of 0.5 was attributed for inclusiveness in Table 2.42, because many different target groups

Table 2.40 Evaluation of the Dutch policy design's pressure on target groups

Indicator	Level of performance	Score
Pressure on target group	No performance: *No policy instruments* adopted	0

Table 2.41 Evaluation of the Dutch policy design's sanctions

Indicator	Level of performance	Score
Sanctions	No performance: *no* existing sanctions	0

Table 2.42 Evaluation of the Dutch policy design's potential for inclusiveness

Indicator	Level of performance	Score
Inclusiveness	Medium performance: *Half* of the sources of emissions may be targeted by the policy	0.5

comprise the pharmaceutical sector. Among those are the pharmaceutical industry, pharmacists, medical doctors, and patients, which have all been named as potential future target groups in the framework of the Dutch policy debate.

Proportionality: Solution Versus Problem Scale

The proportionality of a future policy design can only be anticipated here. Considering the architecture of Dutch water governance, it is likely that the implementation of a future policy on pharmaceutical micropollutants takes place on the jurisdictional level of Water Boards. Such implementation would be advantageous where emissions display regional differences, so that policy design could be tailored to these variations. However, numerous micropollutants, including human and veterinary pharmaceutical residues, are likely to exist on the national scale, considering that the Netherlands is a very densely populated and cultivated country. While the problem scale might be national, the solution may exist at the Water Board level. Hence, the problem scale is likely to be larger than the solution level for some substances. Building on this reasoning, a medium-level performance for the indicator proportionality was estimated for a future policy project (score of 0.5 in Table 2.43).

Directness: Dispersion of Decision-Making, Financing, and Implementation

No matter how a future policy will be designed, its implementation is likely to be a task of the water boards since the latter are responsible for wastewater treatment and for issuing discharging permits. The adoption and funding of micropollutants policies are likely to come from another body because of missing majorities in water boards and resource shortages. Due to the involvement of several authorities responsible for deciding, funding, and implementing, and a probable lack of supervision of implementation, the potential for a direct policy design was estimated as low here (score of 0.25 in Table 2.44).

Table 2.43 Evaluation of the Dutch policy design's potential for proportionality

Indicator	Level of performance	Score
Proportionality	Medium performance: Mostly, problem scale and solution level are likely to *match*, but for substances detected on the national scale, the defined solution level by Water Board basin could be *smaller* than the problem scale	0.5

Table 2.44 Evaluation of the Dutch policy design's potential for directness

Indicator	Level of performance	Score
Directness	Low performance: *Implementation* is passed on to *Water Boards*, the task of *decision-making* and *financing* is likely to be taken over by *another entity*	0.25

Table 2.45 Evaluation of the Dutch policy design's bindingness

Indicator	Level of performance	Score
Bindingness	No performance: *no* existing legal document	0

Table 2.46 Performance of the Dutch policy design after the policy comprehensiveness index

Indicators of the policy comprehensiveness index	Scores
Pressure on target groups	0
Sanctions	0
Inclusiveness	0.5
Proportionality	0.5
Directness	0.25
Bindingness	0
Overall index score	0.2 (low)

Bindingness and Type of Policy Document

Lastly, the score for bindingness was estimated at zero for now (score of 0 in Table 2.45). Although the policy process has shown some output through the adoption of pilot projects for the reduction of pharmaceutical pollution in water bodies, these measures remain voluntary and sporadic so far.

Summary of Index Results for the Dutch Policy Design on Micropollutants

The index result shown in Table 2.46 reveals that the Dutch policy design has not comprehensively addressed the issue of micropollutants so far (score of 0.2). Since the Dutch policy remains a project, it can compel neither target groups (pressure 0, sanctions 0), nor implementers (directness 0.25) or the state (bindingness 0) to act. Nevertheless, the proposed ideas on how to reduce veterinary and pharmaceutical micropollutants have the potential to rather effectively and efficiently reduce pharmaceutical micropollutants in waters on behalf of potential emitters (inclusiveness 0.5) by quite well adapting problem scale and solution level (proportionality 0.5).

2.3.7 Summary of the Comprehensiveness of Policy Designs on Micropollutants

Table 2.47 summarizes the assessment about policy designs' degree of comprehensiveness with regard to reducing micropollutants in waters as elaborated above. According to the policy comprehensiveness index, Switzerland possesses the comparably most comprehensive policy design. Germany's design displays a medium degree of comprehensiveness. A medium/low rating can be assigned to France and a low one to the Netherlands.

Table 2.47 Performances of the studied policy designs after the policy comprehensiveness index

Indicators	CH	G	F	NL
Pressure on target group	0.75	0.5	0.25	0
Sanctions	0.75	0	0	0
Inclusiveness	0.5	0.25	0.75	0.5
Proportionality	1	0.75	0.5	0.5
Directness	0.75	0.75	0.25	0.25
Bindingness	0.75	0.625	0.25	0
Policy comprehensiveness index	0.75 (high)	0.48 (medium)	0.34 (medium/low)	0.2 (low)

The ranking of the policy designs confirms that Germany, France, and the Netherlands lag behind the policy innovator of Switzerland. This dichotomy is a reflection of Swiss policy design effectively contributing to the reduction of micropollutants in surface waters, and therefore being considered comprehensive. The German policy design, on the contrary, relies on monitoring the quality of surface waters, which neither reduces emissions nor improves water quality in the first place. Nevertheless, the German policy design has the potential to effectively improve water quality if further pollution reduction measures are adopted. Both the French and Dutch policy designs remain vague, and hence, improvements of water quality in these countries are less certain.

The ranking of the policy designs also reflects the fact that the four Rhine countries focus on different aspects of the policy problem and have diverse approaches to address aquatic micropollutants. The Swiss approach consists of a confinement of the problem to insufficient wastewater treatment, at least initially. Hence, the amended Waters Protection Act and Ordinance introduce technical standards, which require selected wastewater treatment plants to filter 80% of micropollutants from their wastewater. The Swiss end-of-pipe approach can be characterized as particularly pragmatic and strategically clever: pragmatic, because an entire range of industrial agricultural, household, and pharmaceutical substances can potentially be eliminated through improved wastewater treatment; and strategic, because the regulation of wastewater treatment might be a comparably feasible endeavor in a political environment favorable to clean water. The focus of the Swiss policy design on one aspect of a more complex phenomenon successfully resulted in a comprehensive policy design.

The other three Rhine countries, in contrast, feared the costs of investments in new wastewater filtering technology and therefore have not yet followed the Swiss example. With a larger territory, more inhabitants, and a different sewage treatment structure, France and Germany argued that, compared to Switzerland, a much higher number of (small) treatment plants would need investments. A fourth treatment step for wastewater plants would be a particularly costly project, which has so far lacked the necessary political support (interviews with French Ministry of

Ecology (DEB) October 18, 2013, German State of Rhineland-Palatinate (RLP) April 17, 2012, German State of Bavaria (BAY) April 26, 2012).

Compared to the Swiss end-of-pipe approach, the German EQN approach was evaluated here as less comprehensive. The German policy design must be understood in the context of federalism, where the federal state sets concentration limits, and the Länder adopt pollution reduction measures where necessary to comply with the limit. The adopted EQN in the German Surface Water Ordinance focuses on the control of concentration levels of agricultural and industrial emissions. This way, point-source as well as diffuse pollution is considered. Pharmaceutical residues in surface waters are not listed and remain unregulated in the timeframe of this study because of the opposition of the German Länder.

With regard to the French approach, in particular, micropollutants are framed as comprehensively as possible, when considering all types of pollutants, sectors, and levels. The National Government adopted a holistic strategic orientation in the Micropollutants Plan, a non-legally binding policy document, and proposed policy measures for all potential sources of emissions. Of all the propositions, the national Government has to-date mainly focused on commissioning research on micropollutants in order to set the foundations for future policies. The reticence of the French national Government may seem surprising considering that France is well known for its centralism. However, France is highly decentralized when it comes to water governance. The adoption of measures for the reduction of micropollutants in waters has consequently been a task of Water Agencies thus far, which often work on a case-by-case basis.

The Dutch situation is similar to the French one in that the Dutch Water Boards, to date, have adopted some pilot measures. The national policy process has not yet resulted in a concrete policy output. Nonetheless, the Dutch case is the only one with a clear focus concerning the risks arising from human and veterinary pharmaceutical residues in waters. Source-directed as well as end-of-pipe policy solutions have been proposed, but also rejected.

In conclusion, the EU member states France, Germany, and the Netherlands display a less-comprehensive policy design for the reduction of micropollutants than does Switzerland. One reason for this lag lies in the EU membership itself. Clearly, certain policy issues are best dealt with on the EU level. However, EU membership seems to provide false incentives if member states claim policy action on behalf of the EU, but then fail to achieve consensus on a comprehensive EU policy design. The case of micropollutants is an example of a situation where the EU regulation, here the WFD, provides for a general framework and leaves concrete pollution reduction measures to the member states. The member states, in turn, wait for further action on the EU level and in the meantime adopt incomprehensive policy designs at best. It remains to be seen whether future national measures will go beyond research and case-by-case decisions, and propose effective ways of reducing micropollutants in surface waters.

References

Abegglen, C., & Siegrist, H. (2012). Mikroverunreinigungen aus kommunalem Abwasser. Verfahren zur weitergehenden Elimination auf Kläranlagen. *Umwelt-Wissen* (Vol. 1214, pp. 210). Bern: Bundesamt für Umwelt (BAFU).

Altmann, D., Schaar, H., Bartel, C., Schorkopf, D. L., Miller, I., Kreuzinger, N., et al. (2012). Impact of ozonation on ecotoxicity and endocrine activity of tertiary treated wastewater effluent. *Water Research, 46*(11), 3693–3702.

Arellano-Gault, D., & Vera-Cortés, G. (2005). Institutional design and organisation of the civil protection national system in Mexico: The case for a decentralised and participative policy network. *Public Administration and Development, 25*(3), 185–192.

Bach, M., & Frede, H.-G. (2012). Trend of herbicide loads in the river Rhine and its tributaries. *Integrated Environmental Assessment and Management, 8*(3), 543–552.

BAFU. (2012). *Verursachergerechte Finanzierung der Elimination von Spurenstoffen im Abwasser—Änderung des Gewässerschutzgesetzes. Auswertung der Vernehmlassung von April–August 2012.* Bern: Bundesamt für Umwelt.

Baumgartner, F., & Jones, B. (1991). Agenda dynamics and policy subsystems. *The Journal of Politics, 53*(4), 1044–1074.

Bercu, J., Parke, N., Fiori, J., & Meyerhoff, R. (2008). Human health risk assessments for three neuropharmaceutical compounds in surface waters. *Regulatory Toxicology and Pharmacology, 50*(3), 420–427.

Birkland, T. (2010). *An introduction to the policy process: Theories, concepts, and models of public policy making* (3rd ed.). Armonk NY: M.E. Sharpe.

Bressers, H. (2004). Implementing sustainable development: How to know what works, where, when and how. In W. Lafferty (Ed.), *Governance for sustainable development: The challenge of adapting form to function.* Cheltenham: Edward Elgar.

Bressers, H., & Huitema, D. (2000). What the doctor should know: Politicians are special patients. The impact of the policy-making process on the design of economic instruments. In M. S. Andersen, & R.-U. Sprenger (Eds.), *Market-based instruments for environmental management* (pp. 67–88). Cheltenham: Edward Elgar.

Bressers, H., & O'Toole, L. (1998). The selection of policy instruments: A network-based perspective. *Journal of Public Policy, 18*(3), 213–239.

Bressers, H., & O'Toole, L. (2005). Instrument selection and implementation in a networked context. In P. Eliadis, M. Hill, & M. Howlett (Eds.), *Designing government: From instruments to governance* (pp. 132–153). Montreal, Kingston: McGill-Queen's University Press.

Burkhardt-Holm, P., Peter, A., & Segner, H. (2002). Decline of fish catch in Switzerland Project Fishnet: A balance between analysis and synthesis. *Aquatic Sciences, 64*(1), 36–54.

Christopoulos, D., & Ingold, K. (2015). Exceptional or just well connected? Political entrepreneurs and brokers in policy making. *European Political Science Review, 7,* 475–498.

Clarke, B., & Smith, S. (2011). Review of 'emerging' organic contaminants in biosolids and assessment of international research priorities for the agricultural use of biosolids. *Environment International, 37*(1), 226–247.

Crawford, S., & Ostrom, E. (1995). A grammar of institutions. *The American Political Science Review, 89*(3), 582–600.

Cunningham, V., Perino, C., D'Aco, V., Hartmann, A., & Bechter, R. (2010). Human health risk assessment of carbamazepine in surface waters of North America and Europe. *Regulatory Toxicology and Pharmacology, 56*(3), 343–351.

Dahl, R., & Lindblom, C. (1953). *Politics, Economics and Welfare.* Chicago: The University of Chicago Press.

Doern, B., & Phidd, R. (1983). *Canadian public policy: Ideas, structure, process* (2nd ed.). Michigan: University of Michigan.

Doern, B., & Wilks, S. (1998). *Changing regulatory institutions in Britain and North America.* Toronto: University of Toronto Press.

Dye, T. (1976). *Policy analysis: What governments do, why they do it, and what difference it makes.* Alabama: University of Alabama Press.

Edelenbos, J., Van Schie, N., & Gerrits, L. (2010). Organizing interfaces between government institutions and interactive governance. *Policy Sciences, 43*(1), 73–94.

EEA. (2011). *Hazardous substances in Europe's fresh and marine waters.* European Environmental Agency: An overview.

Eliadis, P., Hill, M., & Howlett, M. (2005). *Designing government. From instruments to governance.* Montreal, Kingston: McGill-Queen's University Press.

Esmark, A. (2009). The functional differentiation of governance: Public governance beyond hierarchy, market, and networks. *Public Administration, 87*(2), 351–370.

Falkner, G., Treib, O., Hartlapp, M., & Leiber, S. (2005). *Complying with Europe: EU Harmonisation and Soft Law in the Member States.* Cambridge: Cambridge University Press.

Fischer, M. (2012). *Entscheidungsstrukturen in der Schweizer Politik zu Beginn des 21. Jahrhunderts.* Glarus, Chur: Rüegger.

Foster, C., & Plowden, F. (1996). *The state under stress: Can the hollow state be good government?.* Berkshire: Open University Press.

Gälli, R., Ort, C., & Schärer, M. (2009). Mikroverunreinigungen in den Gewässern. Bewertung und Reduktion der Schadstoffbelastung aus der Siedlungsentwässerung. *Umwelt-Wissen Nr. 0917.* Bern: Bundesamt für Umwelt.

Gibbs, D., Jonas, A., & While, A. (2002). Changing governance structures and the environment: Economy–environment relations at the local and regional scales. *Journal of Environmental Policy & Planning, 4*(2), 123–138.

Götz, C., Abegglen, C., McArdell, C., Koller, M., Siegrist, H., Hollender, J., et al. (2010a). Mikroverunreinigungen. Beurteilung weitergehender Abwasserreinigungsverfahren anhand von Indikatorsubstanzen. *GWA Gas, Wasser, Abwasser, 90*(4), 325–333.

Götz, C., Kase, R., & Hollender, J. (2010b). *Mikroverunreinigungen - Beurteilungskonzept für organische Spurenstoffe aus kommunalem Abwasser. Studie im Autrag des BAFU.* Dübendorf: Eawag.

Grabosky, P. (1995). Counterproductive regulation. *International Journal of the Sociology of Law, 23,* 347–369.

Gunningham, N., Grabosky, P., & Sinclair, D. (1998). *Smart regulation: Designing environmental policy.* Oxford: Clarendon Press.

Gunningham, N., & Sinclair, D. (1991). Regulatory pluralism: Designing policy mixes for environmental protection. *Law and Policy, 21*(1), 49–76.

Gunningham, N., & Young, M. (1997). Toward optimal environmental policy: The case of biodiversity conservation. *Ecology Law Quarterly, 24,* 243–298.

Hajkowicz, S. (2006). Multi-attributed environmental index construction. *Ecological Economics, 57*(1), 122–139.

Hill, M., & Hupe, P. (2009). *Implementing public policy: An introduction to the study of operational governance.* Thousand Oaks: Sage.

Hollender, J., Singer, H., & McArdell, C. (2008). Polar organic micropollutants in the water cycle. In P. Hlavinek, O. Bonacci, J. Marsalek, & I. Mahrikova (Eds.), *Dangerous pollutants (xenobiotics) in urban water cycle* (pp. 103–116). Dodrecht: Springer.

Hood, C. (1986). *The tools of government.* Chatham, N.J.: Chatham House Publishers.

Hood, C. (2007). Intellectual obsolescence and intellectual makeovers: Reflections on the tools of government after two decades. *Governance-an International Journal of Policy and Administration, 20*(1), 127–144.

Howlett, M. (1991). Policy instruments, policy styles, and policy implementation. *Policy Studies Journal, 19*(2), 1–21.

Howlett, M. (2000). Managing the "hollow state": Procedural policy instruments and modern governance. *Canadian Public Administration, 43*(4), 412–431.

Howlett, M. (2004). Beyond good and evil in policy implementation: Instrument mixes, implementation styles, and second generation theories of policy instrument choice. *Policy and Society, 23*(2), 1–17.

Howlett, M. (2005). What is a policy instrument? Tool, mixes, and implementation styles. In P. Eliadis, M. Hill, & M. Howlett (Eds.), *Designing government. from instruments to governance* (pp. 31–49). Montreal, Kingston: McGill-Queen's University Press.

Howlett, M. (2009). Governance modes, policy regimes and operational plans: A multi-level nested model of policy instrument choice and policy design. *Policy Science, 42*, 73–89.

Howlett, M. (2011a). *Designing public policies: Principles and instruments.* New York: Routledge.

Howlett, M. (2011b). *Revisiting policy design: The rise and fall (and rebirth?) of policy design studies.* Paper Presented at the General Conference of the European Consortium for Political Research (ECPR), Reykjavik, Iceland, July, 11.

Howlett, M. (2014). From the 'old' to the 'new' policy design: Design thinking beyond markets and collaborative governance. *Policy Sciences, 47*(3), 187–207.

Howlett, M., & Giest, S. (2012). The policy-making process. In E. Araral, S. Fritzen, M. Howlett, M. Ramesh, & X. Wu (Eds.), *Routledge handbook of public policy* (pp. 17–28). Oxon: Taylor & Francis.

Howlett, M., & Ramesh, M. (1995). *Studying public policy: Policy cycles and policy subsystems.* Toronto, New York: Oxford University Press.

Howlett, M., & Ramesh, M. (2003). *Studying public policy: Policy cycles and policy subsystems.* Oxford: Oxford University Press.

Howlett, M., Ramesh, M., & Perl, A. (2009a). *Studying public policy: Policy cycles and policy subsystems.* Oxford: Oxford University Press.

Howlett, M., & Rayner, J. (2006). Convergence and divergence in 'new governance' arrangements: Evidence from European integrated natural resource strategies. *Journal of Public Policy, 26*(2), 167–189.

Howlett, M., & Rayner, J. (2007). Design principles for policy mixes: Cohesion and coherence in 'new governance arrangements'. *Policy and Society, 26*(4), 1–18.

Howlett, M., Rayner, J., & Tollefson, C. (2009b). From government to governance in forest planning? Lessons from the case of the British Columbia Great Bear Rainforest initiative. *Forest Policy and Economics, 11*(5–6), 383–391.

Hupe, P. (2011). The thesis of incongruent implementation: Revisiting Pressman and Wildavsky. *Public Policy and Administration, 26*(1), 63–80.

Hysing, E. (2009). From government to governance? A comparison of environmental governing in swedish forestry and transport. *Governance-an International Journal of Policy and Administration, 22*(4), 647–672.

IAWR. (2007). *Position der IAWR und IAWD zu spurenstoffen in den gewässern.* International Association of Water Works in the Rhine Basin.

ICPR. (2003). *Upstream. Outcome of the Rhine action programme.* Koblenz: International Commission for the Protection of the Rhine.

ICPR. (2010a). *Evaluation report for medicinal products for human use* (Vol. Report number 182e). Koblenz: International Commission for the Protection of the Rhine.

ICPR. (2010b). *Evaluation report for odoriferous substances* (Vol. Report number 194e). Koblenz: International Commission for the Protection of the Rhine.

ICPR. (2010c). *Evaluation report on biocidal products and anti-corrosive agents* (Vol. Report number 183e). Koblenz: International Commission for the Protection of the Rhine.

ICPR. (2010d). *Evaluation report radiocontrast agents.* (Vol. Report number 187e). Koblenz: International Commission for the Protection of the Rhine.

ICPR. (2010e). *Our common objective: Living waters in the Rhine catchment.* Koblenz: International Commission for the Protection of the Rhine.

ICPR. (2010f). *Strategy for micro-pollutants—Strategy for municipal and industrial wastewater* (Vol. Report number 181e). Koblenz: International Commission for the Protection of the Rhine.

ICPR. (2011a). *Evaluation report estrogens* (Vol. Report number 186e). Koblenz: International Commission for the Protection of the Rhine.

ICPR. (2011b). *Report on contamination of fish with pollutants in the catchment area of the rhine ongoing and completed studies in the Rhine states (2000–2010)* (Vol. Report number 195e). Koblenz: International Commission for the Protection of the Rhine.

ICPR. (2012a). *Evaluation report complexing agents* (Vol. Report number 196e). Koblenz: International Commission for the Protection of the Rhine.

ICPR. (2012b). *Evaluation report on industrial chemicals* (Vol. Report number 202e). Koblenz: International Commission for the Protection of the Rhine.

ICPR. (2012c). *Strategy for micro-pollutants integrated assessment of micro-pollutants and measures aimed at reducing inputs of urban and industrial wastewater* (Vol. Report number 201e). Koblenz: International Commission for the Protection of the Rhine.

Ingold, K. (2007). The influence of actors' coalition on policy choice: The case of the Swiss Climate Policy. In T. Friemel (Ed.), *Applications of social network analysis.* UVK: Constance.

Ingold, K. (2008). *Analyse des mécanismes de décision: Le cas de la politique climatique suisse.* Zürich and Chur: Rüegggger Verlag.

Ingold, K. (2011). Network Structures within Policy Processes: Coalitions, Power, and Brokerage. Swiss Climate Policy. *Policy Studies Journal, 39*(3), 435–459.

Ingold, K. (2014). How involved are they really? A comparative network analysis of the institutional drivers of local actor inclusion. *Land Use Policy, 39,* 376–387.

Jann, W., & Wegrich, K. (2014). Phasenmodelle und Politikprozesse: Der Policy-Cycle. In K. Schubert & N. Bandelow (Eds.), *Lehrbuch der Politikfeldanalyse* (pp. 97–132). München: Oldenbourg Wissenschaftsverlag.

Johnson, A., Jürgens, M., Williams, R., Kümmerer, K., Kortenkamp, A., & Sumpter, J. (2008). Do cytotoxic chemotherapy drugs discharged into rivers pose a risk to the environment and human health? An overview and UK case study. *Journal of Hydrology, 348*(1–2), 167–175.

Jordan, A., Wurzel, R., & Zito, A. (2003). Comparative conclusions—'New' environmental policy instruments: An evolution or a revolution in environmental policy? *Environmental Politics, 12* (1), 201–224.

Jordan, A., Wurzel, R., & Zito, A. (2005). The rise of 'new' policy instruments in comparative perspective: Has governance eclipsed government? *Political Studies, 53*(3), 477–496.

Jordan, A., Wurzel, R., & Zito, A. (2013). Still the century of 'new' environmental policy instruments? Exploring patterns of innovation and continuity. *Environmental Politics, 22*(1), 155–173.

Keeney, R., & Raiffa, H. (1993). *Decisions with multiple objectives: Preferences and value trade-offs.* Cambridge: Cambridge University Press.

Kingdon, J., & Thurber, J. (2011). *Agendas, alternatives, and public policies.* New York: Longman.

Knill, C. (2006). Implementation. In J. Richardson (Ed.), *European Union: Power and policy-making* (pp. 351–376). London, New York: Routledge.

Knill, C., & Lenschow, A. (2003). Modes of regulation in the governance of the European Union: Towards a comprehensive evaluation. *European Integration Online Papers, 7*(1), 4–15.

Knill, C., & Tosun, J. (2012). *Public policy: A new introduction.* New York: Palgrave Macmillan.

Kortenkamp, A., Faust, M., Scholze, M., & Backhaus, T. (2007). Low-level exposure to multiple chemicals: Reason for human health concerns? *Environmental Health Perspectives, 115*(S-1), 106–114.

Lasswell, H. (1956). *The decision process: Seven categories of functional analysis.* College Park: University of Maryland Press.

Lasswell, H. (1958). *Politics: Who gets what, when, how. With postscript (1958).* New York: Meridian Books.

Lemos, M. C., & Agrawal, A. (2006). Environmental governance. *Annual Review of Environment and Resources, 31*(1), 297–325.

Linder, S., & Peters, G. (1984). From social theory to policy design. *Journal of Public Policy, 4*(3), 237–259.

Linder, S., & Peters, G. (1989). Instruments of government: Perceptions and contexts. *Journal of Public Policy, 9*(1), 35–58.

Lockwood, B. (2004). How robust is the Kearney/Foreign policy globalisation index? *World Economy, 27*(4), 507–523.

Lowi, T. (1964). American business, public policy, case-studies, and political theory. *World Politics, 16*(04), 677–715.

Lowi, T. (1972). Four systems of policy, politics and choice. *Public Administration Review, 32*(4), 298–310.

Mayntz, R., & Scharpf, F. (1995). *Gesellschaftliche Selbstregulierung und politische Steuerung* (Vol. 23, Schriften des Max-Planck-Instituts für Gesellschaftsforschung Köln). Frankfurt am Main: Campus Verlag.

McGonigle, D., Harris, R. C., McCamphill, C., Kirk, S., Dils, R., Macdonald, J., et al. (2012). Towards a more strategic approach to research to support catchment-based policy approaches to mitigate agricultural water pollution: A UK case-study. *Environmental Science & Policy, 24,* 4–14.

Metz, F., & Ingold, K. (2014). Sustainable wastewater management: Is it possible to regulate micropollution in the future by learning from the past? *A Policy Analysis. Sustainability, 6*(4), 1992–2012.

Mostafa, F., & Helling, C. (2002). Impact of four pesticides on the growth and metabolic activities of two photosynthetic algae. *Journal of Environmental Science and Health, Part B, 37*(5), 417–444.

Müller, M. S. (2011). *Polar organic micro-pollutants in the River Rhine: Multi-compound screening and mass flux studies of selected substances.* Berlin: Eawag, Technische Universität Berlin Dübendorf.

Niemeijer, D. (2002). Developing indicators for environmental policy: Data-driven and theory-driven approaches examined by example. *Environmental Science & Policy, 5*(2), 91–103.

Ostrom, E. (1990). *Governing the commons. The evolution of institutions for collective actors.* Cambridge, New York: Cambridge University Press.

Ostrom, E. (2009). *Understanding institutional diversity.* Princeton, Oxford: Princeton University Press.

Pape, J. (2009). *Domestic driving factors of environmental performance: The role of regulatory styles in the case of water protection policy in France and the Netherlands.* Konstanz: Universität Konstanz.

Peters, G. (2013). *American public policy: Promise and performance* (9th ed.). Thousand Oaks: CQ Press.

Peters, G., & Hoornbeek, J. (2005). The problem of policy problems. In P. Eliadis, M. Hill, & M. Howlett (Eds.), *Designing government.* Montreal, Kingston: McGill-Queen's University Press.

Pollitt, C., Talbot, C., Caulfield, J., & Smullen, A. (2006). *Agencies: How governments do things through semi-autonomous organizations.* Basingstoke: Palgrave Macmillan.

Pressman, J., & Wildavsky, A. (1984). *Implementation* (3rd ed.). Berkeley, Los Angeles: University of California Press.

Provan, K., & Kenis, P. (2008). Modes of network governance: Structure, management, and effectiveness. *Journal of Public Administration Research and Theory, 18*(2), 229–252.

Reungoat, J., Escher, B., Macova, M., & Keller, J. (2011). Biofiltration of wastewater treatment plant effluent: Effective removal of pharmaceuticals and personal care products and reduction of toxicity. *Water Research, 45*(9), 2751–2762.

Richardson, S., & Ternes, T. (2011). Water analysis: Emerging contaminants and current issues. *Analytical Chemistry, 83*(12), 4614–4648.

Rowney, N., Johnson, A., & Williams, R. (2009). Cytotoxic drugs in drinking water: A prediction and risk assessment exercise for the thames catchment in the United Kingdom. *Environmental Toxicology and Chemistry, 28*(12), 2733–2743.

RÜS. (2010). *Rheinüberwachungs-Station Weil am Rhein. Jahresbericht 2010.* Weil am Rhein: Monitoring Station Weil am Rhine, Umweltministerium Baden-Württemberg, Bundesamt für Umwelt BAFU, Amt für Umwelt und Energie Basel-Stadt.

Rüthers, B., Fischer, C., & Birk, A. (2011). *Rechtstheorie mit Juristischer Methodenlehre* (6th ed.). München: C.H. Beck.

Sabatier, P. (2007). *Theories of the policy process.* Boulder: Westview Press.

Sabatier, P., & Jenkins-Smith, H. (1993). Policy Change and Learning: An Advocacy Coalition Approach. Boulder: Westview Press.

Sacher, F., Ehmann, M., Gabriel, S., Graf, C., & Brauch, H.-J. (2008). Pharmaceutical residues in the river Rhine-results of a one-decade monitoring programme. *Journal of Environmental Monitoring, 10*(5), 664–670.

Sager, F. (2009). Governance and Coercion. *Political Studies, 57*(3), 537–558.

Salamon, L. (2000). The new governance and the tools of public action: An introduction. *Fordham Urban Law Journal, 28*(5), 1611–1674.

Salamon, L. (2002). *The tools of government: A guide to the new governance.* Oxford, New York: Oxford University Press.

Schneider, A., & Ingram, H. (1988). Systematically pinching ideas: A comparative approach to policy design. *Journal of Public Policy, 8*(1), 61–80.

Schneider, A., & Ingram, H. (1993). Social construction of target populations: Implications for politics and policy. *The American Political Science Review, 87*(2), 334–347.

Schneider, V., & Janning, F. (2006). *Politikfeldanalyse: Akteure, Diskurse und Netzwerke in der öffentlichen Politik.* Wiesbaden: VS Verlag für Sozialwissenschaften.

Schwarzenbach, R., Escher, B., Fenner, K., Hofstetter, T., Johnson, A., Von Gunten, U., et al. (2006). The challenge of micropollutants in aquatic systems. *Science, 313*(5790), 1072–1077.

Sciarini, P., Fischer, A., & Nicolet, S. (2004). How Europe hits home: Evidence from the Swiss case. *Journal of European Public Policy, 11*(3), 353–378.

Sedlak, D., Gray, J., & Pinkston, K. (2000). Peer reviewed: Understanding microcontaminants in recycled water. *Environmental Science and Technology, 34*(23), 508A–515A.

Shelton, D. (2006). Normative hierarchy in international law. *The American Journal of International Law, 100*(2), 291–323.

Smith, K. (2002). Typologies, taxonomies, and the benefits of policy classification. *Policy Studies Journal, 30*(3), 379–395.

Touraud, E., Roig, B., Sumpter, J., & Coetsier, C. (2011). Drug residues and endocrine disruptors in drinking water: Risk for humans? *International Journal of Hygiene and Environmental Health, 214*(6), 437–441.

UBA (2014). *Antibiotika und Antiparasitika im Grundwasser unter Standorten mit hoher Viehbesatzdichte* (Vol. Report number 27, March 2014). Dessau-Rosslau: Umweltbundesamt.

Valiente Moro, C., Bricheux, G., Portelli, C., & Bohatier, J. (2012). Comparative effects of the herbicides chlortoluron and mesotrione on freshwater microalgae. *Environmental Toxicology and Chemistry, 31*(4), 778–786.

Varone, F. (1998). *Le choix des instruments des politiques publiques. Une analyse comparée des politiques d'efficience énergétique du Canada, du Danemark, des Etats-Unis, de la Suède et de la Suisse.* Bern: Paul Haupt Verlag.

Vedung, E. (1998). Policy Instruments: Typologies and Theories. In M.-L. Bemelmans-Videc, R. Rist, & E. Vedung (Eds.), *Carrots, sticks & sermons: Policy instruments and their evaluation* (pp. 21–58). New Brunswick, NJ: Transaction Publisher.

Von Neumann, J., & Morgenstern, O. (1944). *Theory of games and economic behavior*. Princeton: Princeton University Press.

Weible, C. (2007). An advocacy coalition framework approach to stakeholder analysis: Understanding the political context of California marine protected area policy. *Journal of Public Administration Research and Theory, 17*(1), 95–117.

WHO. (2008). *Guidelines for drinking-water quality*. World Health Organization.

WHO. (2012). *Global assessment of the state-of-the-science of endocrine disruptors*. World Health Organization.

Wilson, J. (1974). *Political organizations*. Princeton: Princeton University Press.

Wilson, J. (1986). *American government: Institutions and policies* (3rd ed.). Lexington, MA: D.C. Heath.

Chapter 3
Water Policy Networks—The Structural Perspective

3.1 Policy Networks

There are five main theoretical perspectives on policy design that can be distinguished (John 2013), which focus on either macro-, meso-, or microlevel explanations, or on explanations exogenous to the political realm:

(1) *Institutional* approaches focus on macro-level explanations by analyzing how the institutional architecture of a political system (polity) shapes policy design (e.g., North 1990; Weaver and Rockman 1993; Immergut 1992).

(2) *Group* and *network* approaches are meso-level and put forward that groups of actors, or networks of relationships between actors, impact policy design (e.g., Heclo 1978; Marsh and Rhodes 1992).

(3) *Actor* approaches focus on microlevel explanations to policy design. They claim that actors' interests explain why certain policies are chosen over others (e.g., the rational choice approaches in Olson's (1965) classic work *The logic of collective action* or in Ostrom's (1990) work *Governing the Commons*).

(4) *Idea-based* approaches also follow a microlevel perspective. They put forward that ideas, concerning solutions to policy problems, can gain influence and thus may impact policy design [e.g., King's (1973) work on ideas or Hall's (1993) work on learning].

(5) Lastly, some analysts advance *socioeconomic* factors outside the political system. They argue that factors *exogenous* to the political system strongly influence public actors and policy design. Examples include the idea of focusing events by Birkland (1997) or Lindblom's (1977) work on markets and politics.

According to John (2013), the five perspectives can be combined by differentiating between constraints and causes of policy action. While macro- and meso-level factors promote or constrain policy action, individual actors are seen as the real causes for policymaking according to John. Actors are considered change

© Springer International Publishing AG 2017

F. Metz, *From Network Structure to Policy Design in Water Protection*,
Springer Water, DOI 10.1007/978-3-319-55693-2_3

agents who are at the root of policy action. Institutions, networks, and socioeconomic factors, on the other hand, limit or promote policy action of actors. According to John (2013), looking for explanations of policy design should start with an actor-focus and gradually integrate different levels of explanation.

Among the actor-centered perspectives, again, three main theoretical approaches can be distinguished (for an overview see Knill and Tosun 2012, pp. 5–9). The so-called *rationalists* conceive of policymaking as a *problem-solving activity*, where actors weigh different potential solutions in order to rationally choose the most adequate one for solving a given problem (Lasswell 1956; Birkland 2010). The *incrementalists*, on the other hand, stress that actors do only rarely have the capacities (e.g., time, intellectual abilities) to gather perfect information. With limited information, rational choices of policy actors seem impossible. Hence, incrementalists adhere to the idea of bounded rationality and conceive of policy as a result of interaction processes among actors who possess differing pieces of information (see Charles Lindblom's famous article *The science of muddling through* from 1959). Incrementalists also put forward the notion that policy actors choose the option that causes the least opposition, which usually represents a small (incremental) change to the status quo and label this strategy *partisan mutual adjustment* (Wildavsky 1964; Berry 1990; Breunig et al. 2010). The third approach conceives of policymaking as a *power game* in which different groups of actors compete for influence. Finally, powerful parts of society manage to impose their preferred policy design on weaker societal groups (Baxter-Moore 1987; Kitschelt and Wilkinson 2007).

In comparison, the rationalist approach describes how policy should be made and therefore can be considered *normative*. The incrementalist and the power approach to policymaking can both be characterized as *positivist* since both approaches aim at describing how policy actually is made (as opposed to how it should be).

Policy process theories

The aforementioned conceptions of policymaking as a power game or as a mutual adjustment demonstrate that the policymaking process is key to explaining policy design. By acknowledging that the policymaking process largely impacts the design of the policy output, a large body of literature on the policy process has emerged. Its beginnings are routed in the *policy cycle* model (Lasswell 1956, 1971; Easton 1965), which conceptualizes policy processes as a series of five consecutive phases (Jann and Wegrich 2014, p. 106):

(1) *Problem definition* is a preliminary stage to the policy process during which a societal problem is made public and its root causes and effects are discussed or framed.

(2) *Agenda setting* is an early phase of the policy process, in which policy actors decide to which societal problems they are going to pay serious attention in the next phase(s) of the policy cycle.

(3) *Policy formulation and adoption* involves exploring the various options available for addressing a policy problem and coming to a decision. More precisely, policy formulation is a process during which different available policy designs are put forward and negotiated. The negotiation process ends with policy adoption, also termed *decision making*, which refers to accepting one policy design and rejecting all other potential options. Hence, policy design is considered the output of policy formulation. The decision-making phase is not identical to the entire policymaking process, but both terms are often falsely used as synonyms (Howlett and Ramesh 2003).
(4) In the *implementation* stage of the policy process, an adopted policy design is carried out.
(5) *Evaluation* involves an assessment of the implemented policy design with the potential consequence of policy termination or reformulation.

This cyclical model, which conceptualizes policy processes as a series of stages (therefore also called *stages heuristic*), was criticized as an over-simplistic textbook approach, which largely neglects the empirical reality where stages overlap or are skipped. Moreover, the policy cycle model is unable to make causal claims or to explain the substance of policy design, and it neglects the role of actors involved in the process. Still, it served the purpose of drawing scholars' attention to the procedural element of policymaking, i.e., temporal activities that are interrelated and worth further studying (Howlett and Giest 2012, p. 17).

Most importantly, the *stages heuristics* inspired policy process theories, such as famously summarized in the book by Paul Sabatier and Christopher Weible (Sabatier and Weible 2014, first edition in 1991). Comprehensive theories that combine several theoretical perspectives on policymaking (institutions, networks, actors, ideas, and exogenous factors) include the Advocacy Coalition Framework by Sabatier and Jenkins-Smith (1993), the Multiple Streams Framework by Kingdon (1984), and the Punctuated Equilibrium Framework by Baumgartner and Jones (1993). *Policy process theories* owe their label to the focus on politics, i.e. the conception of policymaking as a process shaped by institutional, network, and actor variables as well as idea-based and socio-economic variables.

The *Advocacy Coalition Framework* (ACF) can be classified as a combination of an actor-, ideas-, network-based, and socioeconomic perspective. According to the ACF, actors form coalitions by coordinating their policy action with others who share similar beliefs (Ingold 2007; Sabatier and Jenkins-Smith 1993; Weible and Sabatier 2007). Thanks to these coordinative efforts, two to four advocacy coalitions form during a policy process, each with its own idea about policy content. Coalitions compete for dominance in a subsystem in order to carry out their preferred policy design. Actors' values, or *beliefs* as the ACF terms it, are stable over time and can only be changed through exogenous shocks (such as a nuclear accident, financial, or economic crises). Hence, coalitions are stable until a large socioeconomic event changes the coalition pattern and enables policy change. The ACF explains stability with belief systems, but has to resort to external factors to explain policy change. Some authors also put forward that the ACF does not consider the rising constraints for actors through institutional structures (John 2013).

Kingdon's *Multiple Streams* analysis combines actor-, idea-, institution-based, and socioeconomic perspectives (Kingdon and Thurber 2011; Kingdon 1984). Policy formation is considered the result of the simultaneous interplay among three processes, so-called *streams*: problems, policies, and politics. When (a) a policy problem enjoys public attention, (b) policy proposals advocating change are made by policy entrepreneurs in the policymaking process, and (c) election results, media coverage, or the public mood are favorable to the proposed policy change, a window of opportunity opens for policy change. Only if the three streams occur simultaneously, conditions for policy change are favorable. If institutions are flexible, they can create favorable conditions for change through promoting the free flow of ideas or the integration of diverse policy actors in the policymaking process. Critics argue that the Multiple Streams approach is more descriptive than explanatory with regard to policy change and completely lacks explanation for policy variation (John 2013, p. 160).

The *Punctuated Equilibrium Theory* by Baumgartner and Jones (1993) seeks to explain policy decisions by combining actor-, idea-, network-, institution-based and socioeconomic perspectives. The underlying assumption suggests that policies are stable and change only punctually when stability is interrupted by increased public interests, media coverage, and public action. According to the Punctuated Equilibrium Theory, partial equilibria consist of stable policy monopolies, which are created through the institutionalization of interest groups or bureaucracies where policies change only incrementally. Under specific circumstances, established ideas can be challenged rapidly through a bandwagon effect. This effect arises when policy entrepreneurs, media, and public opinion align their attention to the very same topic, which then diffuses rapidly. The policy monopoly is challenged once the expansion of new ideas and policies is unstoppable. In such cases, partial equilibrium is punctuated and policy change is possible. Networks are important in the Punctuated Equilibrium Theory, too, because changes in the agenda are often associated with new coalitions. Moreover, the theory builds on the assumption that media and public opinion outside of the policymaking process can influence politicians and thereby challenging the existing stability. This public opinion approach contrasts with the many instances where policy makers have shaped agendas from above.

Although these frameworks combine the various theoretical perspectives on policymaking (institutions, groups, actors, ideas, and socioeconomic factors) quite differently, they all draw particular attention to the interactions (over time) between policy actors in the policymaking process in order to discern explanations for policy outputs, variation, or change (Sabatier and Weible 2014, p. 5). The actors participating in the policymaking process are central to all three frameworks. Likewise, the present work is situated within the family of policy process research.

The study focuses on the policymaking process by considering policy networks as snapshots that represent the aggregated result of multi-actor interactions in the policymaking process over time. Actors' involvement in the policymaking process, as well as their interactions and embeddedness into a web of relations, is studied in order to understand the resulting policy designs and variations across countries. Apart from its focus on policy networks (a meso-level variable), this work also maintains a strong connection to microfoundations (microlevel variables). Such a focus can be attributed to actors, defined as collective actors throughout this study, who form the building blocks of networks. Hence, this study begins with an actor-focus by considering actors' attributes (e.g., their values, preferences, and interests). In contrast to a pure actor perspective, the study also accounts for the inter-dependencies between policy actors through its network approach. Actors' ability to carry out policy action is studied as a function of the web of relations in which they are embedded.

3.1.1 The Network Approach to Policymaking

The popularity of the network approach can be attributed to the acknowledgment that a huge quantity and diversity of actors contribute to policymaking today and that contemporary Governments are unable to move unilaterally without incorporating the interests and resources of other societal actors (Bressers and O'Toole 1998, p. 215). Moreover, the network approach fascinates policy analysts, as it exhibits that a policy actor does not only have intrinsic values (or interests, preferences, positions, knowledge, perceptions, resources), but an actor's values are also influenced by the web of relations in which the actor is embedded. As a result of its popularity, a large variety of network approaches have emerged in the literature (for review articles, see: Börzel 1998; Dowding 1995; Adam and Kriesi 2007; Rhodes and Marsh 1992; Ward et al. 2011; Thatcher 1998; Pappi and Henning 1998; Robinson 2006).

Three main uses of the term *policy network* may be distinguished among its differing definitions. First, the term *policy network* has been employed as a *metaphor* to take into consideration the fact that policymaking involves a large number and variety of interlinked actors. Among others, the notions of *iron triangles* (Ripley and Franklin 1984 after Dowding 1995), *issue networks* (Heclo and Wildavsky 1974; Heclo 1978), or *policy communities* (Marsh and Rhodes 1992) emerged to denote differing types of policy networks. Secondly, some authors employ policy networks as an analytical tool or a *method* (Laumann and Knoke 1987; Hennig et al. 2012; Wasserman and Faust 2009; Scott 2000). As a method, social network analysis is able to describe and measure the interactions or inter-dependencies among actors who exchange resources in the policymaking process. In a third body of literature, the policy network approach is considered a *theory* with explanatory power (Marsh 1998; Adam and Kriesi 2007; Provan and Kenis 2008). The underlying assumption of network theory is that not only the attributes of policy actors matter, but most importantly the embeddedness into a social structure helps to understand policy actors' resources. Hence, the patterns of linkages are said to provide a strong explanation for the course and results of policymaking processes.

Lastly, the popularity of the network approach is due to its broadness, combining a macro-, meso-, and microlevel perspective. In fact, many network approaches to policymaking go beyond a pure actor-focus by incorporating structural or institutional factors (macro), which impact the constellation of actors (meso) and, hence, actors' interactions or attributes (micro) (Schneider 2014, p. 259; Mayntz and Scharpf 1995).

Diverse Conceptions of Policy Networks

Previous researchers have applied the policy network approach as a theory, an analytical tool, or a method to visualize and analyze network data mathematically (Börzel 1998; Serdült 2002; Lang and Leifeld 2008). The latter will be explained in Sect. 3.2.2. The underlying assumption of network theorists (Börzel 1998; Howlett 2002; Bressers and O'Toole 1998) is a so-called *logic of appropriateness*, which rests on two ideas (Sager 2009). First, public policies are not only technical, but also social, because they organize social relations between the state and society (Lascoumes and Le Gales 2007). Secondly, the logic of appropriateness states that a policy output, which preserves the existing social order in the policy network, is more likely to be selected. The central claim of Bressers and O'Toole's 1998 article is (p. 220): '[T]he more an instrument's characteristics help to maintain the existing features of the network, the more likely it is to be selected during the policy formation process.' Put differently, the policy output reflects the patterns of social cleavages of the policymaking process (Linder and Peters after Howlett and Ramesh 2003).

As an analytical tool, the network approach uncovers these social cleavages by describing what happens in the 'black box' between the input and the output of a policymaking process (Easton 1965; Immergut 1998). Policy networks can be understood as a map of the policymaking process, which provides a synthetic overview of the involved actors, their roles, and positions as well as relations among each other (Dowding 1995).

In addition to the diverse conceptions of policy networks, there are numerous empirical studies that portray the existence and characteristics of policy networks (Fischer 2013; Ingold 2008; Bressers et al. 1995; Leifeld and Schneider 2012; Fischer 2012; Knoke et al. 1996). So far, theoretical as well as empirical research failed to demonstrate in which way policy networks are relevant to policy design (Marsh 1998; O'Toole 1997; Robinson 2006). It remains unclear under which conditions networks enhance or reduce the efficiency and legitimacy of policy outputs. Scholars who studied the explanatory value of policy networks fall short of formulating hypotheses, which systematically link network structures with policy outputs (Crona and Bodin 2006; Sciarini 1996; Marsh and Rhodes 1992; Fischer 2012; Sandström and Carlsson 2008; Ingold 2011; Lubell and Fulton 2007; Knoke et al. 1996; Marin and Mayntz 1991). Consequently, existing scientific literature does not give satisfactory answers regarding the explanatory value of the network approach to the selection of policy designs. Nevertheless, in order to understand the essence of policymaking in a networked context, it is essential to know the impact of different network structures on policy design. The present study seeks to address this research gap by formulating hypotheses linking network structures to policy design (see Chap. 4).

3.1.2 Characteristics of Policy Networks

There is a large body of empirical studies in the policy process literature, which generated a deep understanding of the characteristics of policy networks. A first element characterizing policy networks is the *diversity of policy beliefs* and preferences in the network. According to the Advocacy Coalition Framework (ACF), developed by Sabatier and Jenkins-Smith, policy actors have policy beliefs and preferences (1993), which guide their actions in the policymaking process. Building on these ideas, Bressers and O'Toole (1998) have developed the idea of *belief cohesion* which refers to a consensus between network members about policy beliefs and content, about what issues the network should deal with, and how to best solve them (see also Marsh 1998). The assumption is that it is more likely to come to a policy decision when actors share common worldviews and policy preferences than when objectives fundamentally clash (Kickert et al. 1997).

In quantitative network analysis, the term *cohesion* implies dense relations among members of a social group (Knoke and Yang 2008). More concretely, a cohesive subgroup denotes, in social network analysis, a group of actors connected by many direct and reciprocated ties and disconnected from outsiders. Hence, cohesion captures dense ties, but does not predetermine the type of tie, such as belief similarity.

In order to do combine the two types of uses, the term *belief cohesion* is employed here. Belief cohesion reflects, on the one hand, the notion of cohesive subgroups of network members who, on the other hand, display homogeneity of beliefs. In this work, belief cohesion refers to ties signaling a general agreement among network members about fundamental policy beliefs. The usage of the term *belief cohesion* is compatible with the network literature insofar as cohesive subgroups of network graphs are generally considered to display homogeneity of thought, identity, and behavior (Knoke and Yang 2008, p. 72), which is in line with the meaning of the term belief cohesion here.[1]

With high belief cohesion, non-governmental network members are also more likely to attain support by the Government (which often takes a neutral position), because all actors agree on the need to take policy measures and on the direction of action. Low belief cohesion between policy actors, on the other hand, can lead to

---

[1]Note that sharing beliefs does not imply that actors with similar beliefs hold other direct relational ties aside from their belief similarity. For example, actors who share beliefs do not, by definition, have to collaborate directly.

mutual distrust between the major coalitions, creating opportunities for minority interests or even network outsiders to influence the policymaking process (Marsh and Rhodes 1992). In the same line of thought, other scholars have looked at the variety of actor types participating in the network. Networks, which display heterogeneous actor structures, are labeled as *open*, and homogenous ones as *closed* (Howlett 2002). The openness (and likewise closedness) of a policy network provides information on the diversity of ideas and interests, as well as the network's receptiveness toward new ideas (Adam and Kriesi 2007). Scholars still disagree on the impact of actor diversity on policy outputs. Some argue that the more open a network is, the more resources and constructive ideas there are concerning how to best address a policy problem and, consequently, the more likely a comprehensive solution is to come about (Adam and Kriesi 2007). Other researchers believe that the more open a network is, the less likely a comprehensive solution will materialize, because clashes of interests become more likely (Kickert et al. 1997). Rather than simply looking at actor types, another group of scholars uses actors' beliefs to identify relational conflict lines in policy networks (Ingold 2011; Fischer and Sciarini 2013; Henry et al. 2010; Weible and Sabatier 2005). The underlying rationale is that diverging policy beliefs lead to conflictive network relations (also termed *disagree-ties*), while converging policy beliefs lead to cooperative network relations (also termed *agree-ties*). In summary, the literature indicates that heterogeneous beliefs and conflictual relations imply low belief cohesion. Homogeneous actor beliefs and cooperative relations point to high belief cohesion. Scholars are divided as to which way belief cohesion promotes or hampers comprehensive collective outputs. There is large agreement, however, that actors' beliefs and preferences matter in explaining policy design.

This insight points to the next distinctive element of policy networks, i.e., *interconnectedness*. Interconnectedness refers to the number of contacts among actors in the policymaking process (Bressers and O'Toole 1998). For instance, when policy actors frequently exchange information during a policymaking process or collaborate intensively, a network is considered highly interconnected. The literature puts forward that interconnectedness is decisive in overcoming divergences and reaching a policy decision (Fischer 2015). Actors who maintain some sort of social relation, e.g., information exchange or collaboration, create so-called bonding ties —building trust and social capital (Fischer 2012; Berardo and Scholz 2010; Coleman 1990). Ultimately, interconnectedness promotes actors' organizing or problem-solving capacities as well as their ability to find solutions for policy problems. Hence, with more ties in the network, collective outputs are more likely (Weible and Sabatier 2007). In fact, empirical studies demonstrate that network density is pivotal for reaching policy decisions (Hermans et al. 2013) with the results suggesting that large networks can come to a decision as successfully as small networks do if policy actors are well-connected. Building on this insight, many network typologies are based on the degree to which interconnectedness occurs. Rod Rhodes, for instance, establishes a continuum of network types ranging from highly integrated policy communities at one end to loosely integrated issue networks at the other end (Rhodes 1988).

Another important element of policy networks is the distribution of power in the network. Many typologies build on power as a criterion for distinguishing types of networks (Adam and Kriesi 2007; Fischer 2012; Van Waarden 1992). At the two extremes, power can be distributed across the whole policy network (or groups of actors), or it can be concentrated among a few actors (or one dominant group of actors) (Kriesi et al. 2006). Additionally, some authors include the attribution of power to specific types of actors, such as state actors, political parties, economic associations, or NGOs (Kriesi et al. 2006). Clearly, powerful actors have more leverage in influencing policy outputs according to their preferences.

Power can refer to two different actor attributes. First, power is institutionally defined and granted to (mostly governmental) actors through formal authority to propose or adopt legislation. Secondly, power is a network characteristic. Central actors with many ties, as well as actors who hold a strategic position in the network, may control the flow of information and may have power over other network members. Depending on how the power is distributed in the network, the courses of policy action can turn in one or another direction. Hence, some authors describe networks as a power dependency relationship (Henning 2009; Pappi and Henning 1998, 1999).

Empirical studies suggest a further network phenomenon through their observation that actors' preferences are a strong predictor for interactions and, more precisely, *coalition formation* (Ingold 2011; Henry 2011). ACF scholars like Karin Ingold and Christopher Weible have studied coalition formation in policy networks. They demonstrate that policy actors form coalitions based on shared policy beliefs during a policymaking process in order to pool resources, coordinate their action, and influence policy design (Ingold 2008; Weible and Sabatier 2006). Hence, coalitions are defined by both a high degree of internal belief cohesion and a high degree of internal interconnectedness. ACF scholars could also show that advocacy coalitions compete in order to turn their beliefs and preferences into policy. A further insight is that policymaking processes display different types of coalition structures. Weible et al. (2010) or Fischer (2014, 2015) distinguishes policy networks with a dominant coalition from those with competing coalitions of similar strength. Coalitions are considered dominant when they include a majority of network members as well as powerful actors. According to Fischer (2014), dominant coalitions are more likely to make their preferred policy outputs be adopted. Going one step further by linking these theoretical ideas about dominant coalitions to the quality of policy outputs leads to the following expectations: If the dominant coalition pushes for policy action, it is likely that an encompassing policy solution is agreed upon. If the dominant coalition is against addressing a policy problem, policy design is likely to be weak. In cases of several equally powerful coalitions, the assumption is that coalitions block each other and that little is done to alleviate a policy issue. Hence, coalition structures are decisive in understanding policy design. Policy scholars have linked the different coalition types to the concept of policy subsystems. *Policy subsystems are common organizational forms of policymaking, in which actors of a defined policy field interact to address that field's issues by establishing relatively stable relations with one another* (Baumgartner and

Jones 1991; Thurber 1996). Three types of policy subsystems can be distinguished: *unitary, collaborative, and adversarial subsystems* (Weible et al. 2010; Ingold and Gschwend 2014). While unitary subsystems are characterized by one coalition with high intra-coalition belief similarity and high intra-coalition coordination, collaborative subsystems are defined as two coalitions with intermediate inter-coalition belief similarity and high inter- as well as intra-coalition coordination. Adversarial subsystems are described as competitive through their low inter-coalition belief similarity and coordination, but high intra-coalition coordination (see Fig. 3.1). This distinction is helpful in order to assess whether comprehensive problem solving is a likely outcome of a policymaking process. Theoretically, certain subsystem structures could be more conducive to comprehensive problem solving than others. One can assume that comprehensive policy solutions should be more likely when (a) a dominant coalition forms in an adversarial or a collaborative subsystem and pushes for a comprehensive policy solution, or (b) the one existing coalition in a unitary subsystem pushes for a comprehensive policy solution. As might be expected, a comprehensive policy solution can also be impeded by dominant coalitions or unitary subsystems if they are opposed to a comprehensive policy solution.

Beyond the characteristics that refer to the overall structure of policy networks, network scholars argued that single actors with privileged structural positions are also an important constitutive element of policy networks (Wasserman and Faust 2009). Through their strategic network positions, those key actors are able to induce new impulses to the policymaking process, as well as constrain or promote linkages in the network. As they exert a particular influence on the overall network structure, the actors are able to modify the status quo (Kingdon 1995). The policy process literature distinguishes between two types of exceptional actors, namely policy brokers and policy entrepreneurs (Christopoulos and Ingold 2015; Mintrom and Norman 2009). Both types are assumed to have a decisive influence on the overall decision-making process and policy design, but in distinctive ways. While brokers occupy a mediation role, policy entrepreneurs are considered 'pushers' toward their preferred policy output (Schneider et al. 2011). Neither brokers nor entrepreneurs are conceptualized as individuals or personalities, but rather as roles that organizations occupy through their strategic network position in policymaking processes (Mintrom 2000).

The ACF literature examined and further developed the concept of *brokerage* (Sabatier and Jenkins-Smith 1993).[2] According to the ACF, *policy brokers* exhibit a

---

[2]The concept of brokerage has been used in different streams of literature. The network literature, such as in Burt's (2000) famous work, for example, states that brokers occupy a strategic network position allowing them to bridge structural holes. The idea that brokers are neutral mediators, on the contrary, has its origins in the ACF literature (Sabatier and Jenkins-Smith 1993). Common to both streams of literature is the concept of brokers who hold strategic bridging positions enabling them to spread resources and impact the entire network positively. One such positive impact of brokers is the mediation of compromise as put forward in the ACF literature.

Ideal network structures

Adversarial

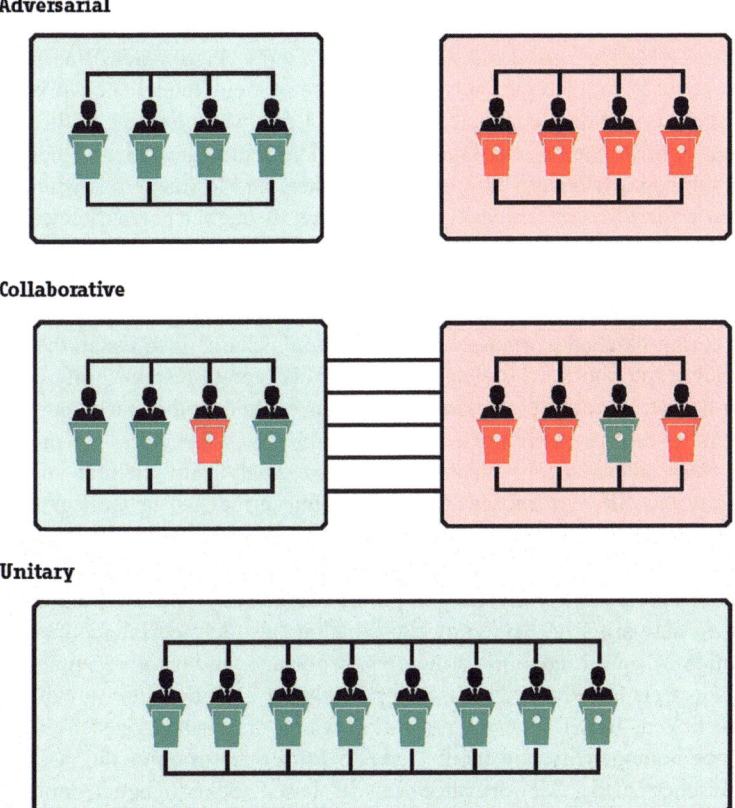

Collaborative

Unitary

Fig. 3.1 Three policy subsystem types

moderate belief system and are therefore located between conflicting coalitions, and connected to actors who are split on diverging beliefs. As a result of their neutral position, brokers take a role of mediators advocating a feasible compromise solution (Ingold and Varone 2012). Brokers take on this role by facilitating communication and promoting trust, thereby helping to prevent blockades and to overcome existing conflicts (Christopoulos and Ingold 2015). Empirical studies approve the pivotal role of policy brokers for policy networks and the production of policy outputs (Henry et al. 2010; Bodin et al. 2006). Ingold (2011), for example, found that brokers have the ability to mediate compromise and enable a policy solution.

While policy brokers are said to help work toward conciliatory policy solutions, *policy entrepreneurs* are considered pushing toward their own preferred policy design. The concept of policy entrepreneurs, as employed here, is mainly rooted in the Multiple Streams Framework (Kingdon and Thurber 2011) and the Punctuated

Equilibrium Theory (Baumgartner and Jones 1993).[3] Entrepreneurs are described as self-interested actors with leadership qualities who know how to take advantage of inefficiencies in the policymaking process in order to promote their own policy preferences (Michael Mintrom and Norman 2009; Ingold and Varone 2012). Entrepreneurs are able to gain this edge because of their strategic network position, which consists of being connected to powerful and well-connected other network members (Christopoulos and Ingold 2015). Their strategic position helps entrepreneurs to capture the attention of policy makers and to frame a demand for their preferred policy design, especially in situations of inefficiencies (Schneider et al. 2011; Christopoulos and Ingold 2015).

Evidently, both concepts are ideal-typical. Ingold and Varone (2012) demonstrated that brokers, too, have some kind of self-interest in promoting negotiations and hindering blockades when their organization's charge is to insure the adoption of a feasible solution for a societal problem. This type of responsibility can be the case for ministries or political parties, for instance. On the other hand, entrepreneurs may be in favor of a compromise if a compromise is in line with their preferences. Nevertheless, those ideal-typical concepts are analytically helpful in order to understand the different impact of actors holding privileged network positions on policymaking processes and on policy design.

In total, seven aspects of policy networks have been put forward by different network scholars, namely diversity of policy beliefs and preferences (labeled belief cohesion), network size, degree of coordination (also termed interconnectedness), distribution of power, coalition structures, brokerage, and entrepreneurship. Those different aspects have been discussed separately for analytical clarity, but in reality they are linked. Belief-cohesive networks, which are characterized by belief and preference homogeneity, are more likely to form one strong coalition (Nohrstedt 2008; Fischer 2015). Tie-formation can be based, next to belief similarity, on power-seeking—meaning that actors have a strong incentive to form ties in order to increase their own level of power (Henry 2011; Ingold and Leifeld 2014; Fischer and Sciarini 2015; Heaney 2014). Moreover, large networks are more likely to be

---

[3]The present definition of entrepreneurs is rooted in policy process theories, i.e., Multiple Streams and Punctuated Equilibrium, but does not claim universality across different bodies of literature. There is a fundamental difference between economic (Schumpeter 1939; Schumpeter and Opie 1934; Leibenstein 1968; Kirzner 1978) and political entrepreneurship (Schneider and Teske 1992), for example. Some parts of the literature on political entrepreneurship are largely inspired by the economic literature and define political entrepreneurs as innovators who aim at profit maximization as in the economic sector (Kuhnert 2001). Other literature puts forward that market mechanisms are different from rules defining the electoral and political realms and therefore, political entrepreneurship should be conceived fundamentally different from the ideas inspired by economic profit maximization (Mintrom 2000; Holcombe 2002; Michael Mintrom and Norman 2009). Even within the literature on political entrepreneurship there exist different concepts: Some define political entrepreneurs as some sort of 'altruistic' actors who solve problems of collective actions in order to supply collective goods (Frohlich and Oppenheimer 1978). Others argue that a "glorifying view of [policy entrepreneurs] as catalysts for change" neglects their inherent self-interests (Christopoulos 2008, p. 3).

less dense, and thus less interconnected, than small networks, because policy actors have only limited resources (money, time, and motivation) to form ties (Ingold 2011).

In summary, the network approach, as used in this work, assumes that network structures can constrain or promote collective decisions. Some network structures seem to be more efficient in producing collective outputs than others. The constellation of actors is considered a decisive element of the networks' capacity to solve societal problems. Which combination and level of network characteristics promote comprehensive problem solving still remains an open empirical question. To reach closer to answering this question, the Chap. 4 assesses in which way belief cohesion and interconnectedness, strong brokerage, and a dominant coalition and entrepreneurs (advocating an encompassing solution) may have a positive effect on the emergence of comprehensive policy designs from a theoretical perspective.

3.2 Network Structure: Data, Method, Operationalization

Social network analysis is a method that enables one to explain tie-formation (Gerber et al. 2013; Henry et al. 2010; Robins et al. 2012; Leifeld and Schneider 2012). Likewise, it is employed here to operationalize network structure. Section 3.2.1 explains the methodology behind network data collection employed for the purpose of this study. Section 3.2.2 then provides a short introduction into social network analysis as a method for the visualization and mathematical analysis of relational data. Lastly, Sect. 3.2.3 explains in detail the different types of analyses applied to the gathered network data in order to operationalize variables capturing different structural network properties.

3.2.1 Data Collection for Policy Network Structure

Surveys generally represent an efficient method for systematically collecting data from a large number of observations. In the case of policy networks, surveys are a particularly pertinent way of gathering relational data, because complete information about interactions is often difficult to access otherwise. In the framework of this study, the majority of actors, i.e., 158 out of 199, were mail-surveyed, and additionally, a subset of actors, i.e., 41 actors, was surveyed through semi-structured personal interviews in order to complement quantitative survey data with qualitative information. Due to limited resources, it was not possible to personally interview all 199 actors. In subsequent chapters, the distinction will be made between survey interviews and preliminary interviews. Preliminary interviews refer to those interviews which were conducted before the survey started, e.g., to pretest questions and gain case knowledge. Survey interviews were conducted to gather the data that is analyzed in the present study.

Gathering data on policy networks is not a simple task, and several preliminary explanations are needed: Sect. 3.2.1.1 defines policy actors, Sect. 3.2.1.2 explains why policymaking processes can be conceptualized as policy networks, and Sect. 3.2.1.3 goes on to explain how to define a network boundary and key actors based on policymaking processes. Sect. 3.2.1.4 then highlights the procedures of collecting network data. Finally in Sect. 3.2.1.5, response rates of the surveys are calculated for each of the four investigated countries.

3.2.1.1 Policy Actors

Rooted in policy process theories is the conception of policy actors as organizations, defined as *collective actors* (as opposed to individual persons), which are active in the policymaking process through their involvement in policy decisions (Schubert and Bandelow 2014, p. 8; Schneider 2014). In modern democracies, a multitude of actors participate in policy decisions, including state as well as non-state actors. Hence, the classic distinction between 'the state', which is solely responsible for policymaking, and 'the society', which is targeted by state-made policies, is not valid anymore (Kriesi 2007; Fischer 2012, p. 44). In Western democracies, state actors typically include the Government, the bureaucracy, the legislature, the courts, and the political parties when elected for positions in the legislature or Governments. Non-state actors are organizations representing the different interests of society, such as political parties (unless in power), environmental, consumer, worker, employer or industry associations, and also scientific research.

Categorizing policy actors into *political-administrative actors, target groups, beneficiaries,* and *third-party members* provides a helpful way of analyzing these groups (Knoepfel et al. 2011, pp. 66–77). *Political-administrative* actors are public authorities responsible for drafting and implementing public policies. *Target groups* are those that have been identified as the cause or contributors to a societal problem. As addressees of a policy, target groups must implement the adopted policy and change their procedures or behavior accordingly. The *beneficiaries*, on the contrary, are those actor groups who suffered previously from the negative effects of the collective problem. Such a group profits from the introduction of the new policy as their social, economic, or environmental situation should be improved. *Third-party groups* are not the direct target of a public policy and, yet, their situation changes significantly with the introduced policy. They may either be unintended beneficiaries of the introduced policy in cases where the policy changes have positive consequences for them; or their situation can be impaired. In such cases, third-party groups may be considered 'collateral targets' of an introduced policy.

Many policy analysts have adopted an actor-centered perspective and have conceptualized policymaking as the collective result of decisions and actions of multiple actors who interact with one another (for an overview, see Schneider 2014, p. 259). Actors are said to form positions and strategies based on their values, preferences or, interests, and are constrained by the polity, i.e., institutionalized obligations or resources, in order to reach their policy aim (Schubert and Bandelow 2014, p. 1; Mayntz and Scharpf 1995). This microfoundation not only focuses on the number and types of actors participating in policymaking processes, but also on the constellation of actors. Hence, the notion of *policymaking in a networked context* emerged (Bressers and O'Toole 1998).

3.2.1.2 Conceptualizing Policymaking Processes as Policy Networks

Numerous studies demonstrate that policymaking is the result of interactions between multiple actors during policymaking processes (Klijn 1996, p. 90). The policy cycle model (Lasswell 1956, 1971; Easton 1965) conceptualizes policymaking processes as a series of five consecutive phases (Jann and Wegrich 2014, p. 106): (1) problem definition, (2) agenda setting, (3) policy formulation and adoption, (4) implementation, and (5) evaluation. Although the stages heuristics have been criticized for their analytical weaknesses, they still helped to draw scholars' attention to the procedural element of policymaking. In fact, policymaking processes are composed of a sequence of events, which are points in time where a contribution is made to reach a commonly supported policy decision in order to address a collective problem (Knoke 1994, p. 285). Thus, policy processes include all events that occur between the emergence of a collective problem and the evaluation of policy adoption and implementation (Easton 1965; Knoke 1994). This sequence of events can involve hundreds of actors from the executive, the administration, the legislature, parties, interest groups, or science (Sabatier 2007, p. 3; Kriesi 2007, p. 263). Even though a state's political system assigns decision-making power to defined governmental bodies, no single actor has enough steering capacity to determine the result of the policymaking process. State and non-state actors are mutually dependent on each others' resources, such as information or support (Pappi and Henning 1998, 1999; Knoke et al. 1996). Hence, the notion emerged that policymaking processes take place in a networked context (Bressers and O'Toole 1998, 2005). According to Bressers and O'Toole, *networked* means that the policy-relevant actors operate in a matrix of inter-dependencies, rather than as autonomous units or as a consequence of a predefined hierarchy (2005, p. 140). To understand the inter-dependencies and interactions between actors, many scholars have relied on network analysis. From a network approach, policymaking is conceptualized as the result of a multi-actor process of inter-dependent relations. Network scholars have translated the described theoretical concepts (policy process, actors, and inter-dependencies) into network terms as follows:

- A policy process, which can span over 10 years from agenda setting to evaluation (Sabatier 2007, p. 3; Easton 1965), is conceptualized as a policy network (Pappi and Henning 1999; Knoke et al. 1996; Ingold 2008; Fischer 2012).
- State and non-state actors who participate in the policy process are the nodes in the network.

- Actors who exchange resources, such as information or support, form the ties in a network. Present and absent ties between actors reflect the existing inter-dependencies (Burt 2000, 2009).

The present study constrains its analyses to those events occurring between the emergence of a policy problem on the political agenda and the adoption of a policy design. Hence, the underlying policy networks represent snapshots of actors' interactions in stages preliminary to implementation.

3.2.1.3 Definition of Network Boundary and Key Actors

The network boundaries were delimited based on Knoke and Laumann's concept of policy domain[4] that includes all actors 'concerned with formulating, advocating, and selecting courses of action to solve that domain's problem' (Knoke 1994, p. 279). The main criteria for delineating the network boundary are, thus, participation as well as having a particular interest in problem-solving activities (Börzel 1998). Applied to this study, the network boundary includes all state and non-state policy actors who were involved or have a stake in the policymaking processes on aquatic micropollutants previously described in Sect. 2.3. Each country's policy network is considered a snapshot representing the aggregated result of multi-actor interactions in the specific policymaking processes over time (Lubell et al. 2014). The time frame for the policymaking processes varies between the countries under study. More specifically, the Swiss policymaking process on the amendment of the Waters Protection Act and Ordinance is analyzed for the period between 2007 and 2013. The German process adopting the Surface Water Ordinance is restricted to the time frame between 2008 and 2011 here. The analysis of the French process adopting the Micropollutants Plan concentrates on the period between 2009 and 2013. The Dutch policymaking process on pharmaceutical micropollutants is assessed since its emergence in 2001 until the end of this study's observation period in 2013. All actors concerned with the previously mentioned policymaking processes in the stated time frames are considered part of the policy network in this study.

In Sect. 3.2.1.1, *policy actors* are defined as collective actors (as opposed to an individual person), including state actors with formally assigned regulatory competences, and non-state actors who contribute to the creation or implementation of policies (Knoepfel et al. 2011; Schneider 2014).

Within the network boundary, the key actors were determined by applying Laumann's et al. (1983, pp. 22–24) decisional, positional, and reputational approach for complete networks, which has been approved in many empirical studies (Ingold 2008, 2011; Fischer 2012). The aim is to compile a complete list of the network population, which demands substantial prior knowledge of the policymaking process by the investigator (Marsden 2005).

---

[4]In this work, I use the word *policy field* rather than *policy domain*.

With the *decisional approach*, actors 'that participate in the making or influencing [of] the collectively binding policy decision' are listed (Knoke 1994, p. 280). The *positional approach* locates all those actors holding key roles in the political (sub)system, e.g., elected or executive positions in state bodies or major economic, societal, or political non-state organizations. Finally, the *reputational approach* identifies all those actors who are believed by knowledgeable observers to have the actual or potential power to 'move and shake' the policy domain (Knoke 1994, p. 280).

To apply the *decisional method* in practice, the aforementioned four policy-making processes were reconstructed by means of document analysis[5] and in-person interviews. During in-person interviews preliminary to the survey (see Annex 1 for a full list of interviews), top-level state officials who followed the policymaking process closely were asked to verify whether policymaking processes were correctly retraced. A total of 12 preliminary interviews[6] took place between 2011 and 2014. Those interviewed included: three Swiss representatives of the Department for Water and the Swiss Water Association; seven German represen-tatives of federal and state ministries for the environment, and one representative of the German Institute of Hydrology; one French representative of the Department of Water; and a Dutch representative of the Water Board Vechtstromen. Where nec-essary, interviewees were asked for missing documents about the policymaking processes, e.g., meeting protocols or attendance lists. Since the policymaking processes lasted several years where numerous events took place, the processes were subdivided into the following phases, which are generally part of every pol-icymaking process: trigger, concept phase, elaboration of the draft proposal, con-sultation, finalization of the legal draft, parliamentary phase, and implementation. Actors participating twice in different phases or twice in the same phase have been retained. The decisional method has been applied to the four countries as follows:

- Switzerland: A total of 156 actors were identified, who participated in the policymaking process at least once, after excluding all individual persons, companies, wastewater treatment plants, and communities, whose interests are represented by diverse associations. As explained above, out of the 156 actors, only those who participated twice in different phases or twice in the same phase were retained. As such, a total of 45 actors were determined through the deci-

---

[5]Documents analyzed include records of public hearings, parliamentary debates, legal drafts, texts or other policy documents on micropollutants, governmental letters, or information found on Web sites of the lead administrations for micropollutants policy.

[6]Preliminary interviews refer to those interviews which were conducted before the survey started, e.g., to pretest questions and to gain case knowledge. I distinguish preliminary from survey interviews. Survey interviews are those which were conducted for gathering the data analyzed in the present study.

sional approach (see Annex 2).[7] The openness of the policymaking process to all kinds of policy actors is remarkable. All actor categories, i.e., political, economic, and societal actors, participated in the policymaking process.

- Germany: In Germany, a total of 47 actors participated in the policymaking process at least once (see Annex 3). Of note is that only very few actors participated twice in the process.[8]
- France: Compared to Switzerland, many fewer actors, namely 28, participated twice in different phases, or twice in the same phase of the policymaking process (see Annex 4). One should note that agricultural organizations did not participate in the elaboration process of the Micropollutants Plan, except through the National Water Committee (CNE), the 'water parliament' that votes on every water-related legal project in France. One reason for the absence of agricultural organizations could be that there is the *Plan Ecophyto*, which specifically targets micropollutants from agriculture.
- The Netherlands: In the Netherlands, 24 actors were identified through the decisional approach (see Annex 5). As the policymaking process focused on micropollutants from pharmaceuticals, it is not surprising that the pharmaceutical and the health sector participated in the policymaking process. In the other countries, these two actor categories did not play a huge role. It is noteworthy that agricultural or economic (non-pharma) organizations, as well as environmental and consumer associations, did not participate in the Dutch policymaking process. Two engineering bureaus (KIWA, TAUW) took part in the process, but no longer exist in the same form. These bureaus were not considered for this study because of difficulties in locating a contact person who had previously worked for the company in the framework of the policymaking process on micropollutants.

Secondly, actors with key roles (*positional approach*) in the political system governing water quality and micropollutants were retained. This includes national- or lower-level governmental administrations or agencies responsible for water quality, elected political actors such as parties, the parliament, or parliamentary

---

[7]In fact, 73 actors were identified through the decisional approach. Out of the 73 actors, 27 were excluded, i.e., the Department of Foreign Affairs, as well as the 26 Swiss cantons. The Department of Foreign Affairs was not involved in the underlying policy process because the issue of micropollutants did not fall into its portfolio of competences. Water-related international issues are dealt with in the Department of Environment, and hence, the Department did not release a statement in consultations internal to the bureaucracy. In the latter case, cantonal associations representing the Cantons on the federal level were surveyed, rather than individual cantonal Governments.

[8]An environmental NGO called Deutsche Umwelt-Aktion e.V. was identified through the decisional approach since it was consulted. However, the NGO reported in a telephone call that it did not participate in the process and was therefore not considered here.

commissions, the key environmental, industrial, agricultural, worker and consumer associations, drinking water or wastewater associations, and science (universities, laboratories, and consultancies). The following actors have been identified via the positional method in the four countries:

- Switzerland: Even if already identified through the decisional approach, all federal departments were singled out through positional reasoning, because by default all departments are consulted in the inter-departmental consultation procedure as members of the bureaucracy. Those departments that do not have an obvious relationship to environmental or water-related issues, and did not release a statement, were deleted from the list (Department of Homeland Security, Department of Foreign Affairs; the latter, because water-related international issues are dealt with in the Department of the Environment). In a federal state like Switzerland, the cantons always have a say in policymaking. In this study, cantonal associations, rather than individual cantonal Governments, were surveyed because the associations represent cantonal interests on the national level. In general, the parliament plays an important role in amending laws. However, the whole parliament cannot be regarded as an actor with a common position. Alternatively, the environmental committees of both parliamentary chambers are considered here, because a common position has been worked out in the committees. Finally, the biggest Swiss political parties with seats in parliament were identified through the positional approach, as well as the biggest associations. In total, the following 12 actors were identified through the positional approach in addition to the decisional approach: Department of Air Protection and Chemicals of the Federal Office for the Environment (BAFU-CHEM), Federal Office for Spatial Planning (ARE), Federal Office for Agriculture (BLW), Federal Office for Energy (BFE), Nation Council's Committee on the Environment, Spatial Planning and Energy (UREKN), Christian Democratic People's Party (CVP), Green Liberal Party of Switzerland (GLP), Civil Democratic Party (BDP), Swiss Municipalities Association (SGV), Swiss Employers' Association (SAV), Swiss Trade Association (SGEWV), and Swiss Farmers' Association (SBV).
- Germany: Two German actors were singled out through the positional approach: the German Federal Institute of Hydrology (BfG) and the Federation of German Consumer Organizations (VZBV). The former in particular is a key organization when it comes to water protection. Some actors with potentially crucial roles in the German political architecture were not taken into account here. For instance, the German Chancellery (in German: Bundeskanzleramt) was not considered here because the preliminary interviews revealed that the Chancellery had not played a major role in this particular policy process. Furthermore, the parliamentary committees of the Council of Constituent States (in German: Bundesrat) were excluded because the Council of Constituent States is bound to the instructions of the Länder. Thus, directly surveying the German Länder, rather than the Council of States, is necessary in order to capture the interests of

the Länder.[9] Therefore, the five Länder, which are part of the Rhine watershed, were included in addition to two Länder working groups on water, which assemble all 16 German states. The ordinance did not have to be adopted by the lower house of the German Parliament (Bundestag), and therefore, the legislature, as well as parties on the national level, did not play a key role in this policy process.

- France: In addition to those actors participating in the policymaking process, 17 actors were identified because of the significant role they play. Among those actors are two national public authorities, i.e., the National Water Committee (CNE) and the French Environment and Energy Agency (ADEME). Moreover, regional and local governmental actors play an important role with regard to water quality policy, i.e., the prefects of water basins (DELEGBASSIN), the Local Territorial Authority (DDT), the Assembly of the French Departments (ADF), the Basin Committee, and Local Water Commissions (CLE). Additionally, a number of non-governmental actors achieved a crucial position in French policymaking with regard to water quality policy, including the World Wide Fund For Nature France (WWF), Friends of the Earth France, Chemical Industries Union (UIC), Assembly of French Chambers of Commerce and Industry (CCI), EDF Energy (EDF), Association of French Paper Industries (COPACEL), Federation of Mechanical Engineering Industries (FIM), Farmers Association (CHAMBREAGRI), Federal Union of Consumers (UFC), and Consumers, Housing and Well-being Association (CLCV).

Some French actors were not retained despite their potentially momentous roles. All those actors who were only identified through the positional approach, but did not participate to a significant degree in the policymaking process, and were not identified through the reputational approach either, were not added to the list. Examples include interministerial initiatives (Mission interministerielle de l'eau MIE, Comité interministeriel pour l'Environnement, Commissariat général au Développement durable CGDD), green NGOs (Société national de protection de la nature), water associations (L'Office International de l'Eau OIEAU, Association Scientifique et Technique pour l'Eau et l'Environnement ASTEE, Cercle français de l'eau), business associations (Union nationale des producteurs de granulats, Associations des Industriels pour la Protections de l'Environnement ALSAPE), agricultural organizations (Fédération nationale de la propriété privée rurale, Association nationale des industries alimentaires), societal associations (UNAF), the legislature, as well as political parties, and finally some governmental bodies (Mission inter-services de 'eau MISE, Services départementales de l'ONEMA, Commissions locales de l'eau du schéma d'aménagement et de gestion des eaux, and Association française des établissements publics territoriaux de bassin). Greenpeace is confirmedly one of

---

[9]Telephone interview with Bureau of the Council of Constituent States (Bundesrat) on Jan 27, 2014.

the most pivotal environmental NGOs in France, but was not added to the list due to their self-reported disengagement in politics on micropollutants.[10]

- The Netherlands: 20 actors were identified through the positional approach in addition to the decisional method. Among those includes the Netherlands Enterprise Agency (RVO), which composes part of the Ministry of Economic Affairs. RVO potentially plays an important role in water protection policy since it supports new types of businesses, such as those developing environmental and sewage treatment technologies. In addition, there are two national-level advisory or discussion groups in the Netherlands (National Executive Talk Water, Advisory Committee on Water), where high-ranking personalities initiate discussions about new issues such as micropollutants. Moreover, several ministries play a key role in water protection. Prior to the administrative reform in 2010, two ministries with water-related tasks existed, namely the Former Ministry of Transport and Water Management (V&W), and the Former Ministry of Housing, Spatial Planning, and the Environment (VROM). Both were added to the list of actors because the policy process had begun well before the reform. Hence, it is reasonable to assume that the ministries played out significantly in the policy process throughout their existence. During the reform, the former Ministry of Agriculture, Nature and Food Quality (LNV) was affiliated with the Ministry of Economic Affairs. Prior to 2010, however, the LNV potentially played a crucial role in agricultural micropollutants policies and was therefore considered in this study. On the regional level, the Association of Dutch Provinces plays a pivotal role in ground water protection as well as spatial planning. The Association of Dutch Municipalities is also a key player with regard to water protection, since the municipalities are responsible for rainwater runoff, urban ground water, and collecting sewage from households. The legislature (parliamentary committees of both chambers) holds an important role in the Dutch political system and was therefore added to the actor list. Agriculture as a large source of pollution, particularly in the Netherlands, has led the Dutch Federation of Agriculture and Horticulture (LTO) to be identified an important actor, as well as the three biggest economic and water-related associations (Confederation of Netherlands Industry and Employers, Association of the Dutch Chemical Industry VNCI, Industrial Water User Association VEMW). The concern over antibiotics has been discussed in the Dutch policy discourse, and therefore, the Dutch Working Party on Antibiotic Policy (SWAB) was identified as a potentially significant actor. The Rioned Foundation is a center of expertise in sewer management and urban drainage and was therefore also identified as important actor pertinent to this study. Lastly, most environmental and consumer associations were singled out through the positional approach (Cooperative Fishery Organization, Society for the Preservation of Nature in the Netherlands, World Wide Fund For Nature the Netherlands (WWF), and the Consumer Association).

---

[10]Telephone interview with Greenpeace France on Oct 28, 2013.

Thirdly, a list containing all the actors identified through the decisional and positional approach was then presented to top-level state officials responsible for the legal draft proposal (*reputational approach*). These persons followed the policy process closely and thus are knowledgeable of the involvement of actors. In 12 in-person interviews preliminary to the survey (see Annex 1), the top-level state officials were asked to identify those actors on the list that played a role in the policymaking process and to name those who might be missing. Finally, the network population consists of all those actors identified by one of the three approaches.[11] The following actors were identified by the reputational method in the four countries, in addition to the decisional and positional method:

- Switzerland: The following five actors were identified in three interviews: Association of Cantonal Chemists of Switzerland (VKCS), Lab'Eaux (LABEAUX), Swiss Fishery Association (FISCH), Swiss Cosmetics and Detergent Association (SKW), and Université de Lausanne (UNIL).
- Germany: Eight preliminary interviews were conducted, and only one missing actor, the Agrochemical Association (IVA), was identified in a survey interview.[12]
- France: In a preliminary interview, two additional actors were pointed out. A regional environmental organization, which is particularly active in the Rhine basin, i.e., Alsace Nature (ALSACENATURE); and the National Innovation Centre for Sustainable Development and Environment in Small Companies (CNIDEP), a public research institute that develops environmentally friendly solutions for small businesses.
- The Netherlands: A total of 5 additional actors were identified during interviews and therefore added to the list. While a preliminary interview revealed that the political parties had not played a major role in the field of micropollutants, survey interviews pointed to the engagement of three political parties in the field of micropollutants (D66, PvdA, and Groenlinks). Moreover, interviews revealed that an environmental NGO, the Foundation House of the Earth (STICHTINGHA), was very active in sensitizing negative environmental effects of pharmaceutical use. Lastly, the engineering company Grontmij was selected, due to their *Grontmij Report* concerning the potential costs of upgrading all Dutch treatment plants with a fourth treatment technology.

An overview over the final target survey population is provided in Table 3.1.

---

[11]The policy network literature does not clearly state whether one of the three approaches is sufficient enough to identify actors as members of a policy network, or whether only those actors qualify as members of the network that are identified simultaneously by all three approaches. It is not clear whether a key actor must be identified by one, two, or all three approaches, or whether a hierarchy exists with regard to them.

[12]Interview with State of Hesse (HES) on Mar 13, 2014.

Table 3.1 Overview of survey populations

Actor type	Number of actors			
	CH	G	F	NL
National governmental	13	7	6	13
Lower-level governmental	7	9	12	3
Elected	9	0	2	9
Environmental	4	7	5	4
Business	10	14	10	10
Consumer, social	3	4	2	1
Water	5	5	2	4
Science	11	4	9	5
Total (N)	62	50	48	49

3.2.1.4   Questionnaire Design, Mixed-Mode Survey, and Contact Procedures

The survey conducted in this research project aimed at understanding the role of actors in the underlying policymaking processes concerning aquatic micropollutants, i.e., actors' degree of involvement in the policy processes, their policy beliefs, and collaboration patterns. The survey was a mixed-mode survey, including Web, paper, and electronic e-mail questionnaires surveying only those policy actors who are concerned and who participated in the national policymaking process on aquatic micropollutants in the four countries under investigation, Switzerland, Germany, France, and the Netherlands (see Sect. 3.2.1.3 for an explanation of actor identification). The ensuing paragraphs first explain questionnaire design and then go on to describe the procedure of data gathering. The content of the questionnaires that is relevant to this study is exposed in Sect. 3.2.3 (see Annexes 6–9 for full questionnaires).

Questionnaire design

The questionnaire was developed with the aid of literature on survey methods, in order to improve measurement validity (Krosnick 1990), optimize scales and labels on rating points (Saris et al. 2010; Prüfer et al. 2003), create incentives to complete and return the questionnaire, and gain knowledge on pretesting methods (Schaeffer and Presser 2003). Moreover, the questionnaire was developed based on a workshop on survey design in social network analysis with Mark Lubell, Karin Ingold, and Philip Leifeld (March 1, 2012) at the University of Berne. Further literature on gathering network data was used (Wasserman and Faust 2009; Scott 2000; Borgatti et al. 2013).

Pretests

The Swiss questionnaire was the first of the four to be sent out and was therefore intensively pretested. Twenty-two persons, i.e., researchers from sociology, political science, and water research, as well as top-ranked bureaucrats and water professionals, reviewed the questionnaire. All other questionnaires (G, F, and NL)

consisted of the exact same questions and were pretested in an interview preliminary to the survey. With the time and field experience underway, minor improvements were made to the design of single questions.

The questionnaire's language was German in the case of Switzerland and
Germany, French in the case of France, and English in the case of the Netherlands.
An English survey would not constitute a problem in the Netherlands according to
Dutch project partners from the University of Twente's Department of Governance
and Technology for Sustainability, as well as a representative from the water board
Vechtstromen. Native speakers proofread the German, French, and English
questionnaires.

Overview of questionnaire

The questionnaire contained 13 questions in the case of Switzerland and 12
questions in the case of Germany, France and the Netherlands, which cover the
following topics (see Annexes 6–9 for full questionnaires):

- Participation in a defined policy process on micropollutants (3 questions),
- Network relations (3 questions),
- Policy beliefs and preferences (3 questions),
- Responsibilities in water quality issues (3 questions in the Swiss case/2 ques-
  tions otherwise), and
- International collaborations (1 question).

The exact wording and response options of the survey questions used here are
explained in Sect. 3.2.3. This book refrains from highlighting all the survey
questions that provide background knowledge, but whose responses were not used
to operationalize variables.

Mixed-mode survey

For the purpose of this study, a mixed-mode survey procedure was chosen in order
to secure data from various types of respondents (Dillman et al. 2009; Van Selm
and Jankowski 2006). The mixed-mode survey was applied to the present study
such that about ten actors were interviewed in each country, and all other actors
received a paper or online mail survey. Actors were interviewed who hold a key
role regarding micropollutants, and who have been named in a pretest interview as
particularly important. In total, 41 out of 199 surveyed actors were interviewed.
Complementing mail surveys along with interview surveys has several advantages:
Some of the survey respondents were high-ranking political leaders whose
responses were crucial for the success of the present study. At the same time,
however, it was likely that high-ranking political leaders do not make the time to fill
out online or paper-and-pencil surveys. An interview, on the other hand, is a more
appropriate and promising mode of interaction with high-ranking officials. Hence,
the first purpose of in-person interviews was to ensure survey responses from
important policy actors. Another advantage of mixed-mode surveys is that interviews allow for gathering qualitative background knowledge throughout the

Table 3.2 Types of survey responses

	CH	G	F	NL
In-person interview	11	8	12	10
Paper/electronic	36	23	1	5
Online	–	–	5	–

conversations. Non-structured interview segments were allotted room by asking additional, open questions on the policy itself, the political system, water-related institutions, the policy process, and roles, and competences of policy actors. These qualitative insights help with understanding and contextualizing the analyzed quantitative data gathered through structured survey segments. A mixed-mode survey also offers the advantage of different modes being offered to the respondents. In fact, in many instances, the respondents themselves chose the mode through which they wished to take the survey. Some actors were asked for an interview but preferred to fill out a paper or electronic questionnaire; others refused the paper or electronic version and requested an in-person or telephone interview. Table 3.2 provides an overview of the survey response modes.

Potential disadvantages of mixed-mode surveys are that respondents might produce inconsistent answers based on the mode in which they are surveyed (Check and Schutt 2011, p. 178). The present survey uses the same question structures, response choices, and instructions across the different survey modes in order to reduce the likelihood of divergent response, and to ensure data compatibility. Interviewees were asked the exact same questions during face-to-face interviews as survey respondents of paper or online questionnaires. Based on this methodology, the same information exists for all actors, and data can be analyzed quantitatively over a larger number of observations. Additionally, the likelihood of mode-related response differences was reduced by offering only a small number of response choices to respondents for each question (Check and Schutt 2011).

Survey contact procedures

Prior to launching the survey, each sampled policy actor was contacted in order to identify the responsible senior representative of the respective organization for the topic of aquatic micropollutants. The survey was then personally addressed to the identified contact. In the case of paper surveys, full survey packets were sent out containing an official cover letter summarizing the research project, a 12-page questionnaire, and a self-addressed postage-paid reply envelope. One week before the paper questionnaire was due, and in the two consecutive weeks, all non-responders received an e-mail reminder with an electronic version of the questionnaire attached. All those who did not return the paper survey after two e-mail reminders were personally contacted by telephone. In the case of in-person interviews, the identified responsible person received an official letter asking for an interview in an e-mail attachment. In the case of France, the paper survey was replaced by an online survey implemented in Qualtrics (2013). An e-mail invitation was sent out that included a summary of the project and the link to access the survey's Web site.

Table 3.3 Time frame for the surveys

	CH	G	F	NL
Sending out	April 2013	January 2014	October 2013	April 2014
Closure	July 2013	May 2014	May 2014	August 2014

The French survey took about 8 months (see Table 3.3), while the other surveys were closed after about 5 months. The low response rate to the online survey was a factor affecting the lengthier time frame of the French survey. In order to compensate this low response rate, actors were recontacted and asked for in-person interviews, due to the improved responsiveness of in-person conversations.

3.2.1.5 Response Rates

Each response was coded by applying the standard coding scheme of the American Association for Public Opinion Research (AAPOR 2011). Each individual response or non-response was recorded as displayed in Table 3.4. Questionnaires where more than half of the questions were completely filled in are coded as *completion*. Partial completion refers to surveys where less than half of the questions were filled in. *Refusal or break-off* are cases in which the respondent clearly stated that they did not wish to participate, or returned a blank questionnaire. If a respondent was unable to take the survey for health reasons, and no other person was able or willing to replace the respondent, the code *unable for health reasons* applies. The code *not mailed* describes cases where it was impossible to locate a contact person for an organization, and consequently, the survey was not mailed. When a respondent never returned the questionnaire, it was coded as *nothing returned*. A mailed survey that never reached the person eligible to fill it out was coded as *unable to reach*. Finally, some actors self-reported to not have been involved in the policy process

Table 3.4 Combined disposition summary (Web, paper, interview)

Dispositions (AAPOR code)	CH	G	F	NL
Completion >50% (1.0)	44	27	17	14
Partial completion $\leq$ 50% (1.2)	3	2	1	1
Refusal or break-off (2.112, 2.12)	4	8	10	5
Unable for health reasons (2.32)		2		
Not mailed (3.11)	1	1	10	1
Nothing returned (3.19)	1	3	4	12
Unable to reach (3.17)				3
Ineligible (4.1)	5	5	1	8
Nonexistent (4.6)				5
Duplicate (4.9)	4	2	5	
Total	62	50	48	49
Response rate	89.2%	68.4%	45.5%	50.5%

on micropollutants and were coded as *ineligible*. In the Netherlands, an encompassing administrative reform and budget cuts led the Government to dissolve and merge several ministries and agencies in the years 2007 and 2010. These cases were coded as *nonexistent*. Nevertheless, these ministries or agencies played a role in the policy process when they were still in existence and are therefore part of the network population. The code *duplicate* refers to cases where the actors themselves reported to have a common position with another actor. This code also refers to actors that represent different groups or levels of the same policy actor, e.g., different departments in the same organization. The full actor list for each country is provided in Annexes 10–13.

The response rate formula and key is given below (AAPOR 2011). Using the AAPOR's response rate formula 4, the survey response rate is 89.2% for Switzerland, 68.4% for Germany, 45.5% for France, and 50.5% for the Netherlands.

$$RR4 = \frac{I+P}{(I+P)+(R+NC+O)+e(UH+UO)}$$	Response Rate 4 (RR4) is the number of completely plus partially returned questionnaires divided by the total number of eligible questionnaires sent out. The eligible questionnaires sent out include the number of completely and partially returned questionnaires, the number of non-responses (refusal and break-off, non-contacts, others) plus all cases of unknown eligibility. RR4 calculates response rate by estimating what proportion of cases of unknown eligibility is actually eligible (e) (AAPOR 2011).
RR = Response rate I = Complete survey (1.0) P = Partial survey (1.2) R = Refusal and break-off (2.112, 2.12) NC = Non-contact	O = Other (2.32) UH = Unknown (3.11, 3.17, 3.19) UO = Unknown, other e = Estimated proportion of cases of unknown eligibility that are eligible.

3.2.2 Social Network Analysis

Social network analysis has been used as a method in political sciences since the 1980s and has grown in popularity ever since (Lang and Leifeld 2008). The success of this type of analysis can be attributed to many political phenomena involving complex dependencies for which relational analysis provides particularly suitable explanations (Hennig et al. 2012; Lang and Leifeld 2008). Power dependencies typically emerge in politics, which can be 'decoded' using social network analysis. Social network analysis can account for the division of power and labor among state and non-state actors without predefining a hierarchical structure (Fischer 2012, p. 44; Kenis and Schneider 1991, p. 34). By accounting for relational dependencies, social network analysis provides powerful explanations for political phenomena including policy outcomes.

Policy networks are *social networks* defined as structures composed of a set of actors and their relations (Hennig et al. 2012). *Policy actors* are collective actors, i.e., formal organizations, such as political parties or associations (Knoke 1990). A *relation* depicts one particular type of connection between a pair of actors, i.e., a *dyad*. A huge variety of relations can occur among social actors. For most analytical purposes, each type of relation needs to be represented in a separate network, even with the same set of actors (Knoke and Yang 2008, p. 45). Relations can either be *directed* or *undirected*. Directed relations are represented through lines with arrowheads pointing from the sender to the receiver (Knoke and Yang 2008, p. 6). In a policy network, directed relations can be, among others, power, hierarchical, or financial dependencies; reporting duties; advice seeking; or knowledge transfer. Some relations such as dialogues or conversations are by definition mutual, which means that they exist only when both actors maintain the connection. Collaboration is another example for a mutual relation, as it is impossible for an actor to collaborate with another actor when the other refuses to. Mutual relations can be represented as lines without arrowheads indicating a symmetric, undirected tie (Knoke and Yang 2008, pp. 6–9).

Policy networks can be depicted as *sociograms*, also defined as graphs or network diagrams, which are two-dimensional diagrams composed of points and lines. The points in sociograms represent policy actors and are also called *vertices* or *nodes*. They often carry the label of the organization's name for which they stand. Lines, on the contrary, represent the ties among actors and are also termed *edges* or *arcs*. The precise placement of nodes and the length of ties are usually arbitrary in sociograms (Knoke and Yang 2008, pp. 45–49). Therefore, a visual representation of sociograms provides only a first insight into the underlying data structure. The application of mathematical concepts from graph theory allows complementing visual representation with a larger set of powerful, quantitative analysis in order to rigorously examine the underlying social structure (Borgatti et al. 2013).

Social network data is conventionally stored as a tabular display taking the form of an *actor* × *actor* matrix, also called *sociomatrix* or *adjacency matrix*. An adjacency matrix is a matrix in which the rows and columns represent actors, and the sequence of actors is identical across rows and columns. The cell contents display information about the presence, absence, or intensity of ties between a defined pair of actors (Borgatti et al. 2013, p. 18; Scott 2000). Two nodes are termed *adjacent* when a tie exists between them. By convention, row actors are the senders of ties, while column actors are receivers. Another convention is the notation of a cell entry as x_{ij}, where x_{46}, for example, denominates the tie that the fourth row actor sends to the sixth column actor. Information about the presence or absence of this relation is stored in the cell entry, where the entry $x_{ij} = 1$ signals that the actor i sends a tie to actor j; $x_{ij} = 0$ illustrates the absence of a tie. In symmetric networks x_{ij} equals x_{ji} ($x_{ij} = x_{ji}$), which means that the tie i sends to j is reciprocated by a tie from j to i (Knoke and Yang 2008, pp. 40–50; Wasserman and Faust 2009, pp. 150–157). Not all network members need to have direct links to all other actors, because network analysis accounts for present and absent ties. The analysis of complete graphs, where all possible ties are present, is in many cases less informative or revealing than the analysis of incomplete graphs where actors fall into separate clusters, cliques, or components, for example. Actors without any connection to others are called *isolates* in social network analysis, while those with only one tie are termed *pendants* (Borgatti et al. 2013, p. 14).[13]

Researchers analyzing social networks have a number of network measures at their disposal to quantitatively analyze relational structures on the graph-level, dyad-level, or node-level. A commonly applied graph-level measure is network density, which captures the degree of present ties compared to the total amount of possible ties, if all actors were connected. Reciprocity is a typical dyad-level measure and captures the tendency of nodes to reciprocate ties. Popular node-level measures are centrality measures, i.e., nodal degrees, which quantify graph theoretical ideas about actors' activity, importance, or power. Other examples of commonly applied network measures include structural equivalence, blockmodeling, or the analysis of cycles or geodesics (Wasserman and Faust 2009; Scott 2000). Those network measures that are used for the empirical analysis in this study will be explained in more detail in the next chapter.

In conclusion, researchers can use social network analysis to study individual and systemic levels of analysis. On the individual level, social network analysis allows for studying the degree to which actors' attitudes, beliefs, voting behavior, or policy preferences are influenced by their social relations. On the network level, network analysis can reveal variations in social structure along with their consequences. The structural patterns of ties influence systemic performance and, therefore, can be used to explain collective outputs. In summary, the quantitative analysis of social relations takes account of the claim that 'relations matter' to collective performance and provides powerful explanations for policy research in addition to individual attributes or institutional contexts.

3.2.3   Operationalization of Policy Network Structure

A key task of every empirical researcher is to adequately translate theoretical concepts into measurable (or categorical) variables in order to inspect whether a theory is supported by observation (Hardy and Bryman 2004). Capturing a theoretical concept adequately is no simple task, for instance when necessary data is difficult or impossible to gather.

Translating theory into verifiable indicators is termed the *operationalization* of variables. Operationalization includes, first, how a theoretical concept is measured in the data-gathering process. Secondly, operationalization concerns the way in

---

[13]Technically, pendants are only those actors who have one tie toward others exhibiting more than one tie (Borgatti et al. 2013, p. 14).

which the gathered data is arranged, i.e., on which scales it is coded. Thirdly, operationalization illustrates which data analysis method is employed in order to analyze the coded data. Hence, the operationalization of variables provides information about (a) measurement in the data-gathering process, (b) coding of the data, and (c) method of data analysis. Scholars must describe these steps carefully in order to make a study replicable for other researchers. To this end, the following chapter explains how policy network structures were translated into measurable variables and how data was gathered, coded, and finally analyzed.

Section 3.2.3 draws attention to several concepts characterizing the structure of policy networks, and based on these concepts, Sect. 4.1 formulates different hypotheses for the impact of structural network properties on policy design. Hereafter, these theoretical concepts are translated into measurable indicators. Sections 3.2.3.1–3.2.3.6 explain one-by-one how the studied network characteristics are operationalized.

3.2.3.1 Belief Cohesion

Belief cohesion (see Sect. 3.1.2 for a detailed explanation), as employed in this work, refers to the consensus between actors about policy content. To capture the level of consensus between actors, the present study relies on information about actors' beliefs. Following the ACF definition, beliefs are deeply rooted values (Sabatier and Jenkins-Smith 1993). The ACF literature establishes a three-tiered hierarchical concept of beliefs by broadly distinguishing stable *deep core* and *policy core beliefs* from less stable *secondary aspects*. While deep core beliefs are described as fundamental values spanning over several policy fields, core beliefs apply to a specific policy field (e.g., environmental, social, or economic policy). In this study, beliefs refer to policy core beliefs regarding the policy issue of micropollutants in waters. Relevant policy core beliefs include actors' level of concern for the aquatic environment. Actors with high environmental concern are more likely to approve policy measures for the reduction of micropollutants in waters than are actors with a low level of environmental concern. Another relevant belief here is actors' risk averseness, since many uncertainties remain regarding micropollutants (Metz and Ingold 2014). Actors with high risk averseness are likely to favor reduction measures because of potential risks for humans and the environment. Actors with low risk averseness, on the contrary, are likely to neglect the necessity of adopting policy measures.

For the purpose of this study, actors' beliefs, i.e., environmental concern and risk averseness, were surveyed by asking respondents for their level of agreement with the five following statements (the process of data collection was previously explained in Sect. 3.2.1.4):

Please indicate your organization's level of agreement with the following statements:

1. *Measures should address the sources of pollution.*
2. *Measures should be end-of-pipe.*
3. *Precautionary measures should be taken to reduce potential risks for humans and the environment (precautionary principle).*
4. *It is reasonable to wait with policy measures until the impact of micropollutants is fully understood.*
5. *Policy measures should aim at completely eliminating micropollutants in waters.*

Among the above-mentioned statements, numbers 1, 2, and 5 test for different aspects of environmental concern, while statements 3 and 4 test for risk averseness.

Response options for statements 1 through 4 ranged from *strongly agree* to *strongly disagree* on a four-point scale (Prüfer et al. 2003). For statement 5, the response options included: *completely* (meaning: policy measures should aim at completely eliminating micropollutants), *largely*, *only a few substances*, or *not at all*.

In all statements, a four-point scale was applied, because all surveyed actors participated in decision-making processes and are therefore assumed to hold an opinion on the policies targeted at micropollutants. Middle—or 'I don't know'–options were excluded since they may be a way for respondents to avoid the cognitive work that is necessary for generating optimal answers (Krosnick 1990, p. 559). With the four-point scale, respondents are encouraged to openly report their organizations' policy beliefs and to provide well-informed answers even if time-consuming and resource-intensive. Answers to each of the five statements were coded between 1 (strongly disagree) and 4 (strongly agree).

All five statements are employed to operationalize belief cohesion as one variable. Consequently, belief cohesion refers here to networks where actors display similar levels of environmental concern and risk averseness. Using the five statements on actors' beliefs, a *dissimilarity matrix* was constructed. A dissimilarity matrix is an *actor × actor* matrix where each cell describes the level of belief dissimilarity between two actors. The higher the values, the more dissimilar actors' beliefs are. The lower the values, the more similar actors' beliefs are. Distances are calculated for each pair of actors on all five surveyed statements. The matrix is obtained by calculating the Manhattan distance d between each pair of actors p and q for all five surveyed statements i using the software UCINET, version 6.352 (Borgatti et al. 2002). In the formula, p_i denotes the coded value of the answers the actor p gave to each of the five survey statements:

$$d(p,q) = \sum_{i=1}^{5} |p_i - q_i|$$

The Manhattan distance is therefore, the absolute difference between the scores for individual statements. For instance, if one actor obtains a score of 4 and another actor a score of 1 for a statement, they share a distance of 3 for that statement.

Using this methodology, distances for all five statements are added together in order to generate one dissimilarity matrix for the whole network.

Since belief cohesion concerns belief similarity, the dissimilarity matrix was transformed into a *similarity matrix* $s(p, q)$ by subtracting each dissimilarity value d (p, q) from the maximum dissimilarity value $\max(d)$ (Leifeld and Schneider 2012) using the statistical computing environment R (R Development Core Team 2014, version 3.1.2.):

$$s(p, q) = \max(d) - d(p, q)$$

In similarity matrices, the higher the values, the more similar actors' beliefs are, and the lower the values, the more dissimilar actors' beliefs are.

Finally, the density of the belief similarity matrix was calculated in UCINET for each of the four underlying networks separately. Densities provide information about the amount of present ties compared to the total amount of possible ties, if all actors were connected. Mathematically, density Δ for an undirected graph represents the ratio of the number of present ties, L, to the maximum number possible.

When g stands for the total number of nodes in the network, $g(g - 1)/2$ corresponds to the number of maximum possible ties in undirected graphs. Hence, density Δ is calculated as follows (Wasserman and Faust 2009, p. 101):

$$\Delta = \frac{L}{g(g - 1)/2} = \frac{2L}{g(g - 1)}$$

Here, a tie L signifies that there exists some degree of similarity between two actors, that is, nonzero similarity between them. The density of belief similarity ties, thus, indicates the average level of belief similarity in the overall policy network.

3.2.3.2 Interconnectedness

From a theoretical perspective, interconnectedness refers to the number of contacts among actors in the policymaking process (Bressers and O'Toole 1998). The word *contacts* is quite broad and could point to different types of relations, including information exchanges or negotiations. A widely used relational type in policy network literature is *collaboration* (for instance Kriesi et al. 2006; Fischer and Sciarini 2013; Ingold 2011; Henry et al. 2010). Collaboration is defined as a mutual engagement and may therefore be considered a rather stable, reciprocal relation (Ingold and Gschwend 2014). Engaging in collaboration can be time-consuming for policy actors and might involve travel costs or the appropriation of other resources. Hence, collaboration is considered a sign that actors coordinate their policy action in order to achieve a defined strategic goal. Through collaboration, actors develop policy options, exchange their positions, evaluate alternatives, and finally decide upon policy proposals. Since collaboration constitutes a central element of policymaking processes, I have chosen collaboration ties as a proxy for *contacts among actors in the policymaking process.*

To survey actors' collaboration ties, a complete list of actors participating in the underlying policy process (actor identification was explained in Sect. 3.2.1.3) was presented to each respondent.

Respondents were then asked the following question:

Please check all the actors with whom your organization has closely collaborated during the policy process on micropollutants (time frame and name of the respective process was given).

Close collaboration is defined as discussing new findings, developing policy options, exchanging positions, and evaluating alternatives.

If there are actors missing, please add them to the bottom of the list, and indicate with an 'x' if you closely collaborate.

The survey participants' responses were coded as a dummy variable (0,1) in an *actor × actor* matrix. To compare how closely actors collaborated in the four policy networks under investigation, indicators need to be independent from the network size. The density indicator discussed earlier is unsuited for this purpose, because large networks tend to be less dense than small networks (Friedkin 1981). Being well-connected in large networks demands a higher amount of resource investment per actor than in small networks. Given limited resources for tie-formation, densities therefore tend to be lower in large networks.

To overcome this difficulty, two different parameters were calculated here that are less dependent on network size: *ties per node* and *triangles per tie*. What is called here *ties per node* is equivalent to the term *mean degree* of a node. Mean degree provides information on the number of ties that are sent out, on average, by each node. To calculate mean degree (or ties per node), the total quantity of undirected ties in the policy network is divided by the total quantity of nodes. It can be shown mathematically that mean degree is less dependent on network size than is density: Given a fixed number of ties in a network, network density decreases faster with an increasing amount of network members N, than mean degree. With increasing N, the mean degree of a network will scale as a $1/N$ function, whereas density will scale as a $1/N^2$ function, given a fixed number of network ties. The curve of the $1/N$ function approaches the x-axis more slowly than the curve of the $1/N^2$ function. Hence, mean degree is less exposed to size-related biases than does density.

The second parameter considers triangular patterns in the network. Triadic structures are considered here because they indicate that ego's alters (the contacts of the focal node) collaborate with each other. Triangles in networks correspond to closed social circles. Where such closed triadic clusters are present, collaboration benefits from extensive redundancy and cross-checking (Berardo and Scholz 2010). For *triangles per tie*, first, a triad census is conducted with the *statnet* package in the statistical software environment R (Handcock et al. 2008, version 2014.2.0), and secondly, the total number of undirected triangles (MAN 300) is divided by the number of ties in the network. The result provides an average measure of the number of triangles that each tie is part of. A strong preference for closed triads in a network can be considered a proxy for dense collaboration patters.

3.2.3.3 Coalition Formation

Advocacy coalitions are defined by the ACF literature as groups of policy actors with shared belief systems who coordinate their actions within their coalition in order to pool resources and influence policy design (Zafonte and Sabatier 1998). Hence, coalitions are defined by both a high degree of internal belief cohesion as well as a high degree of internal interconnectedness. These defining characteristics are also employed in the policy network literature, which usually applies a two-step approach to the identification of advocacy coalitions (e.g., Knoke 1990; Kriesi and Jegen 2001; Sciarini et al. 2004; Fischer 2012; Ingold 2008; Knoke et al. 1996): First, actors with shared beliefs become clustered into a block, while actors with distinctive beliefs are grouped in opposite blocks. Secondly, the degree of actors' collaboration within blocks is calculated. To qualify as an advocacy coalition, both criteria, namely within-block belief cohesion and interconnectedness, have to be fulfilled.

To identify blocks of actors with shared beliefs, some studies in the policy network literature have used the subjective perception of actors' agreement with each other as a proxy for joint beliefs (Ingold 2011; Fischer 2015), because within a coalition, policy actors share common beliefs and agree with one another on policy preferences. Across coalitions, on the contrary, actors' beliefs fundamentally clash and there exists disagreement between actors on the direction of policy. To survey actors' agreement and disagreement with other actors, the same, complete list of policy actors was presented to respondents as for the collaboration question (in previous Sect. 3.2.3.2).

Respondents were asked the following question:

Please check all the actors with whom your organization had convergences and/or divergences about policy content during the policy process on micropollutants (time frame and name of the respective process were given).

'Convergence' is defined as agreement on policy content; 'divergence' as disagreement.

If there are actors missing, please add them to the bottom of the list and indicate your convergences and divergences.

Responses were coded in an *actor* × *actor* matrix, where 1 corresponds to a convergence tie between the sender and the receiver and −1 to a divergence tie; 0 is equivalent to a neutral tie. In some rare cases, actors reported to have both convergence and divergence ties, because they agree about some parts of the policy design, but disagree about others. These cases can indicate a traditionally conflicting relationship between two policy actors, where a long history of mediation has led the actors to stress some degree of common ground. Because belief systems are nevertheless divergent, these types of multiplex relations were recoded as −1 (divergence).

To identify blocks of advocacy coalitions, actors who agree with each other are grouped together, and actors who disagree are placed into opposite groups. This grouping can be best achieved by applying the blockmodeling procedure called *balance* or the *Doreian-Mrvar method* in the social network analysis software PAJEK (Batagelj and Mrvar 1996, version 1.23; Batagelj and Mrvar 1998). The balance procedure partitions the actor matrix into blocks such that actors with agree-ties are grouped together, while actors with disagree-ties are placed in opposite blocks (Doreian et al. 2005; Doreian 2008; De Nooy et al. 2005). The procedure continuously rearranges the actor matrix until it reaches a configuration that is closest to a predefined block structure (analysts predefine the number of blocks) with positive within-block ties and negative between-block ties (Fischer 2015). An error term indicates deviations from the ideal arrangement (Doreian and Mrvar 1996, 2009). More concretely, the error term accounts for disagreement ties within blocks and for agreement ties between blocks. The block structure with the smallest error term provides the best fit. For all four policy networks under scrutiny, partitions with two blocks resulted in the smallest error term. This represents a first indication that the underlying policy networks fall into two advocacy coalitions. In some rare cases, policy actors had a tie profile, which allowed for membership in both coalitions. These actors were treated as missings in subsequent analyses.

To evaluate whether the blocks of actors with similar beliefs fully qualify as an advocacy coalition, the degree of coordination within and between blocks was analyzed. To this end, the densities of collaboration ties within and across blocks were calculated (for an explanation of density, see Sect. 3.2.3.1).[14] If a block of actors displays fewer ties internally than it does toward another block of actors, this strongly indicates that such a block of actors is not a real coalition. The block of actors with some degree of belief cohesion does not satisfy the criterion of inter-connectedness and can thus not be called a coalition by virtue of its definition, but rather a loose conglomeration of actors without common strategic goals. By contrast, if a block of actors with similar beliefs exhibits high densities of collaboration ties among its members, this is a strong indication for an advocacy coalition.

After having identified actors' memberships in coalitions, I analyzed the mean beliefs of coalitions in order to gain knowledge of coalitions' general policy orientations and to provide coalitions with a suitable label. For the purpose of this study, five beliefs were surveyed (see Sect. 3.2.3.1). Actors were asked about their level of agreement with a source-directed, end-of-pipe, and preventive approach for the reduction of micropollutants, and about their agreement with the political aim to achieve zero-concentration levels of micropollutants in waters. Response options

---

[14]For the calculations of densities within coalitions, the neutral and well-connected brokers are excluded because their inclusion would impact densities to the advantage of the coalition of which the brokers are a part.

ranged from strongly agree (coded as 4) to strongly disagree (coded as 1) on a four-point scale (Prüfer et al. 2003). The mean of the answers to these four beliefs was calculated for each coalition in order to capture coalitions' pro-environmental policy orientation. Where a coalition attained a high mean value (closer to a score of 4), it was labeled *water quality coalition*. Additionally, survey respondents were asked to indicate their level of agreement with a policy orientation that opposes preventive measures. High mean values of that belief were treated as an indicator for a coalition opposing a pro-environmental policy orientation. These coalitions were labeled *opposing coalitions*.[15]

3.2.3.4 Coalition Structure

Once the formation of coalitions has been identified, the exact coalition structure can be analyzed in more detail. The goal of this step of analysis is twofold: The first aim is to assess whether coalitions display similar levels of power, or whether they fall into dominant or minority coalitions. Secondly, this step of analysis concerns the classification of coalition structures as collaborative, adversarial, or unitary subsystems (Weible et al. 2010; Ingold and Gschwend 2014).

The first of the goals mentioned, i.e., analyzing the power of coalitions, involves the distinction between *dominant coalitions* and *minority coalitions*. Power was measured here by combining three indicators. The first indicator represents the size of coalitions in terms of the number of actors, where larger size is a first indicator for a dominant coalition. The second indicator builds on the notion that a dominant coalition should include actors who are well-connected. I measure connectedness based on the mean of all coalition members' scores of degree centrality (for undirected collaboration ties).

Mathematically, the degree of a node n_i, $d(n_i)$, denotes the number of ties incident with it (Wasserman and Faust 2009, p. 100). Here, the subscript i denotes an index that runs through all nodes within a coalition. The coalition degree, denoted by $\bar{d}_c$, signifies the average of coalition members' individual degrees. When c represents the number of actors in one coalition, the *coalition degree centrality* is calculated as follows:

$$\bar{d}_c = \frac{\sum_{i=1}^{c} d(n_i)}{c},$$

---

[15]The classic opposition between pro-environmental and pro-economic coalitions did not seem valid in the present study, because actors reported diverse motives (aside from economic ones), in order to oppose a pro-environmental policy: While there do not exist any actors openly reporting against 'clean water', opposition is more related to questions of risk averseness (do micropollutants cause problems?), the distribution of costs (who has to pay for reduction measures?), responsibilities (who are the responsible emitters?), and timing (how much environmental protection do we need at this point?).

A third indicator for coalition power is *reputational power*. To gather data about each actor's reputational power, i.e., the degree to which actors perceive others as important in the policymaking process, the aforementioned complete actor list was again presented to survey respondents. Respondents were then asked:

Please check all the actors that have been particularly important in the policy process on micropollutants (time frame and name of the respective process was given) from the point of view of your organization.

If there are actors missing, please add them to the bottom of the list and evaluate their importance.

As in previous examples, responses were coded as a dummy variable (0,1). In contrast to the previous examples, the resulting *actor* × *actor* matrix—the *reputational power matrix*—constitutes a directed graph, where ties between actors are directed from one actor to another. To analyze the way in which reputation is distributed across such a directed network, we have to extend the notion of degree centrality to directional graphs. More concretely, to assess reputational power, *indegree centrality* scores were first calculated for each actor individually and then averaged for coalitions. The indegree $d_I(n_i)$, of a node n_i, is equal to the number of ties a node receives (excluding outgoing ties) and captures a node's popularity (Wasserman and Faust 2009, p. 126). Indegree centrality $\overline{d}_I$ is generally calculated as the number of incoming ties of a node over the total number of nodes in the network. In order to obtain information about the average indegree centrality scores of coalition members, here the number of incoming ties $d_I(n_i)$ of coalition members are summed and then divided by the number of coalition members c as follows:

$$\overline{d}_I = \frac{\sum_{i=1}^{c} d_I(n_i)}{c}$$

Note that the difference between $d(n_i)$ and $d_I(n_i)$ lies in that the former applies to undirected graphs, whereas the latter applies to the indegree of directed graphs.

In summary, a *dominant coalition* can be identified when three conditions are simultaneously met: first, higher scores on degree centrality for collaboration compared to minority coalition(s); secondly, higher scores on reputational indegree centrality than in minority coalition(s). Thirdly, a dominant coalition should include the majority of network members, but if a minority of actors holds more power in terms of degree centrality for collaboration and reputational indegree centrality, then the smaller group can be a dominant coalition. By contrast, coalitions of similar strength exist when groups of similar size form and when those groups display comparable levels of mean degree centralities and mean reputational indegree centralities.

To assess the structure of coalitions, not only dominant and minority coalitions matter, but also the type of subsystem. As Sect. 3.1.3 explains, unitary subsystems are characterized by one single coalition with high intra-coalition belief similarity and coordination; collaborative subsystems, meanwhile, are defined as two coalitions with intermediate inter-coalition belief similarity and high inter- as well as intra-coalition coordination. Adversarial subsystems are described as competitive through their low inter-coalition belief similarity and coordination, but high intra-coalition coordination.

In order to distinguish these coalition structures, one needs information about (a) belief cohesion within and between coalitions and (b) interconnectedness within and between coalitions. The previous Sect. 3.2.3.3 explains how to calculate the latter, namely via mean densities of collaboration ties within and between coalitions. Regarding the former, the data from the aforementioned belief similarity matrices (see Sect. 3.2.3.1) is employed in order to calculate the mean densities of belief similarity ties within and between coalitions. Higher average belief similarity values point to coalitions with comparably good belief cohesion; and lower average values signal that coalition members have a comparably lower degree of belief cohesion. By combining the information about intra- and inter-coalition belief cohesion and interconnectedness, it becomes straightforward to categorize subsystems as collaborative or adversarial. Unitary subsystems, on the contrary, might be more difficult to identify due to the fact that some actors may not belong to the only existing coalition, while still not forming their own coalition. Based on this reasoning, a unitary subsystem can either imply that all actors align as one coalition, which is then characterized by high intra-coalition belief cohesion and interconnectedness; or a unitary subsystem can exist when a vast majority of actors display high belief cohesion and interconnectedness, while a small minority of actors is internally disintegrated. If these disintegrated actors display lower levels of belief cohesion and interconnectedness internally than they do across coalitions, it is a strong indicator that these actors do not form a coalition and thereby remaining unaligned. In such situations, the only existing coalition is best categorized as a unitary subsystem.

3.2.3.5 Brokerage

The ACF literature has advanced the concept of brokerage. A policy broker denotes an actor with moderate belief systems, who mediates between coalitions, and is well-connected to actors from all coalitions (Sabatier and Jenkins-Smith 1993; Ingold and Varone 2012). According to this definition, three conditions need to be satisfied simultaneously in order for an actor to qualify as a broker. First, through their roles as mediators, brokers are expected to lie on many paths between any other pair of nodes in the network. Secondly, brokers hold neutral beliefs, and thirdly, they are equally well-connected to all coalitions.

The mediation role of brokers is best operationalized by the concept of *betweenness centrality* (Ingold and Varone 2012). Betweenness centrality captures interactions in which two nonadjacent actors are only indirectly connected through a third actor. In other words, the 'third actor' lies on the path between the two nonadjacent actors. This 'third actor' exerts some measure of control over the interactions that exist between the two nonadjacent actors, and therefore carries responsibility for the nonadjacent ones (Pitts 1979, p. 507; cited after Wasserman and Faust 2009, p. 189). The analogy to the theoretical concept of a broker is that, by definition, members of opposite coalitions are not directly connected, but can be indirectly connected through the broker. Hence, it should be possible to identify brokers in practice by their exceptionally high betweenness centrality scores. Mathematically, the betweenness centrality for a node n_i is the sum of the probabilities that actor i lies on the path between two actors j and k. This probability is given by the ratio of all geodesics g_{jk}, i.e., the shortest paths linking j and k that contain i, to all geodesics linking j and k that do not contain i (Wasserman and Faust 2009, p. 190; after Freeman 1977, the measure is also called freeman centrality):

$$C_B(n_i) = \sum_{j<k} \frac{g_{jk}(n_i)}{g_{jk}},$$

One should note that all geodesics are assumed to be equally probable and that j and k should not be i.

In addition to exceptionally high betweenness centrality scores, brokers are characterized by their moderate belief system. To identify neutral beliefs (Ingold 2011), I analyzed survey respondents' answers to the questions about their level of agreement to an end-of-pipe approach for the reduction of micropollutants versus a source-directed one (see Sect. 3.2.3.1). This specific data is used here because it appropriately captures actors' level of concern for the aquatic environment. Highly concerned actors are assumed to favor preventive, source-directed measures to keep waters clean, while actors who are more pragmatic are assumed to favor end-of-pipe solutions. Where respondents strongly agreed to either of the approaches, the answer was coded as 4. Where they strongly disagreed, it was coded as 1. A neutral belief system was then attributed to those actors who display the same level of agreement or disagreement to both approaches. If actors agree both with an end-of-pipe and a source-directed approach, actors therefore do not have a particular preference for one or the other and can be considered neutral.

The third condition for actors to qualify as a broker is that their connections are evenly distributed among all coalitions. This criterion was operationalized as the densities of collaboration ties that the broker maintains toward the different coalitions (Beyers and Braun 2014).

If all three conditions are present, i.e., high betweenness centrality scores, neutral beliefs, and good connections to all coalitions, an actor is likely to take the role of a broker in a policymaking process.

3.2.3.6   Entrepreneurship

The concept of entrepreneurship is mainly rooted in two strands of policy literature, the Multiple Streams Framework (Kingdon and Thurber 2011) and the Punctuated Equilibrium Theory (Baumgartner and Jones 1993). According to that literature, entrepreneurs are self-interested actors who push toward their own policy preferences. Moreover, they are connected to other powerful and well-connected network members (Christopoulos and Ingold 2015). Entrepreneurs are likely to be part of the dominant coalition, which in turn are characterized by a high level of internal interconnectedness. In fact, if well-connected network members form dominant coalitions, and entrepreneurs tend to be connected to well-connected others, entrepreneurs are very likely to be part of the dominant coalition. Hence, to assign entrepreneurship to an actor, three conditions need to be fulfilled simultaneously: first, membership in the dominant coalition, secondly, exceptionally high connections to well-connected others, and thirdly, clear (non-neutral) policy beliefs.

Section 3.2.3.4 already explained how dominant coalitions were identified in this study. The membership of an actor in the dominant coalition can simply be looked up whenever the coalition structure of the policy network has previously been established. In cases of two equally strong coalitions, entrepreneurs can be part of both coalitions. The second theoretical idea, namely being connected to well-connected others, is best operationalized by *eigenvector centrality*.

Eigenvector centrality, e_i, of a node i, counts the number of nodes adjacent to a given node and weighs each adjacent node by its centrality (Borgatti et al. 2013, p. 168):

$$e_i = \lambda \sum_j x_{ij} e_j,$$

where λ is a proportionality constant called the eigenvalue, x_{ij} is the adjacency matrix of the network ($x_{ij} = 0$ and $x_{ij} = 1$ represent present and absent ties, respectively), and e_j are the centrality values of the other nodes in the network j.[16]

Whether actors adopted a non-neutral belief system was evaluated through their agreement to either an end-of-pipe or a source-directed approach for the reduction of micropollutants.

Where all three conditions, i.e., exceptionally high eigenvector centrality values, membership in the dominant coalition, and non-neutral beliefs, are fulfilled, there is a strong indication that a policy actor plays the role of an entrepreneur in a given policymaking process.

---

[16]The mathematical reasoning behind the adoption of this measure is based on some of the properties of *eigenvectors*. If an adjacency or network matrix is interpreted as a mapping in vector space, then, given full rank of the matrix, there will exist some vectors that are mapped by this matrix to multiples of themselves. These vectors are called eigenvectors, and the multiples are their associated eigenvalues. The eigenvectors represent node values that remain unaltered by the assumed effects exerted by their neighboring actors or nodes. Thus, they represent a state of equilibrium. In the present context, we will only refer to one eigenvector, $\mathbf{e} = (e_1, e_2, ..., e_i, ... e_N)$, by which we mean the eigenvector associated with the largest eigenvalue. The centrality value of the node i is thus simply the ith entry of the eigenvector $\mathbf{e}$ associated with the largest eigenvalue.

Table 3.5 Operationalization of structural network characteristics—an overview

Theoretical concept	Network measure	Data	Explanation of network measure
Belief cohesion	Density of belief similarity ties (in overall network)	Surveyed beliefs (2-mode data) were transformed into a valued belief similarity matrix (1-mode data)	Density of belief similarity ties indicates the mean of nonzero, valued ties. Zero stands for maximum dissimilar beliefs; nonzero for similar beliefs
Interconnectedness	Mean degree	Collaboration matrix, 1-mode data, coded as 0/1	Mean degree of collaboration ties is the total number of undirected collaboration ties divided by the total number of nodes
	Triangles per tie	Collaboration matrix, 1-mode data	Number of undirected triangles divided by the total number of ties
Coalition formation	Balance procedure to assess coalition membership	Agreement/disagreement matrix, 1-mode data, coded as −1/0/1	Blockmodeling procedure: actors with agreement ties are placed in same block; actors with disagreement ties are placed in opposite blocks
	Density of collaboration ties (within and across coalitions)	Collaboration matrix, 1-mode data, coded as 0/1	Number of present ties in a coalition/across coalitions divided by the maximum possible number of ties in that/across coalition(s)
Coalition structure	Size of coalition	Coalitions as identified via balance procedure	Number of coalition members
	Degree centrality (average for coalitions)	Collaboration matrix, 1-mode data, coded as 0/1	Sum of coalition members' individual degrees divided by number of coalition members
	Indegree centrality	Reputational power matrix, 1-mode data, coded as 0/1	Sum of coalition members' incoming ties divided by number of coalition members
	Density (within and across coalitions)	(a) Valued belief similarity matrix, (b) Collaboration matrix	Density within and across coalitions of (a) belief similarity ties, and (b) collaboration ties

(continued)

Table 3.5 (continued)

Theoretical concept	Network measure	Data	Explanation of network measure
Brokerage	Betweenness centrality	Collaboration matrix, 1-mode data, coded as 0/1	Number of times a broker lies on a path connecting a pair of nodes divided by all geodesics linking that pair of nodes, summed for all pair of nodes in the network
	–	Beliefs regarding end-of-pipe and source-directed measures, Attribute data	Neutral belief system: actor displays same level of agreement or disagreement to end-of-pipe as to source-directed measures
	Density(between broker and coalitions)	Collaboration matrix, 1-mode data, coded as 0/1	Number of present ties between broker and coalitions divided by the maximum possible number of ties between broker and coalitions
Entrepreneurship	–	Coalitions as identified via balance procedure	Membership in dominant coalition
	Eigenvector centrality	Collaboration matrix, 1-mode data, coded as 0/1	Count of the number of nodes adjacent to a focal node, weighing each adjacent node by its centrality
	–	Beliefs: end-of-pipe, source-directed measures, attribute data	Partial policy beliefs

3.2.3.7 Synthesis of Network Characteristics and Their Operationalization

Table 3.5 provides a synthesis of the operationalization of each structural network characteristic as explained above.

3.3 Water Policy Networks Along the Rhine River

Water policy networks aggregate information on the interactions occurring between policy actors during the micropollutants policy processes examined here (see Sects. 2.3.31, 2.3.4.1, 2.3.5.1, and 2.3.6.1). Analyzing those interactions through the study of structural characteristics of policy networks allows one to gain insights into the social mechanisms that govern policymaking within the realm of water protection.

3.3.1 Switzerland

The Swiss policy network reflects the aggregated interactions of policy actors participating in the amendment process of the Waters Protection Act and Ordinance (between 2007 and 2013) (see Sect. 2.3.3.1). The resulting collaboration network is depicted in Fig. 3.2. The colors in the diagram indicate different actor types. On the bottom right are mostly governmental actors (red) representing different ministries and bureaucracies; middle right (petrol blue) are mostly water and cantonal associations responsible for water quality, drinking water, or wastewater treatment; top right (yellow) are mainly scientific actors; top left are environmental associations (turquois); middle left are political parties (pink) and the legislature (dark red); bottom left are industrial and agricultural associations (gray).

One should note that the Department for Water of the Federal Office for the Environment (BAFU-W/UVEK) is particularly central to this collaboration network, as can be seen by its comparably high number of ties. In fact, BAFU-W took the lead in this policymaking process and, among others, drafted the proposals of the Waters Protection Act and Ordinance. Moreover, the representation of the Swiss Cantons (BPUK) is located close to BAFU-W in the network diagram, which reflects the importance of Cantons responsible for implementing the policy design. By contrast, the Consumer Forum (KF) constitutes an *isolate* in the network and indicates that consumer interests are less represented in this policy process, even though save drinking water would be a relevant consumer topic. On the left of the network diagram, one can see that political parties (pink) are well-connected to both chambers of the legislature (UREKS and UREKN). Although the Green Liberal Party of Switzerland (GLP) represents a *pendant* in this diagram, the party exhibited a pivotal role in the policy process by leading the meetings on micropollutants of

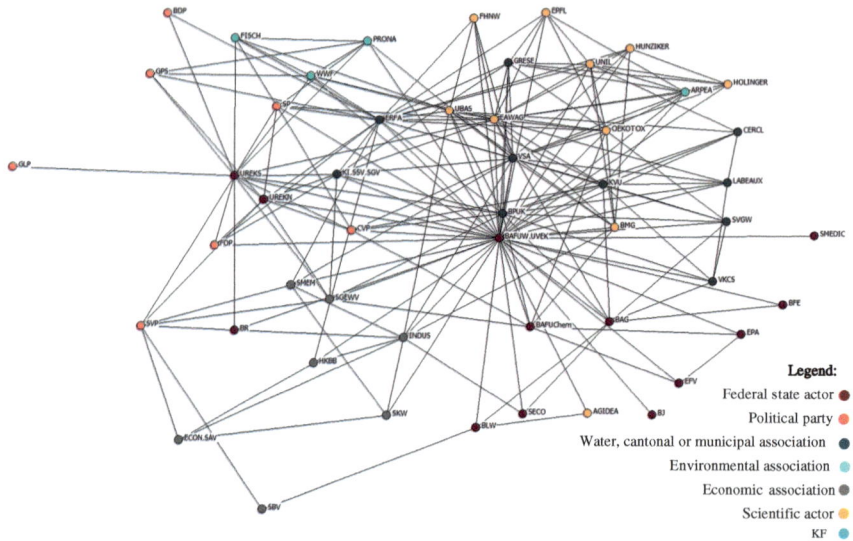

Fig. 3.2 Swiss collaboration network 2007–2013. *Note Circles* represent actors, *colors* represent actor types, labels are actor codes (see Annex 10 for full actor names), *lines* represent collaboration ties, NetDraw's built-in algorithm sets the placement of nodes and length of ties to achieve an optimal visualization

the National Council's Committee on the Environment, Spatial Planning, and Energy (interview with UREKN on May 3, 2013).

Belief cohesion

Belief cohesion refers to a consensus between network members concerning fundamental values, called *policy beliefs*. The beliefs that prevail in this policy issue include environmental concern and risk averseness. Actors who display these beliefs are likely to approve of policy measures for reducing micropollutants in waters, whereas actors who reject such beliefs are likely to disapprove of micropollutants policies.

Swiss policy actors have a 5.92% mean density of belief similarity ties between all actors in the policy network. This statistic indicates that about 6% of all actors hold identical beliefs regarding the regulation of micropollutants, and completely agree about the general need to adopt policy measures as well as ways in which to concretely achieve the reductions of micropollutants in surface waters.

Interconnectedness

Interconnectedness refers to the number of contacts among actors in the policy-making process and is measured as collaboration between actors here. To evaluate the extent to which actors collaborate, two different indicators have been calculated, which attempt function independently of network size, i.e., ties per node and triangles per tie. The result for ties per node, also termed mean degree (score of 3.98), portrays the number of ties each actor maintains on average. Hence, each of the Swiss policy actors has on average almost four contacts. Moreover, the result for triangle per ties illustrates that each tie in the network is involved in 1.54 triangles on average, demonstrating that actors use their ties to form dense relations benefitting from extensive redundancy.

Coalition formation and structure

To identify coalitions, the blockmodeling procedure, called balance, groups network members into one block that share agreement ties; actors with disagreement ties are placed in opposite blocks. The result of the blockmodeling procedure reveals one large coalition with 39 actors and one small coalition with 11 actors in the Swiss case (see Table 3.6). The larger coalition is also more powerful in terms of mean reputational power (12.13 vs. 4.90) and mean degree (9.20 vs. 4.81). Annex 14 displays the complete list of actors' reputational power scores and actors' degrees. The large coalition can be labeled *dominant* because it fulfills all three definition criteria of a dominant coalition as laid down in Sect. 3.2.3.4: It is larger in size and more powerful, as indicated by the higher mean degree centrality scores and mean reputational power scores.

Results for mean beliefs (see Table 3.6) show that the dominant coalition supports a pro-environmental policy with a score of 3.1, where a score of 4 denotes the highest level of support here. Moreover, the dominant coalition strongly disagrees with a contra-environmental policy with a score of 1.4 (Table 3.6), where a score of 1 signifies the highest disagreement. On average, members of the dominant coalition display beliefs that attribute high importance to environmental protection and clean waters and can therefore be labeled *water quality coalition.*

While the water quality coalition is dominant in the Swiss case, its opposition is the minority coalition since its members attain lower mean scores for reputational power (score of 4.90) and mean degree (score of 4.81). The minority coalition displays a policy orientation that disapproves of environmental protection (score of 2.7 for contra-environmental protection in Table 3.6) and only slightly approves of a pro-environmental policy (score of 2.3 in Table 3.6). The minority coalition obtained the label *opposing coalition* here since, on the one hand, its beliefs indicate that the minority coalition does not generally dispute water protection—it slightly approves of a pro-environmental policy; on the other hand, this coalition opposes a policy orientation aiming at very high water protection standards, since its members consider the political goal of aiming for zero concentrations of micropollutants in waters too ambitious (score of 1.5 for belief entitled *aim for zero concentrations* in Table 3.6).

The density of collaboration ties in the water quality coalition is with its 18% density lower than the 25% density within the opposing coalition. With 39 compared to 11 actors, the water quality coalition is larger in size, and reaching high densities is therefore difficult.

Taking into consideration its larger size, the water quality coalition can still be considered well-connected compared to the opposing coalition despite the lower densities. Moreover, the water quality coalition is more cohesive by displaying

Table 3.6 Coalition members and beliefs in the Swiss case

Members of the water quality coalition		Members of the opposing coalition	
Actor type, acronym	**Mean beliefs:**	Actor type, acronym	**Mean beliefs:**
4 ARPEA	Pro-environmental	5 ECON/SAV	Pro-environmental
1 BAFU-CHEM	policy 3.1	1 EFV	policy 2.3
1 BAFU-W/UVEK	Contra-environmental	1 EPA	Contra-environmental
1 BAG	policy 1.4	5 HKBB	policy 2.7
2 BDP	(n = 31, min = 1;	5 INDUS	(n = 8, min = 1;
1 BFE	max = 4)	5 SBV	max = 4)
1 BJ		1 SECO	
1 BLW	*Explanation*	5 SGEWV	*Explanation*
6 BMG	Value of	5 SKW	Value of
3 BPUK	pro-environmental policy	5 SMEM	pro-environmental policy
1 BR	is the mean of the	2 SVP	is the mean of the
3 CERCL	following four surveyed	(n 11)	following four surveyed
2 CVP	beliefs:		beliefs:
6 EAWAG	Source-directed		Source-directed
6 EPFL	approach 3.6,		approach 2.9
3 ERFA	End-of-pipe		End-of-pipe approach 2.4
2 FDP	approach 2.9,		Preventive approach 2.3
6 FHNW	Preventive approach 3.5,		Aim for zero
4 FISCH	Aim for zero		concentrations 1.5
2 GLP	concentrations 2.6		
2 GPS		**Not member of either coalition:**	
3 GRESE		6 AGIDEA	
6 HOLINGER		4 KF	
6 HUNZIKER		(n 2)	
3 KI/SSV/SGV			
3 KVU			
3 LABEAUX			
6 OEKOTOX			
4 PRONA			
1 SMEDIC			
2 SP			
3 SVGW			
6 UBAS			
6 UNIL			
1 UREKN			
1 UREKS			
3 VKCS			
3 VSA			
4 WWF			
(n 39)			

Codes for actor types: 1 = Federal state actor, 2 = Political party, 3 = Water, cantonal, or municipal association, 4 = Environmental or consumer association, 5 = Industrial or agricultural association, 6 = Scientific actor

higher belief similarities than the opposing coalition (6.40 vs. 5.24% mean coalition density of belief similarity ties). Sufficient levels of collaboration and belief similarity within the water quality coalition represent a further indicator of its higher strength compared to the opposing coalition.

Relations between coalitions are characterized by a 4.74% density of belief similarity ties, which is not much less than belief cohesion within the opposing coalition. Collaboration ties across coalitions, on the contrary, are with its 0.41% density much lower than densities within coalitions.

Brokerage

As explained in Sect. 3.2.3.5, an actor qualifies as a broker when three conditions are fulfilled simultaneously: First, brokers lie on many paths between any other pair of nodes in the network, which should translate into high betweenness centrality scores. Secondly, brokers have neutral beliefs, and thirdly, they are equally well-connected to all coalitions.

In the Swiss collaboration network, two policy actors take the role of a broker, because they fulfill all three conditions simultaneously. The first actor who qualifies as a broker is the Department for Water of the Federal Office for the Environment (BAFU-W), which also acts as the lead agency in the policy process on micropollutants. BAFU-W ranks highest among all network members on normalized betweenness centrality with a score of 49.54, compared to a network mean of 1.88 (see Annex 14 for a complete list of actors' normalized betweenness centrality scores). Moreover, BAFU-W has neutral policy beliefs because of its total agreement (score of 4) with both end-of-pipe and source-directed approaches on micropollutants. This result confirms that BAFU-W does not have a particular preference for one or the other and supports its role as neutral mediator. BAFU-W is also well-connected toward both coalitions with 89.5% of all possible collaboration ties toward the water quality coalition, and 72.7% toward the opposing coalition. A situation in which the lead agency also takes a brokerage role has also occurred, for example, in the study by Ingold and Varone (2012). The findings of both studies suggest that the lead agency's role is, among others, to consider all parties that have a stake in the issue and are able to mediate between them.

The other Swiss actor taking the role of a broker is the Council of State's Committee on the Environment, Spatial Planning, and Energy (UREKS), which displays the second highest score of all network members for betweenness centrality, with 9.06, compared to a network mean of 1.88. UREKS became active in the policy process by initiating a motion (Motion 10.3635) that charges the Federal Government to find a financing solution for the upgrade of wastewater treatment plants respecting the *polluter pays principle*. This intervention demonstrates that UREKS mediated between, on the one hand, those who favored policy action on micropollutants, and, on the other hand, those who feared an unequal distribution of costs—typically the Cantons and sewage treatment plants who are responsible for implementing the Swiss policy design. UREKS's broker role is furthermore supported by its neutral beliefs and identical levels of agreement (score of 3) toward end-of-pipe and source-directed approaches on micropollutants. Moreover, UREKS is connected with both coalitions: 39.5% of all possible collaboration ties are present toward the water quality coalition and 18.2% toward the opposing coalition.

Entrepreneurship

Section 3.2.3.6 specified that entrepreneurship can be assigned to actors who simultaneously fulfill three conditions: first, membership in the dominant coalition; secondly, exceptionally high connections to well-connected others through high eigenvector centrality scores; and thirdly, clear (non-neutral) policy beliefs.

In total, four actors take the role of entrepreneurs in the Swiss network. The Swiss Federal Institute of Aquatic Science and Technology (EAWAG) attains the highest normalized eigenvector centrality score with 39.12 compared to a network mean of 16.07 (see Annex 14 for a complete list of actors' eigenvector centrality scores) and a preference for end-of-pipe (score of 4) compared to source-directed approaches (score of 3). EAWAG's central role in the policymaking process emanates from its research on (a) the detection of micropollutants in surface waters and (b) advanced wastewater treatment technologies for the elimination of micropollutants from sewage water. With this research agenda, EAWAG placed the issue of micropollutants on the political agenda and informed policy actors about one potential pollution reduction measure.

Moreover, the Sewage Treatment Plants in Large Cities Initiative (ERFA) obtains the second highest normalized eigenvector centrality score with 38.69 compared to the network mean of 16.07. ERFA strongly agrees with source-directed approaches (score of 4), but disapproves of end-of-pipe approaches (score of 1). With its activity in the policymaking process, ERFA attempted to bring a source-directed approach into focus and to avert a pure end-of-pipe policy orientation that mainly affects large wastewater treatment plants.

The Swiss Water Association (VSA) displays a normalized eigenvector centrality score of 38.59 compared to the average of 16.07. VSA has a preference for source-directed (score of 4) compared to end-of-pipe approaches (score of 3). An interviewee from the VSA reported that the protection of waters represented the ultimate aim of the association (interview with VSA on May 2, 2014). To that end, source-directed measures are considered most effective. Nevertheless, the VSA also supports end-of-pipe measures, as these contribute to water protection as well.

Finally, the Conference of Cantonal Directors of Construction, Planning and Environmental Protection (BPUK) attains an eigenvector centrality score of 30.82 compared to the average of 16.07. BPUK supports source-directed measures (score of 3), but opposes end-of-pipe approaches (score of 2). BPUK's preferences mirror its role as the representation of the Swiss Cantons, which have been entrusted with implementing, and at least partly financing, the Swiss end-of-pipe policy. In order to avert additional duties from Cantons, BPUK took a source-directed stance.

All entrepreneurs approve of source-directed approaches, which indicates that they hold beliefs in favor of environmental protection. The mentioned policy actors are members of the dominant water quality coalition and therefore fulfill all three criteria defining entrepreneurs.

Coalition structure in Switzerland

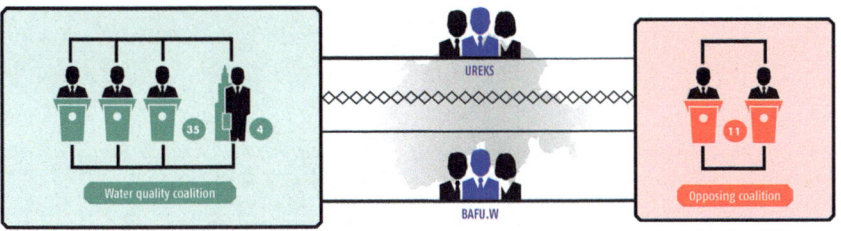

Legend

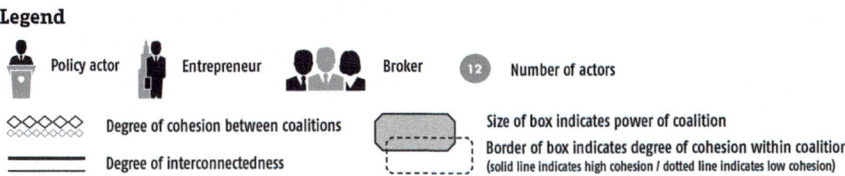

Fig. 3.3 Swiss network diagram

Swiss network diagram

To illustrate the network structure presented above in a more intuitive manner, the Swiss collaboration network is presented schematically in Fig. 3.3. In the diagrammatic network, a dominant and a minority coalition with the respective number of members are shown by means of two squares. The size of the squares indicates a coalition's degree of power. Large squares indicate higher levels of power than small squares. The thickness of lines symbolizes the degree of interconnectedness within and across coalitions, or toward the brokers. Thick lines symbolize a high level of interconnectedness, while a low degree is depicted with thin lines. Specific symbols for brokers and entrepreneurs indicate their existence and strategic position in the network. A high degree of belief cohesion within coalitions is depicted by thick box lines around coalitions, while a high degree of belief cohesion across coalitions is illustrated by thick black rhombs across coalitions.

3.3.2   Germany

The German policy network aggregates the interactions that survey respondents reported toward other policy actors during the adoption process of the Surface Water Ordinance (2008–2011) (see Sect. 2.3.4.1). Figure 3.4 of the German collaboration network displays, on the bottom right, environmental associations (turquois) and water associations (petrol blue) responsible for drinking water or wastewater treatment. On the top right, the diagram shows mostly Länder (pink) and a few federal state actors (dark red), while most federal state actors representing

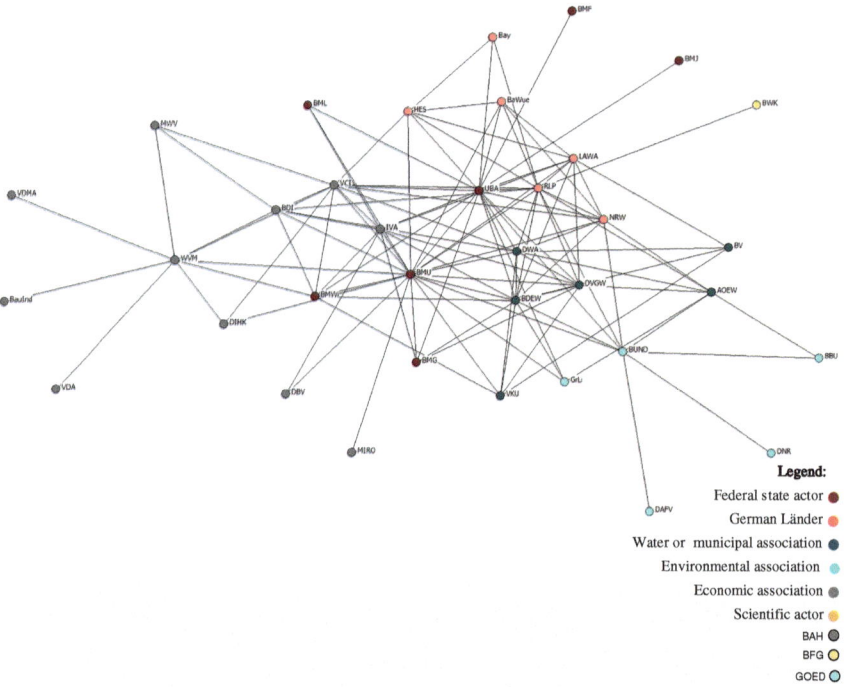

Fig. 3.4 German collaboration network 2008–2011. *Note Circles* represent actors, *colors* represent actor types, labels are actor codes (see Annex 11 for full actor names), *lines* represent collaboration ties, and NetDraw's built-in algorithm sets the placement of nodes and length of ties to achieve an optimal visualization (Color figure online)

different ministries (dark red) are located in the center. Industrial and agricultural associations (gray) are displayed on the left. Scientific actors (yellow) play only a minor role.

The diagram illustrates that there are two particularly well-embedded policy actors: the Federal Ministry for the Environment (BMU), which has the lead in this policy process on micropollutants, and the Federal Environmental Agency (UBA), which supports the Ministry with scientific knowledge. Around this center are located the water associations (petrol blue), which represent influential actors in German water governance; the German Länder (pink), which used to be solely responsible for water protection policy before the federal reform transferred competences to the federal Government; and economic associations (gray), which are generally influential in Germany. Less well-connected to the central actors BMU and UBA are environmental associations (turquois) and scientific actors. The diagram suggests that the latter play a minor role in this network. However, UBA and German water associations hold scientific divisions and represent science here to some degree. One should note that political parties did not become involved in the policymaking process.

Belief cohesion

There is a 2.98% of mean density for belief similarity ties among the German policy actors. These results reveal that about 3% of German actors have identical beliefs regarding the regulation of micropollutants and agree about the general need to act politically.

Interconnectedness

The 39 German policy actors form on average about 3 ties (2.97 ties per node), and each tie is involved in a little less than 1.5 triangles (1.47 triangles per tie).

Coalition formation and structure

The results of the blockmodeling procedure called *balance* exhibited two coalitions Table 3.7), one being slightly more powerful in terms of size (19 vs. 17 actors), reputational power (7.63 vs. 5.88), and mean degree (7.10 vs. 5.59). The complete list of actors' reputational power scores and actors' degrees is displayed in Annex 15. The larger coalition can be labeled *water quality coalition,* since it comprises those actors with beliefs favorable to stricter environmental norms. Table 3.7 displays coalitions' mean beliefs. The water quality coalition strongly supports a pro-environmental policy (score of 3.15 in Table 3.7) and strongly objects to non-action (score of 1.3 for contra-environmental policy in Table 3.7). These results indicate that members of the water quality coalition favor high environmental protection standards, and consider aquatic micropollutants enough of a risk to require immediate policy action. The smaller of the two coalitions can be termed *opposing coalition,* because on the one hand, its members favor non-action (score of 2.5 for contra-environmental policy in Table 3.7); on the other hand, coalition members' slight approval of a pro-environmental policy orientation indicates that these actors do not fully oppose water protection (score of 2.5 in Table 3.7), but rather very high standards and precipitous policy action.

The water quality coalition is more cohesive by displaying higher belief similarities than the opposing coalition (4.32 vs. 2.32% mean coalition density of belief similarity ties). Of note is that both coalitions are equally well-connected with 18.4% present ties of all possible connections.

Belief cohesion across coalitions is with 2.43% density of belief similarity ties less strong than the belief cohesion within the water quality coalition and comparable to belief cohesion within the opposing coalition. Collaboration across coalitions is with 0.87% of tie density less strong than collaboration within coalitions.

Brokerage

There are two actors—the Federal Ministry for the Environment (BMU) and the Federal Environmental Agency (UBA)—that fulfill the three criteria defining policy brokers: First, they attain the highest betweenness centrality scores of all network members (27.26 and 16.13 against a network mean of 2.74; see Annex 15 for a complete list of actors' normalized betweenness centrality scores). Secondly, BMU and UBA are well-connected to both coalitions: BMU has a 55.6% density of belief

Table 3.7 Coalition members and beliefs in the German case

Members of the water quality coalition		Members of the opposing coalition	
Actor type, acronym	**Mean beliefs:** Pro-environmental policy 3.15 Contra-environmental policy 1.3 (n = 15, min = 1, max = 4)	Actor type, acronym	**Mean beliefs:** Pro-environmental policy 2.5 Contra-environmental policy 2.5 (n 10, min = 1, max = 4)
3 AOEW		5 BAH	
2 Bay		5 BauInd	
4 BBU		2 BaWue	
3 BDEW	*Explanation*	5 BDI	*Explanation*
6 BfG	Value of pro-environmental policy is the mean of the following four surveyed beliefs: Source 3.7 End-of-pipe 2.6 Preventive 3.7 Zero concentrations 2.9	1 BML	Value of pro-environmental policy is the mean of the following four surveyed beliefs: Source 3.1 End-of-pipe 2.6 Preventive 2.4 Zero concentrations 2
1 BMG		1 BMWi	
1 BMJ		6 BWK	
1 BMU		5 DBV	
4 BUND		5 DIHK	
3 BV		3 DWA	
4 DAFV		5 IVA	
4 DNR		5 MIRO	
3 DVGW		5 MWV	
4 GrLi		2 RLP	
2 HES		5 VCI	
2 LAWA		5 VDA	
2 NRW		5 WVM	
1 UBA		(n 17)	
3 VKU		**Not member of either coalition:**	
(n 19)		1 BMF	
		4 GOED	
		5 VDMA	
		(n 3)	

Codes for actor types: 1 = Federal state actor, 2 = German Länder, 3 = Water or municipal association, 4 = Environmental or consumer association, 5 = Industrial or agricultural association, 6 = Scientific actor

similarity ties toward the water quality coalition and 70.7% toward the opposing coalition. UBA has 61.1% of all possible collaboration ties toward the water quality coalition and 41.2% toward the opposing coalition. With regard to the third criterion—a neutral beliefs system—UBA equally agrees to source-directed and end-of-pipe approaches for the reduction of micropollutants (agreed somewhat to both in the survey). BMU has a slight preference for end-of-pipe approaches (agreed somewhat) compared to source-directed ones (disagreed somewhat). BMU can still be considered a broker, because of an only small difference of the level of agreement to both approaches, and at the same time, exceptionally high betweenness centrality scores and collaboration ties toward both coalitions.

BMU with the support of UBA together were responsible for a draft proposal of the Surface Water Ordinance. Their brokerage role was necessary in order to draft an ordinance that passes the policymaking process. In this regard, a representative of the UBA reported during an interview for the underlying investigation that, among others, the integration of the German Länder was important, since Länder have veto points in the Federal Council of States enabling them to block a proposal that neglects their interests (interview with UBA on Apr 17, 2014). To circumvent such a failure, BMU and UBA engaged in brokerage.

Entrepreneurship

In the German case, entrepreneurs exist in both coalitions, because there are two opposing, almost equally powerful coalitions. In total, each coalition includes four entrepreneurs, as indicated by their exceptionally high eigenvector centrality scores:

- Entrepreneurs in the water quality coalition: Federal Association of Energy and Water Industry (BDEW) with a normalized eigenvector centrality of 40.61 against a network mean of 17.34, German Technical and Scientific Association for Gas and Water (DVGW) with an eigenvector of 32.54/17.34, Common Working Group on Water of the Federal Government and States Governments (LAWA) with an eigenvector of 32.57/17.34, State of North Rhine-Westphalia (NRW) with an eigenvector of 33.341/17.34 (see Annex 15 for a complete list of actors' normalized eigenvector centrality scores).
- Entrepreneurs in the opposing coalition: German Association for Water, Wastewater and Waste (DWA) with an eigenvector of 34.11/17.34, Agrochemical Association (IVA) with an eigenvector of 31.68/17.34, State of Rhineland-Palatinate (RLP) with an eigenvector of 41.584/17.34, German Chemical Industry Association (VCI) with an eigenvector of 33.213/17.34.

All eight entrepreneurs have a clear preference either for source-directed or end-of-pipe approaches for the reduction of micropollutants, which indicates their clear policy beliefs and affirms their entrepreneurship.

Of note is that entrepreneurs reflect the interests vested in the issue of micropollutants in Germany: First, water associations engage in entrepreneurship in both coalitions. DVGW, for example, participates in water quality coalition, due to its responsibility in the supply of drinking water, and therefore advocates for high water protection standards. By contrast, DWA, which mainly represents the wastewater treatment sector, is a member of the opposing coalitions, because high environmental quality norms could force operators of sewage plants to invest in costly treatment technologies. A high-ranking state official from the BMU reported during an interview (interview on Apr 1, 2015) that despite DWA's mission to contribute to water protection through sewage treatment, the association has repeatedly disapproved of high environmental standards in the past.

Secondly, the entrepreneurship of the German Länder highlights their particular stake in micropollutants policies. LAWA, the representation of the Länder on the federal level, as well as the State of North Rhine-Westphalia, constitutes members of the water quality coalition, while the State of Rhineland-Palatinate belongs to the opposing coalition. During an interview, a representative of the BMU reported that the German Länder were divided on the issue of micropollutants (interview on Apr 3, 2012). A further interviewee from the State of North Rhine-Westphalia (interviews on Mar 27, 2012 and Mar 17, 2014) explained that micropollutants ranked high on the political agenda in that state, and therefore the state also sought to push the issue onto the national agenda. The interviewee from Rhineland-Palatinate (interview on Mar 13, 2014) justified its opposing stance, first, by the fact that many small sewage treatment plants exist in the state and their technological upgrade would therefore be particularly expensive; secondly, due to the intensive agricultural activities in the state, it would be difficult to comply with high standards of water protection.

Finally, the industrial sector, represented by the association of the chemical industry VCI and the agrochemical association IVA, engages in entrepreneurship in the opposing coalition. Despite its critical stance, the representative of the VCI emphasized during an interview that water protection matters to their members, and therefore, industry has optimized production processes (interview on Apr 29, 2014).

German network diagram

Figure 3.5 shows the just presented network structure in a more intuitive manner. The network diagram depicts two almost equally powerful coalitions with equally good interconnectedness within coalitions (thick lines), but low interconnectedness across coalitions (thin line). Belief cohesion within the water quality coalition is stronger (thick box lines around coalition) than within the opposing coalition (dashed box lines around coalition) or across coalitions (gray thin rhombs across coalitions). The strategic positions and amount of brokers and entrepreneurs are also illustrated.

Coalition structure in Germany

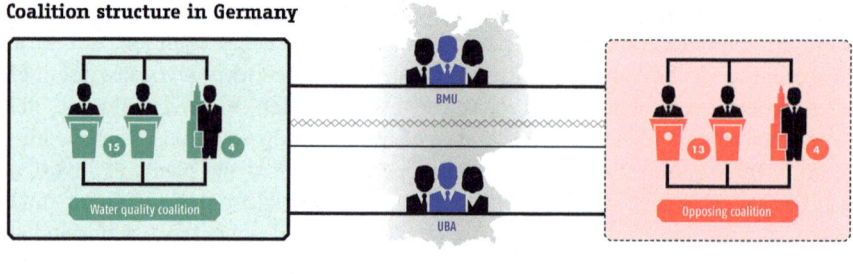

Legend

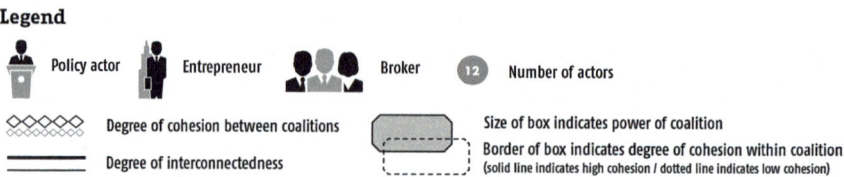

Fig. 3.5 German network diagram

3.3.3   *France*

The French policy network reflects the surveyed interactions between policy actors who contributed to the elaboration of the Micropollutants Plan (2009–2013) (see Sect. 2.3.5.1). A sociograph of the French collaboration network can be shown in Fig. 3.6. In the center of the network diagram are governmental actors (dark red) and water associations (petrol blue). The most central actor is the Department of Water and Biodiversity of the Ministry of Ecology (DEB), which has the lead in drafting the Micropollutants Plan. Located around this center are regional or local authorities at the right (pink), including Water Agencies (AGENCE), which take responsibility for water protection policy in France, and which are particularly central to the network when considering their tie profile. Moreover, scientific actors (yellow) are located at the top of the network diagram. In comparison, French scientific actors outnumber those in the German or Swiss networks. This observation further highlights that the French approach to micropollutants centers around improving scientific knowledge about the existence and effects of micropollutants in waters. Economic associations (gray) are located at the middle left of the diagram, and environmental and consumer associations at the bottom left (turquois).

Belief cohesion

Belief cohesion concerns the degree to which actors share fundamental values regarding a policy issue. The findings for the French network reveal a mean density

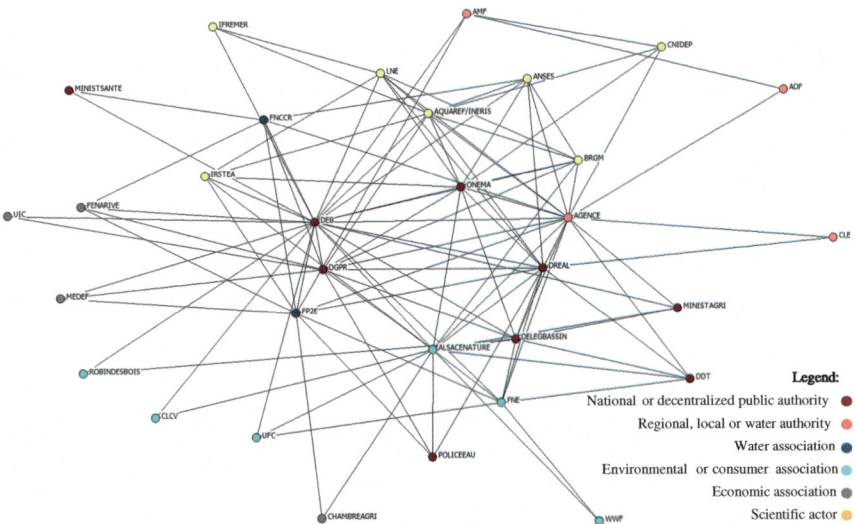

Fig. 3.6   French collaboration network 2009–2013. *Note Circles* represent actors, *colors* represent actor types, labels are actor codes (see Annex 12 for full actor names), *lines* represent collaboration ties, and NetDraw's built-in algorithm sets the placement of nodes and length of ties to achieve an optimal visualization (Color figure online)

of belief similarity ties of 1%, which denotes that 1% of the network members hold identical beliefs regarding the regulation of micropollutants. Those policy actors agree about the need to address micropollutants as a policy issue, as well as about policy content, compared to the vast majority of actors who disagree on the topic.

Interconnectedness

The results for interconnectedness indicate that French actors form 3.5 ties on average, meaning that French actors tend to collaborate with 3.5 alters on the issue of micropollutants. Moreover, each tie is involved in 1.44 triangles. Triangles intend to measure the degree to which actors form dense and redundant relations.

Coalition formation and structure

Moreover, the blockmodeling procedure termed *balance*, which partitions actors by their agree/disagree tie profiles, reveals that one coalition with 21 members developed in the French policy process (Table 3.8). The 10 remaining actors who are not part of this coalition do not seem to form a cluster of their own, but instead loosely assemble around the existing coalition. This formation can be attributed to the fact that an advocacy coalition is defined (in theory) by intra-coalition belief cohesion and interconnectedness, but here, the density of collaboration ties among the 10 remaining actors (13.3%) is lower than the density across coalitions (14.5%). Likewise, belief cohesion is lower among the 10 remaining actors (density of 0.27% for belief similarity ties) than across coalitions (density of 0.72% for belief similarity ties). Both results indicate that the remaining actors form ties and agree with actors from the coalition, more than among each other. As a result, there appears to be one main coalition and some actors loosely scattering around the main coalition without common strategic goals. For an overview of coalition members, see Table 3.8.

With such a coalition structure, it is not surprising that the main coalition scores are higher than the remaining actors taken together on reputational power (6.67 vs. 3.40) and mean degree (8.18 vs. 4.80). Annex 16 displays the complete list of actors' reputational power scores and degrees. Collaboration within the main coalition attains a density of 25.80%. Belief cohesion, measured as the density of belief similarity ties, reaches 1.60%.

The main coalition was labeled *water quality coalition* due to its high agreement (score of 3.3) to a pro-environmental policy and to its high disagreement (score of 1.4) with a contra-environmental policy. The remaining actors agree slightly less (score of 3) with a pro-environmental policy orientation and disagree a little less (score of 1.7) with a contra-environmental policy orientation than the water quality coalition. The similarity of beliefs between the remaining actors and the main coalition further confirms that these actors do not represent an opposing coalition with its own political stance thus far. Moreover, the moderate answers of the remaining actors suggest that they are indifferent toward the issue of micropollutants and therefore not engaged enough to ally into a coalition of their own.

This coalition structure must be understood in the context of the French Micropollutants Plan, which represents a first initiative by the French Government

Table 3.8 Coalition members and beliefs in the French case

Members of the water quality coalition		Remaining actors	
Actor type, acronym	**Mean beliefs:**	Actor type, acronym	**Mean beliefs:**
2 AGENCE	Pro-environmental	2 ADF	Pro-environmental
4 ALSACENATURE	policy 3.3	2 AMF	policy 3
6 ANSES	Contra-environmental	5 CHAMBREAGRI	Contra-environmental
6 AQUAREF/INERIS	policy 1.4	2 CLE	policy 1.7
6 BRGM	(n = 14, min = 1,	1 DDT	(n = 3, min = 1,
4 CLCV	max = 4)	1 DELEGBASSIN	max = 4)
6 CNIDEP		3 FNCCR	
1 DEB	*Explanation*	3 FP2E	*Explanation*
1 DGPR	Value of	5 MEDEF	Value of
1 DREAL	pro-environmental	5 UIC	pro-environmental
5 FENARIVE	policy is the mean of the	(n 10)	policy is the mean of the
4 FNE	following four surveyed		following four surveyed
6 IFREMER	beliefs:		beliefs:
6 IRSTEA	Source 4		Source 4
6 LNE	End-of-pipe 2.8		End-of-pipe 2
1 MINISTAGRI	Preventive 3.6		Preventive 3.3
1 MINISTSANTE	Zero concentrations 3		Zero concentrations 3
1 ONEMA		**Not member of either coalition:**	
4 ROBINDESBOIS		POLICEEAU	
4 UFC		(n 1)	
4 WWF			
(n 21)			

Codes for actor types: 1 = National or decentralized public authority, 2 = Regional, local, or water authority, 3 = Water association, 4 = Environmental or consumer association, 5 = Industrial or agricultural association, 6 = Scientific actor

to become active on the issue. In the main coalition, those actors—including governmental, scientific, environmental, and consumer organizations—engaged in particular whose mandate attributes high importance to water protection. The remaining actors include lower levels of Government, as well as water, agricultural, and industrial associations, which could potentially oppose micropollutants policies, but do not engage in strong opposition as long as the Micropollutants Plan remains a non-legally binding document without immediate consequences to them. Hence, the non-binding character of the French Micropollutants Plan may explain why only one coalition formed in the policymaking process studied here.

Brokerage

The French data points to the Department of Water and Biodiversity of the Ministry of Ecology (DEB) as having the highest normalized betweenness centrality of all network members. DEB receives a score of 33.85 against a network mean of 2.99 (see Annex 16 for a complete list of actors' normalized betweenness centrality scores). DEB is furthermore well-connected toward both blocks of actors with a 90% density of collaboration ties toward the water quality coalition and a 60% density toward the remaining actors. Nevertheless, DEB does not qualify as a typical broker due to its partiality toward measures at the source (strongly agreed in the survey) compared to end-of-pipe solutions (disagreed somewhat in the survey) for the reduction of micropollutants. DEB's preference is, in fact, reflected in the source-directed policy orientation of the Micropollutants Plan. From a theoretical perspective, brokers should be neutral and display a moderate belief system, and hence, the French DEB is not a typical broker. Note that DEB will be listed below under *brokerage* in order to enable comparison across cases, and to assess the consequences of missing brokerage on policy design.

Entrepreneurship

There are five entrepreneurs in the main coalition with exceptionally high normalized eigenvector centrality scores, compared to a network mean of 21.17: Water Agencies (AGENCE) with an eigenvector of 47.69, Alsace Nature (ALSACE) with an eigenvector of 39.73, the National Reference Laboratory for Water Monitoring (AQUAREF) with an eigenvector of 33.44, the Regional Authority for the Environment, Spatial Planning and Housing (DREAL) with eigenvector of 41.45, and the National Agency for Water and Aquatic Environments (ONEMA) with an eigenvector of 38.78 (see Annex 16 for a complete list of actors' normalized eigenvector centrality scores). All entrepreneurs are advocates of the protection of the aquatic environment, as can be seen by their strong agreement with a source-directed approach for the reduction of micropollutants, and their lower support for an end-of-pipe approach (agreed somewhat in the survey).

Actors who oppose micropollutants policies may not invest resources in entrepreneurship as long as the Micropollutants Plan remains a non-legally binding policy document without immediate consequences to them. So far, entrepreneurship has mainly been a matter of importance to governmental actors with a mandate in water protection policy, as shown by the fact that four out of five entrepreneurs are governmental. For instance, water agencies constitute specialized public authorities in the French system of water governance and are entrusted with water protection. Furthermore, the laboratory AQUAREF and the National Agency ONEMA were both created by the French Government in 2007 as a response to the need for improved knowledge regarding the state of the aquatic environment. DREAL corresponds to the regional representations of the French national environmental ministry. In conclusion, entrepreneurship relies mainly on governmental engagement, which may indicate that other parts of society have thus far expressed indifference toward the issue of micropollutants.

French network diagram

The French network structure is summarized in a more intuitive manner in Fig. 3.7. The network diagram displays one coalition and remaining actors, who scatter around the main coalition.

Thick lines indicate that actors within the main coalition are well-connected, while thin lines among the remaining actors signify that they collaborate even less among each other than with actors from the main coalition. Belief cohesion is generally weak in the network as symbolized by the dashed box line around the

Coalition structure in France

Fig. 3.7 French network diagram

coalition, together with an absence of rhombs across coalitions. The strategic positions and number of brokers and entrepreneurs are also illustrated. DEB is located within the main coalition due to results demonstrating that it is not a typical broker lying between blocks.

3.3.4   The Netherlands

The Dutch policy network displayed in Fig. 3.8 is an aggregation of the surveyed collaborative interactions between actors participating in the debate on policy measures for the reduction of pharmaceutical micropollutants (2001–2013) (see Sect. 2.3.6.1). The network diagram displays state actors (dark red) in the center and periphery of the network; water associations (petrol blue) responsible for drinking and wastewater are located at the center and bottom left; political parties (pink) are displayed at the bottom right; most industrial associations (gray) top right, and scientific actors top left (yellow). Environmental associations (turquois) play no major role in the network. Particular to the Dutch network diagram is that there is not just one, best-connected actor around which the network builds, but rather several central actors, namely the Ministry for Infrastructure and Environment (IenM), the Union of Water Boards (UvW), the Association of Dutch Drinking Water Companies (VEWIN), and the National Institute for Public Health and the Environment (RIVM). These multiple centers reflect the fact that there is no single actor who has taken the lead in the issue of pharmaceutical micropollutants in the Netherlands thus far.

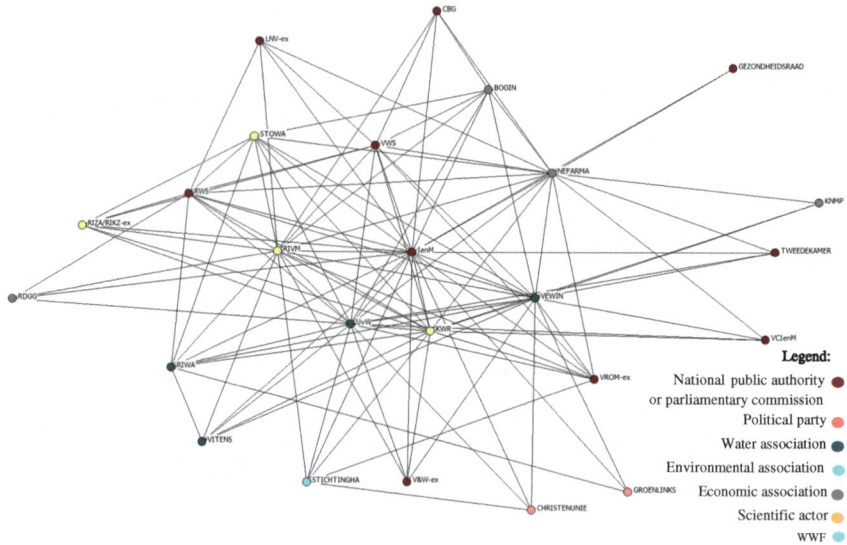

Fig. 3.8 Dutch collaboration network 2001–2013. *Note Circles* represent actors, *colors* represent actor types, labels are actor codes (see Annex 13 for full actor names), *lines* represent collaboration ties, and NetDraw's built-in algorithm sets the placement of nodes and length of ties to achieve an optimal visualization (Color figure online)

Belief cohesion

Dutch actors display a mean density of belief similarity ties of 2.63%. According to these results, 2.63% of Dutch actors share the same values regarding micropollutants policies.

Interconnectedness

In the Dutch collaboration network, each actor has on average 4.34 ties and each tie is involved in 2.16 triangles on average. The latter in particular indicates that actors use their ties to form dense relations benefitting from extensive redundancy.

Coalition formation and structure

The result of the blockmodeling procedure, called *balance*, which relies on agree-/disagree-ties here, reveals a coalition structure with two blocks of actors composed of 10 and 16 network members (see Table 3.9). It is noteworthy that the smaller coalition is still more powerful, because it is better connected than the larger coalition—with 51.1%, compared to 15.2% density of collaboration ties—and includes the powerful actors with 7.50 compared to 6.94 mean reputational power, and 10.7 compared to 7.44 mean degrees (see Annex 17 for a list of reputational power scores and mean degrees of all actors). Belief cohesion within this smaller, but more powerful coalition is with its 5.74% higher than the 1.19% density of belief similarity ties in the larger coalition. Hence, the coalition, which is bigger in size, is still weaker when considering parameters such as connectedness and power.

Table 3.9 Coalition members and beliefs in the Dutch case

Members of the opposing coalition		Members of the water quality coalition	
Actor type, acronym	**Mean beliefs:** Pro-environmental policy 2.8 Contra-environmental policy 2.1 (n = 8, min = 1, max = 4) *Explanation:* Value of pro-environmental policy is the mean of the following four surveyed beliefs: Source 3.4 End-of-pipe 2.4 Preventive 2.2 Zero concentrations 2.3	Actor type, acronym	**Mean beliefs:** Pro-environmental policy 3.4 Contra-environmental policy 1.7 (n = 6, min = 1, max = 4) *Explanation:* Value of pro-environmental policy is the mean of the following four surveyed beliefs: Source 3.8 End-of-pipe 2.8 Preventive 3.5 Zero concentrations 3.4
5 BOGIN 6 KWR 5 NEFARMA 5 RDGG 6 RIVM 3 RIWA 1 RWS 3 VEWIN 3 VITENS 4 WWF (n 10)		1 CBG 2 CHRISTENUNIE 1 GEZONDHEIDSRAAD 2 GROENLINKS 1 IenM 5 KNMP 1 LNV-ex 6 RIZA/RIKZ-ex 4 STICHTINGHA 6 STOWA 1 TWEEDEKAMER 3 UvW 1 V&W-ex 1 VCIenM 1 VROM-ex 1 VWS (n 16)	

Codes for actor types: 1 = National public authority or parliamentary commission, 2 = Political party, 3 = Water association, 4 = Environmental or consumer association, 5 = Industrial or agricultural association, 6 = Scientific actor

The large, but less powerful coalition is labeled *water quality coalition* because its mean beliefs show that coalition members highly support a pro-environmental policy orientation with a mean score of 3.4 and oppose a contra-environmental policy orientation with a mean score of 1.7. The powerful coalition supports a pro-environmental policy only moderately, with a score of 2.8, and demonstrates a higher support for a contra-environmental policy orientation than the water quality coalition with a score of 2.1. These results suggest that the opposing coalition obstructs the activities of the water quality coalition for different reasons: Some object to any type of policy intervention on micropollutants; others favor more encompassing policy action for the reduction of micropollutants.

Compared to interconnectedness within coalitions, the results for interconnectedness across coalitions show that the two coalitions are well-connected. Members from both coalitions have a comparably good degree of exchange with a 34.7% density of collaboration ties across coalition. Belief cohesion across coalitions, on the contrary, is with a 2.84% density of belief similarity ties weaker than within the opposing coalition, but still higher than within the water quality coalition. The Dutch situation is somewhat similar to the French, where one coalition is internally more divided than across coalitions, and where collaboration is weaker internally than across coalitions. In the Dutch case, the water quality coalition is internally divided and disconnected. This lack of activity shows that coalition members do not coordinate in order work toward a concrete strategic goal regarding the issue of pharmaceutical micropollutants.

Brokerage

There is one actor taking the role of a broker in the Dutch network by fulfilling three criteria simultaneously. The first criterion for the identification of brokers is a location on many paths between any other pair of nodes in the network, measured as betweenness centrality. The Ministry for Infrastructure and Environment (IenM) scores highest on betweenness centrality with 18.86, compared to a network mean of 2.41 (see Annex 17 for a complete list of actors' normalized betweenness centrality scores). Secondly, results suggest that IenM has a neutral belief system as it indicated total agreement (score of 4) with both policy orientations on micropollutants, i.e., source-directed and end-of-pipe. The third criterion underlying IenM's role as mediator is its connection to both coalitions, with 90% of all possible collaboration ties toward the opposing coalition, and 86% toward the water quality coalition.

IenM's main role in the policymaking process has thus far been to react to inquiries made by the Dutch parliament (see Sect. 2.3.6.1; Table 2.39). This passive role of the Ministry was mentioned by the interviewee from the Union of Water Boards, who argued that a more active role and initiation of policy action would do justice to IenM's organizational mandate (interview on Apr 10, 2014). During an interview, the representative of the IenM, on the other contrary, highlighted the changing role of Government, which was to refrain from top-down intervention by the state (interview on Apr 10, 2014) and instead to promote bottom-up initiatives on behalf of society or the economy. IenM reported that one of its main tasks was to connect involved actors, which is in fact a typical brokerage task.

Entrepreneurship

Lastly, four entrepreneurs were identified in the opposing coalition and one in the water quality coalition, which shows that the opposing coalition includes more actors who are particularly active in the policy process on micropollutants.[17]

- Entrepreneurs in the opposing coalition: Watercycle Research Institute (KWR) with a normalized eigenvector centrality of 40.67 against a network mean of 24.53, Association for Innovative Medicines in The Netherlands (NEFARMA) with an eigenvector centrality of 38.46/24.53, National Institute for Public Health and the Environment (RIVM) with an eigenvector centrality of 44.49/24.53, and Association of Dutch Drinking Water Companies (VEWIN) 42.32/24.53 (see Annex 17 for a complete list of actors' eigenvector centrality scores).
- Entrepreneur in the water quality coalition: Union of Water Boards (UvW) with an eigenvector centrality of 45.18/24.53.

---

[17]It was mentioned before that entrepreneurs are likely to be part of the dominant coalition. As there is a larger coalition and a more powerful coalition in the Dutch case, entrepreneurs are taken into consideration in both blocks of actors.

Coalition structure in the Netherlands

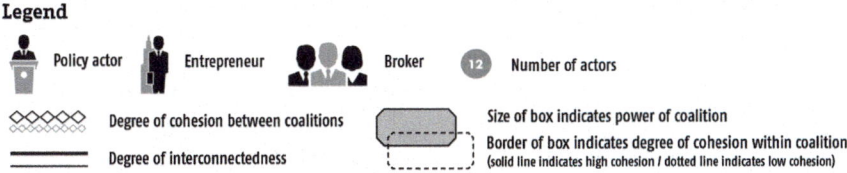

Legend

Fig. 3.9 Dutch network diagram

Aside from their exceptionally high eigenvector centralities compared to other network members, all entrepreneurs display clear policy preferences regarding the reduction of micropollutants. The entrepreneur in the water quality coalition, UvW, displays a strong preference for source-directed approaches (strongly agreed in the survey) for the reduction of micropollutants, compared to end-of-pipe approaches (disagreed somewhat in the survey). In the Dutch water governance system, water boards take responsibility for water protection policy and wastewater treatment, which also explains entrepreneurship of their union UvW.

The entrepreneurs of the opposing coalition are divided: NEFARMA, which represents the Dutch pharmaceutical industry, strongly opposes source-directed and somewhat opposes end-of-pipe approaches. These preferences show that NEFARMA is generally against policies for the reduction of pharmaceutical micropollutants in waters. An interview with a representative of NEFARMA revealed that the association questions the risks arising from pharmaceutical micropollutants in waters and therefore calls for research as opposed to policy action (interview on Apr 28, 2014).

The other three entrepreneurs in the opposing coalition support measures for the reduction of micropollutants. Among those are two actors, VEWIN and KWR, representing the drinking water sector. KWR constitutes a research institute, which focuses on drinking water topics; it agrees strongly with source-directed policies and end-of-pipe approaches and thus stresses the need to take policy action in general. VEWIN, the association for the Dutch drinking water suppliers, clearly favors source-directed policies (strongly agrees) to an end-of-pipe policy orientation (strongly disagrees).

RIVM is a public research institute that investigates the risks for humans arising from pharmaceutical micropollutants in the aquatic environment. More than the other three Rhine riparian countries, the Netherlands relies on surface waters for drinking water purposes, and therefore, questions of public health play a crucial role in the policymaking process on micropollutants. RIVM agrees somewhat with both end-of-pipe and source-directed policy orientations, which indicates that the institute attributes significance to the policy field in general.[18]

Dutch network diagram

In Fig. 3.9, the network structure is presented in an intuitive diagram. A small, but more powerful coalition (larger square), is well-connected (through thick lines) compared to a less powerful coalition (smaller square), which is larger in terms of members. Thick lines within the opposing coalition indicate good collaboration, while thin lines among coalition members of the water quality coalition stand for weak collaboration. Likewise, good belief cohesion within the opposing coalition is illustrated through thick box lines; weak belief cohesion within the water quality coalition through dashed box lines; and weak belief cohesion across coalitions through gray rhombs. Lastly, the strategic positions of the broker and entrepreneurs are easy to identify in the diagram.

3.3.5 Comparison of Network Structures

While the previous chapter presented the structures of the Swiss, German, French, and Dutch policy networks on micropollutants separately, the next chapter takes a comparative perspective. Here, each structural network property, i.e., belief cohesion, interconnectedness, coalition structures, brokerage, and entrepreneurship, is separately compared between the four countries. Only through comparison can one deem results as 'high' or 'low'.

3.3.5.1 Belief Cohesion

When comparing belief cohesion in the four cases (see Table 3.10), the Swiss network clearly scores highest with a 5.92% density of belief similarity ties among actors compared to a 2.98% density for German actors, 1.00% for French actors, and 2.63% for Dutch actors. The ensuing scores of Germany and the Netherlands are less than half the size of that of the Swiss indicating that the Swiss level of belief cohesion can be considered high. In comparison with the Swiss case, the German degree of belief cohesion is evaluated as good, but is at the edge of a medium–low degree of belief cohesion, because it is with 2.98% density of belief similarity ties,

---

[18]The results for KWR and RIVM could also point to a neutral position. Since KWR and RIVM do not have exceptionally high betweenness centrality scores—as by definition brokers do—they do not qualify as brokers here.

Table 3.10 Comparison of belief cohesion

Network characteristic	Operationalization SNA	Results			
		CH	G	F	NL
Belief cohesion	Mean density of belief similarity	5.92% (min = 0; max = 14)	2.98% (min = 0; max = 11)	1.00% (min = 0; max = 9)	2.63% (min = 0; max = 13)

only slightly higher than in the Dutch case with 2.63%. A low degree of belief cohesion was attributed to the Dutch network, followed by a very low level of belief cohesion in the French case (1.00%).

These results indicate high consensus between Swiss actors regarding policy content. German actors seem to disagree more about the design of policies toward micropollutants, but are in consensus about the general need to act politically. The beliefs of French and Dutch actors, on the contrary, fundamentally clash with regard to the need and design of policies toward micropollutants.

One should note that the consensus democracies Switzerland, Germany, and the Netherlands also display higher levels of belief cohesion than the majoritarian democracy France.

3.3.5.2 Interconnectedness

In general, large networks tend to be less dense than small ones, because actors have restricted resources for tie-formation. In the underlying cases, the Swiss network includes the largest number of members with 52 actors, followed by Germany with 39, France with 32, and the Netherlands with 26. Because of these differences in size, indicators for interconnectedness were calculated, i.e., *ties per node* and *triangles per tie*, which attempt to be less size-dependent than network density (see Table 3.11). For ties per node, i.e., mean degree, the Dutch network is —with an average of 4.34—by far the one that is best interconnected, followed by the Swiss network with 3.98 ties per node, the French network with 3.5, and lastly the German network with 2.97. When considering the parameter triangles per tie, the Dutch network still scores highest with 2.16, but the difference to the Swiss (1.54 triangles per tie), German (1.47), and French (1.44) cases is less pronounced. The Dutch network is the smallest in terms of numbers of actors, which might

Table 3.11 Comparison of interconnectedness

Network characteristic	Operationalization SNA	Results			
		CH	G	F	NL
Network size	Total number of nodes	52	39	32	26
	Total number of ties (undirected)	207	116	112	113
Interconnectedness	Ties per node	3.98	2.97	3.5	4.34
	Triangles per tie	1.54	1.47	1.44	2.16

denote that *triangles per tie* are a better indicator for size-independent interconnectedness, than are ties per node. Nevertheless, the Dutch network consistently scores highest on both parameters and is therefore considered highly interconnected here. The Swiss network, which attains the next highest scores, is also the largest one in terms of network members. Hence, the level of interconnectedness was evaluated as good in the Swiss case. The degree of interconnectedness is lowest in Germany and France, and although the ranking changes depending on the parameter, both networks' interconnectedness can be considered medium to low in this context.

One should note that the networks in the smaller countries, i.e., the Netherlands and Switzerland, are better connected than in the comparably larger states Germany and France. While this observation is most likely not related to country size, it may be linked to the macro-political system and pronounced consensual democratic traditions in these countries promoting the exchange of policy actors (Kriesi and Trechsel 2008; Lijphart and Crepaz 1991).

3.3.5.3 Belief Cohesion and Interconnectedness in Combination

When considering both dimensions, belief cohesion and interconnectedness, one can categorize the four-policy network in a more ideal-typical fashion, as shown in Table 3.12. While nuanced analysis was the primary goal of the previous steps, information here is treated in a more condensed fashion in order to enable analytical clarity, as well as to contrast the main characteristics of the four policy networks to one another.

The detailed empirical results reveal that the Swiss network is characterized by high belief cohesion and sufficient interconnectedness, which allows for classifying the Swiss network into the upper right category of the typology shown in Table 3.12. The French network scored low in both measures and can therefore be placed in the opposite category than the Swiss network in the typology. In the German case, a good level of belief cohesion combines with a medium degree of

Table 3.12 Categorization of the policy networks on two dimensions

		Belief cohesion	
		Low	High
Interconnectedness	High	NL Belief cohesion: 2.63% mean density of belief similarity Interconnectedness: 2.16 triangles per tie	CH Belief cohesion: 5.92% mean density of belief similarity Interconnectedness: 1.54 triangles per tie
	Low	F 1.00% mean density of belief similarity Interconnectedness: 1.44 triangles per tie	G Belief cohesion 2.98% mean density of belief similarity Interconnectedness: 1.47 triangles per tie

interconnectedness, which best suits the lower right category of the typology. The almost opposite combination characterizes the Dutch network, i.e., high interconnectedness and low belief cohesion, and therefore fits into the upper left category of the typology.

3.3.5.4   Coalition Structure

The policy network literature distinguishes three types of coalition structures, labeling them as unitary, collaborative, and adversarial subsystems (Weible et al. 2010; Ingold and Gschwend 2014). While unitary subsystems are characterized by one single coalition with high intra-coalition belief similarity and coordination, collaborative subsystems are defined as two coalitions with intermediate inter-coalition belief similarity, and high inter- as well as intra-coalition coordination. Adversarial subsystems are described as competitive through their low inter-coalition belief similarity and coordination, and high intra-coalition coordination. When applying these categories to the Swiss case, empirical results shown in Table 3.13 reveal a coalition structure that is best described as adversarial-collaborative.

Table 3.13 Comparison of coalition structure

Network characteristic	Operationalization SNA	Results			
		CH	G	F	NL
Coalition structure	Size of balance clusters, coalition 1 (c1), coalition 2 (c2)	39 (c1) 11 (c2)	19 (c1) 17 (c2)	21 (c1) 10 (c2)	10 (c1) 16 (c2)
Within-coalition collaboration	Density of collaboration ties within coalition 1 (c1) and 2 (c2)	18% (c1) 25% (c2)	18.4% (c1) 18.4% (c2)	25.8% (c1) 13.3% (c2)	51.1% (c1) 15.2% (c2)
Across-coalition collaboration	Density of collaboration ties across coalitions	0.41%	0.87%	14.5%	34.7%
Power of dominant coalition	Mean degree Mean reputational indegree	9.20 12.13	7.10 7.63	8.18 6.67	10.70 7.50
Power of minority coalition	Mean degree Mean reputational indegree	4.81 4.90	5.59 5.88	4.80 3.40	7.44 6.94
Within-coalition belief cohesion	Density of Manhattan belief similarity ties within c1 and c2	6.40% (c1) 5.24% (c2)	4.32% (c1) 2.32% (c2)	1.60% (c1) 0.27% (c2)	5.74% (c1) 1.19% (c2)
Across-coalition belief cohesion	Density of Manhattan belief similarity ties across coalitions	4.73%	2.43%	0.72%	2.84%

Switzerland represents the only country with one dominant and one minority coalition. The dominant coalition is not only particularly powerful (n = 39, 9.20 mean degree, 12.13 mean reputational power) compared to the minority coalition (n = 11, 4.81 mean degree, 4.90 mean reputational power), but also in comparison with the other three cases (G: 7.10 mean degree, 7.63 mean reputational power; F: 8.18 mean degree, 6.67 mean reputational power; NL: 10.70 mean degree, 7.50 mean reputational power). Moreover, there is a very low degree of interconnectedness across coalitions in the Swiss case (0.41% density of collaboration ties) compared to the other networks (G: 0.87%, F: 14.5%, NL: 34.7%), which points to an adversarial setting. Simultaneously, there is a high level of belief similarity within (c1: 6.40%, c2: 5.24% density of belief similarity ties) and across coalitions (4.73% density of belief similarity ties) in the Swiss network, which is typical for a collaborative subsystem. Taking these two insights into account, i.e., low collaboration across coalitions and high belief cohesion within and across coalitions, the label adversarial-collaborative fits best to describe the Swiss coalition structure. Germany is a more ideal-typical example of an adversarial coalition structure, with two opposing coalitions displaying a low inter-coalition belief similarity (2.43% density of belief similarity ties) and collaboration (0.87% density of collaboration ties), but high intra-coalition collaboration (18.4% density of collaboration ties). The two opposing coalitions are almost equal in size (19 vs. 17 actors) and density (18.4%), while the dominant coalition scores higher on power (7.10 vs. 5.59 mean degree; 7.63 vs. 5.88 mean reputational power) and belief cohesion (4.32% vs. 2.32% density of belief similarity ties).

From a comparative perspective, the Swiss and German coalition structures are more similar to each other than to the French and Dutch coalition structures. Similarities result from a clear coalition structure comprised of two opposing clusters in the Swiss and German policy networks, which is not the case for France and the Netherlands. In the latter networks, one group of actors is internally more divided than across coalitions and is internally less well-interconnected than across coalitions, showing that coalition structures are characterized less by a divide between opposing coalitions, and more by a weaker and a stronger group of actors where 'weak' is defined as poorly interconnected with high belief dissimilarities. Furthermore, Swiss and German actors exhibit a higher tendency to ally themselves into coalitions than do French or Dutch actors, considering that coefficient estimates for within-coalition collaboration are significant and positive in the Swiss and German ERG models, while they are insignificant or negative for the French and Dutch data.

It is of note here that the lower tendency of French and Dutch actors to form coalitions can be related to their specific subsystem structures. The French and Dutch system of water governance relies on task-specific jurisdictions and water agencies that institutionally incorporate integrated water resources management. An integrated approach to resources management provides strong incentives for cooperation and may conflict to some degree with the creation of opposing coalitions. Hence, the lower tendency for coalition formation in the French and Dutch case could result from their integrated approach to water management.

Taking a closer look at the French coalition structure shows that only one coalition formed, and therefore, the French network fits best into the category of a unitary subsystem, although it does not represent a typical example of such a subsystem. While unitary subsystems are typically characterized by high inter-connectedness and belief similarities among actors, French coalition members, by contrast, display less belief cohesion (c1: 1.60% density of belief similarity ties) than the other studied policy networks. Hence, the label of a unitary subsystem is based on the high level of collaboration among French coalition members only (25.8% density of collaboration ties).

The Dutch case seems to fit best into the category of collaborative subsystems, because of its particularly high density of collaboration ties across coalitions (34.7% density). Unlike the French case, the two Dutch groups of actors appear as separate coalitions, because the clusters are internally connected (c1: 51.1%; c2: 15.2% density of collaboration ties) and belief-cohesive (c1: 5.74%; c2: 1.19% density of belief similarity ties) enough to qualify as a coalition. More precisely, the water quality coalition seems to be divided about the question of pharmaceutical micropollutants (1.19%), while the opposing coalition scores higher on belief similarity (5.74%).

In conclusion, results point to a collaborative-adversarial subsystem in the Swiss case, an adversarial subsystem in the German case, a unitary subsystem in the French case, and a collaborative subsystem in the Dutch case.

3.3.5.5   Brokerage

Table 3.14 shows brokers' acronyms (see Annexes 10–13 for a full list of actor names), their betweenness centrality scores compared to the network mean, and their degree of interconnectedness with actors from both coalitions. The table indicates that all brokers display high densities of collaboration ties with both coalitions, which demonstrates that they collaborate with actors from opposing clusters. Through this strategic position, brokers form bridging ties across structural holes existing between coalitions. From a comparative perspective, brokerage is strongest in Switzerland and Germany, while it is less pronounced in France and the Netherlands. Brokerage is stronger in the Swiss and German cases, because there exist two brokers (CH: BAFU-W, URESKS; G: BMU, UBA), and only one in the Dutch network (IenM), as well as an absence of a real broker in the French one. Moreover, strong brokerage can be deduced from higher betweenness centrality of the Swiss (49.54, 9.06, mean 1.88), German (27.26, 16.13, mean 2.74), and French (33.85, mean 2.99) brokers compared to the Dutch one (18.86, mean 2.41). Hence, brokerage is weak when comparing the betweenness centrality scores of all brokers.

In the French case, there does exist an actor with high betweenness centrality scores (33.85). However, the actor most likely finds difficulty mediating between opponents, considering the low belief cohesion scores (c1: 1.60%; c2: 0.27% density of belief similarity ties) in the network. The higher belief dissimilarities and trenches are among actors, the more difficult it is for the broker to mediate

Table 3.14 Comparison of brokerage

Network characteristic	Operationalization SNA	Results			
		CH	G	F	NL
Brokerage	Betweenness centrality score(mean)	BAFU-W 49.54 UREKS 9.06 (1.88)	BMU 27.26 UBA 16.13 (2.74)	DEB 33.85 (2.99)	IenM 18.86 (2.41)
	Ties of broker toward coalition 1 and 2	BAFU-W: 89.5% (c1) 72.7% (c2) UREKS: 39.5% (c1) 18.2% (c2)	BMU: 55.6% (c1) 70.7% (c2) UBA: 61.1% (c1) 41.2% (c2)	DEB: 90% (c1) 60% (c2)	IenM: 90% (c1) 86% (c2)
Within-coalition belief cohesion	Density of Manhattan belief similarity ties within c1 and c2	6.40% (c1) 5.24% (c2) (min = 0; max = 14)	4.32% (c1) 2.32% (c2) (min = 0; max = 11)	1.60% (c1) 0.27% (c2) (min = 0; max = 9)	5.74% (c1) 1.19% (c2) (min = 0; max = 13)
Across-coalition belief cohesion	Density of Manhattan belief similarity ties across coalitions	4.73%	2.43%	0.72%	2.84%

compromise, and to help overcoming existing conflicts. Comparably higher scores for cross-coalition belief similarities in the Swiss, German, and Dutch cases indicate, on the contrary, that brokers have an easier task to mediate between coalitions and foster agreement.

Brokers seem to have encouraged more of a pro-environmental attitude toward the opposing coalition in the Swiss and German networks, thereby enabling agreement on a policy design that emphasizes the protection of aquatic environments. The Dutch network suggests the opposite: The less powerful water quality coalition might have adopted critical arguments of the opposing coalition through brokerage, which impeded environmental protection.

3.3.5.6 Entrepreneurship

The theoretical argument about entrepreneurship is that it might affect policy design positively because entrepreneurs are well-connected and therefore have the ability to mobilize coalition members and improve interconnectedness within their coalition.

The results displayed in Table 3.15 provide a preliminary indication supporting this theoretical argument.[19] Generally when entrepreneurs exist in a coalition, the interconnectedness within that coalition is comparably high. In the German network, four entrepreneurs exist in both coalitions, and intra-coalition interconnectedness is identical in both coalitions as well (18.4% density of collaboration ties). The French coalition includes five entrepreneurs, and the Dutch dominant coalition four entrepreneurs. In both cases, interconnectedness is particularly high (F: 25.5%; NL: 51.1% density of collaboration ties within c_1), which may partially be attributed to the activity of entrepreneurs. The Swiss case does not represent an exception despite its slightly lower density of collaboration ties (18% within c_1). With 39 actors, the Swiss dominant coalition is by far the largest, and therefore, reaching high densities proves more difficult. In conclusion, all four networks display good levels of interconnectedness in the coalitions in which entrepreneurs are active. Entrepreneurs are an important source of tie-creation to the investigated collaboration networks.

In conclusion, results suggest that entrepreneurs contribute to an overall improved interconnectedness, especially within their coalition. Through their activity, entrepreneurs mobilize other network members take action on a policy issue, which may also facilitate a comprehensive policy design.

---

[19]Section 3.2.3.6 explains why entrepreneurs are likely to be part of the dominant coalition, namely because by definition (a) entrepreneurs are connected to well-connected others and (b) dominant coalitions are composed of well-connected, powerful actors. As there are two almost equally powerful coalitions in the German network, entrepreneurs are considered in both coalitions. In the Dutch network, there exists a larger coalition, as well as a more powerful coalition; therefore, entrepreneurs are taken into consideration in both blocks of actors.

Table 3.15 Comparison of entrepreneurship

Network characteristic	Operationalization SNA	Results				
		CH	G	F	NL	
Entrepreneurship	Eigenvector centrality score (mean)	EAWAG 39.12 ERFA 38.69 VSA 38.59 BPUK 30.82 (16.07)	c1: BDEW 40.61 DVGW 32.54 LAWA 32.57 NRW 33.341 c2: DWA 34.11 IVA 31.68 RLP 41.584 VCI 33.21 (17.34)	AGENCE 47.69 ALSACE 39.73 AQUAREF 33.44 DREAL 41.45 ONEMA 38.78 (21.17)	c1: KWR 40.67 NEFARMA 38.46 RIVM 44.49 VEWIN 42.32 c2: UvW 45.18 (24.53)	
Within-coalition collaboration	Densities of collaboration ties within coalitions 1 (c1) and 2 (c2)	18% (c1) 25% (c2)	18.4% (c1) 18.4% (c2)	25.8% (c1) 13.3% (c2)	51.1% (c1) 15.2% (c2)	
Coalition structure	Size of balance clusters, coalition 1 (c1), coalition 2 (c2)	39 (c1) 11 (c2)	19 (c1) 17 (c2)	21 (c1) 10 (c2)	10 (c1) 16 (c2)	

References

AAPOR. (2011). *Standard definitions: Final dispositions of case codes and outcome rates for surveys* (7th ed.). The American Association for Public Opinion Research.

Adam, S., & Kriesi, H.-P. (2007). The network approach. In P. Sabatier (Ed.), *Theories of the policy process* (pp. 129–154). Boulder: Westview Press.

Batagelj, V., & Mrvar, A. (1996). PAJEK—Program for large network analysis (Version 1.23 ed.).

Batagelj, V., & Mrvar, A. (1998). Pajek-program for large network analysis. *Connections, 21*(2), 47–57.

Baumgartner, F., & Jones, B. (1991). Agenda dynamics and policy subsystems. *The Journal of Politics, 53*(4), 1044–1074.

Baumgartner, F., & Jones, B. (1993). *Agendas and instability in American politics*. Chicago: University of Chicago Press.

Baxter-Moore, N. (1987). Policy implementation and the role of the state. A revisited approach to the study of policy instruments. In R. Jackson, D. Jackson, & N. Baxter-Moore (Eds.), *Contemporary Canadian politics*. Scarborough: Prentice-Hall.

Berardo, R., & Scholz, J. (2010). Self-Organizing policy networks: Risk, Partner Selection, and Cooperation in estuaries. *American Journal of Political Science, 54*(3), 632–649.

Berry, W. (1990). The confusing case of budgetary incrementalism: Too many meanings for a single concept. *The Journal of Politics, 52*(1), 167–196.

Beyers, J., & Braun, C. (2014). Ties that count: Explaining interest group access to policymakers. *Journal of Public Policy, 34*(1), 93–121.

Birkland, T. (1997). *After disaster: Agenda setting, public policy, and focusing events*. Washington D.C.: Georgetown University Press.

Birkland, T. (2010). *An introduction to the policy process: Theories, concepts, and models of public policy making* (3rd ed.). Armonk NY: M.E. Sharpe.

Bodin, Ö., Crona, B., & Ernstson, H. (2006). Social networks in natural resource management: What is there to learn from a structural perspective? *Ecology and Society, 11*(2).

Borgatti, S., Everett, M., & Freeman, L. (2002). *UCINET for Windows: Software for social network analysis*. Harvard: Analytic Technologies.

Borgatti, S., Everett, M., & Johnson, J. (2013). *Analyzing social networks*. London: Sage.

Börzel, T. (1998). Organizing Babylon. On the different conceptions of policy networks. *Public Administration, 76*, 253–273.

Bressers, H., & O'Toole, L. (1998). The selection of policy instruments: A network-based perspective. *Journal of Public Policy, 18*(3), 213–239.

Bressers, H., & O'Toole, L. (2005). Instrument selection and implementation in a networked context. In P. Eliadis, M. Hill, & M. Howlett (Eds.), *Designing government: From instruments to governance* (pp. 132–153). Montreal, Kingston: McGill-Queen's University Press.

Bressers, H., O'Toole, L., & Richardson, J. (Eds.). (1995). *Networks for water policy: A comparative perspective*. London: Frank Cass.

Breunig, C., Koski, C., & Mortensen, P. (2010). Stability and punctuations in public spending: A comparative study of budget functions. *Journal of Public Administration Research and Theory, 20*(3), 703–722.

Burt, R. (2000). The network structure of social capital. In B. Staw & R. Sutton (Eds.), *Research in organizational behavior* (pp. 345–423). Greenwich, CT: JAI Press.

Burt, R. (2009). *Structural holes: The social structure of competition*. Cambridge, MA: Harvard University Press.

Check, J., & Schutt, R. (2011). *Research methods in education*. Thousand Oaks: Sage.

Christopoulos, D. (2008). Political entrepreneurs: Network structure and power. Published online: http://www.researchgate.net/publication/265495932 (accessed on July 9, 2015).

Christopoulos, D., & Ingold, K. (2015). Exceptional or just well connected? Political entrepreneurs and brokers in policy making. *European Political Science Review, 7*, 475–498.

Coleman, J. (1990). *Foundations of social theory*. Cambridge, MA: Harvard University Press.

Crona, B., & Bodin, Ö. (2006). What you know is who you know? Communication patterns among resource users as a prerequisite for co-management. *Ecology and Society, 11*(2).

De Nooy, W., Mrvar, A., & Batagelj, V. (2005). *Exploratory social network analysis with Pajek.* Cambridge: Cambridge University Press.

Dillman, D., Smyth, J., & Christian, L. M. (2009). *Internet, mail, and mixed-mode surveys: The tailored design method* (3rd ed.). Hoboken, New Jersey: Wiley.

Doreian, P. (2008). A multiple indicator approach to blockmodeling signed networks. *Social Networks, 30*(3), 247–258.

Doreian, P., Batagelj, V., & Ferligoj, A. (2005). *Generalized blockmodeling, structural analysis in the social sciences.* Cambridge: Cambridge University Press.

Doreian, P., & Mrvar, A. (1996). A partitioning approach to structural balance. *Social Networks, 18,* 149–168.

Doreian, P., & Mrvar, A. (2009). Partitioning signed social networks. *Social Networks, 31,* 1–11.

Dowding, K. (1995). Model or metaphor? A critical review of the policy network approach. *Political Studies, 43*(1), 136–158.

Easton, D. (1965). *A framework for political analysis.* Englewood Cliffs, NJ: Prentice-Hall.

Fischer, M. (2012). *Entscheidungsstrukturen in der Schweizer Politik zu Beginn des 21. Jahrhunderts.* Glarus, Chur: Rüegger.

Fischer, M. (2013). *Policy network structures, institutional context, and policy change.* Paper presented at the COMPASSS Working Paper 73,

Fischer, M. (2014). Coalition structures and policy change in a consensus democracy. *Policy Studies Journal, 42*(3), 344–366.

Fischer, M. (2015). Institutions and coalitions in policy processes: A cross-sectoral comparison. *Journal of Public Policy, 35*(2), 1–24.

Fischer, M., & Sciarini, P. (2013). *Collaborative tie formation in policy networks: A cross-sector perspective.* Paper presented at the Working Paper, EAWAG Dübendorf, University of Geneva,

Fischer, M., & Sciarini, P. (2015). Unpacking reputational power: Intended and unintended determinants of the assessment of actors' power. *Social Networks, 42,* 60–71.

Freeman, L. (1977). A set of measures of centrality based on betweeness. *Sociometry, 40,* 35–41.

Friedkin, N. (1981). The development of structure in random networks: An analysis of the effects of increasing network density on five measures of structure. *Social Networks, 3,* 41–52.

Frohlich, N., & Oppenheimer, J. (1978). *Modern political economy.* Englewood Cliffs, NJ: Prentice-Hall.

Gerber, E., Henry, A. D., & Lubell, M. (2013). Political homophily and collaboration in regional planning networks. *American Journal of Political Science, 57*(3), 598–610.

Hall, P. (1993). Policy paradigms, social learning, and the state: The case of economic policymaking in Britain. *Comparative Politics, 25*(3), 275–296.

Handcock, M., Hunter, D., Butts, C., Goodreau, S., & Morris, M. (2008). statnet: Software tools for the representation, visualization, analysis and simulation of network data. *Journal of Statistical Software, 24*(1), 1–11.

Hardy, M., & Bryman, A. (2004). *Handbook of data analysis.* Thousand Oaks: Sage.

Heaney, M. (2014). Multiplex networks and interest group influence reputation: An exponential random graph model. *Social Networks, 36,* 66–81.

Heclo, H. (1978). Issue networks and the executive establishment. In A. King (Ed.), *The new American political system.* Washington, DC: American Enterprise Inc.

Heclo, H., & Wildavsky, A. (1974). *The private government of public money.* London: Macmillan.

Hennig, M., Brandes, U., Pfeffer, J., & Mergel, I. (2012). *Studying social networks. A guide to empirical research.* Frankfurt am Main: Campus Verlag.

Henning, C. (2009). Networks of power in the CAP System of the EU-15 and EU-27. *Journal of Public Policy, 29*(2), 153–177.

Henry, A. D. (2011). Ideology, power, and the structure of policy networks. *Policy Studies Journal, 39*(3), 361–383.

Henry, A. D., Lubell, M., & McCoy, M. (2010). Belief systems and social capital as drivers of policy network structure: The case of California regional planning. *Journal of Public Administration Research and Theory, 21*(3), 419–444.

Hermans, F., van Apeldoorn, D., Stuiver, M., & Kok, K. (2013). Niches and networks: Explaining network evolution through niche formation processes. *Research Policy, 42*(3), 613–623.

Holcombe, R. (2002). Political entrepreneurship and the democratic allocation of economic resources. *The Review of Austrian Economics, 15*(2–3), 143–159.

Howlett, M. (2002). Do networks matter? Linking policy network structure to policy outcomes: Evidence from four Canadian policy sectors 1990–2000. *Canadian Journal of Political Science, 35*(2), 235–267.

Howlett, M., & Giest, S. (2012). The policy-making process. In E. Araral, S. Fritzen, M. Howlett, M. Ramesh, & X. Wu (Eds.), *Routledge handbook of public policy* (pp. 17–28). Oxon: Taylor & Francis.

Howlett, M., & Ramesh, M. (2003). *Studying public policy: Policy cycles and policy subsystems*. Oxford: Oxford University Press.

Immergut, E. (1992). *Health politics: Interests and institutions in Western Europe*. Cambridge: Cambridge University Press.

Immergut, E. (1998). The theoretical core of the new institutionalism. *Politics and Society, 26*(1), 5–34.

Ingold, K. (2007). The influence of actors' coalition on policy choice: The case of the Swiss climate policy. In T. Friemel (Ed.), *Applications of social network analysis*. UVK: Constance.

Ingold, K. (2008). *Analyse des mécanismes de décision: Le cas de la politique climatique suisse*. Zürich and Chur: Rüeggger Verlag.

Ingold, K. (2011). Network structures within policy processes: Coalitions, power, and brokerage in Swiss climate policy. *Policy Studies Journal, 39*(3), 435–459.

Ingold, K., & Gschwend, M. (2014). Science in policy-making: Neutral experts or strategic policy-makers? *West European Politics, 37*(5), 993–1018.

Ingold, K., & Leifeld, P. (2014). Structural and institutional determinants of influence reputation: A comparison of collaborative and adversarial policy networks in decision making and implementation. *Journal of Public Administration Research and Theory, published online October 21, 2014*.

Ingold, K., & Varone, F. (2012). Treating policy brokers seriously: Evidence from the climate policy. *Journal of Public Administration Research and Theory, 22*(2), 319–346.

Jann, W., & Wegrich, K. (2014). Phasenmodelle und Politikprozesse: Der Policy-Cycle. In K. Schubert & N. Bandelow (Eds.), *Lehrbuch der Politikfeldanalyse* (pp. 97–132). München: Oldenbourg Wissenschaftsverlag.

John, P. (2013). *Analyzing public policy*. New York: Routledge.

Kenis, P., & Schneider, V. (1991). Policy networks and policy analysis: Scrutinizing a new analytical toolbox. In B. Marin & R. Mayntz (Eds.), *Policy networks—Empirical evidence and theoretical considerations*. Frankfurt am Main: Campus Verlag.

Kickert, W., Klijn, E.-H., & Koppenjan, J. (1997). *Managing complex networks*. London: Sage.

King, A. (1973). Ideas, institutions and the policies of governments: A comparative analysis: Parts I and II. *British Journal of Political Science, 3*(3), 291–313.

Kingdon, J. (1984). *Agendas, alternatives, and public policies*. Boston: Little, Brown.

Kingdon, J. (1995). *Agendas, alternatives, and public policies*. New York: Longman.

Kingdon, J., & Thurber, J. (2011). *Agendas, alternatives, and public policies*. New York: Longman.

Kirzner, I. (1978). *Competition and entrepreneurship*. Chicago: University of Chicago Press.

Kitschelt, H., & Wilkinson, S. (Eds.). (2007). *Patrons, clients and policies: Patterns of democratic accountability and political competition*. Cambridge: Cambridge University Press.

Klijn, E.-H. (1996). Analyzing and managing policy processes in complex networks: A theoretical examination of the concept policy network and its problems. *Administration & Society, 28*(1), 90–119.

Knill, C., & Tosun, J. (2012). *Public policy: A new introduction*. New York: Palgrave Macmillan.

Knoepfel, P., Larrue, C., & Varone, F. (2011). *Politikanalyse*. Stuttgart: UTB GmbH.

Knoke, D. (1990). *Political networks. The structural perspective*. New York: Cambridge University Press.

Knoke, D. (1994). Networks of elite structure and decision making. In S. Wasserman & J. Galaskiewicz (Eds.), *Advances in social network analysis: Research in the social and behavioral sciences* (pp. 274–295). Thousand Oaks: Sage.

Knoke, D., Pappi, F. U., Broadbent, J., & Tsujinaka, Y. (1996). *Comparing policy networks. Labor politics in the U.S., Germany, and Japan*. Cambridge, U.K., New York: Cambridge University Press.

Knoke, D., & Yang, S. (2008). *Social network analysis* (2nd ed.). Los Angeles: Sage.

Kriesi, H.-P. (2007). *Vergleichende Politikwissenschaft: Teil I: Grundlagen - Eine Einführung*. Baden-Baden: Nomos.

Kriesi, H., Adam, S., & Jochum, M. (2006). Comparative analysis of policy networks in Western Europe. *Journal of European Public Policy, 13*(3), 341–361.

Kriesi, H., & Jegen, M. (2001). The Swiss energy policy elite: The actor constellation of a policy domain in transition. *European Journal of Political Research, 39*, 251–287.

Kriesi, H., & Trechsel, A. (2008). *The politics of Switzerland. Continuity and change in a consensus democracy*. Cambridge: Cambridge University Press.

Krosnick, J. (1990). Survey research. *Annual Review of Psychology, 50*, 537–567.

Kuhnert, S. (2001). An evolutionary theory of collective action: Schumpeterian entrepreneurship for the common good. *Constitutional Political Economy, 12*(1), 13–29.

Lang, A., & Leifeld, P. (2008). Die Netzwerkanalyse in der Policy-Forschung: Eine theoretische und methodische Bestandsaufnahme. In F. Janning & K. Toens (Eds.), *Die Zukunft der Policy-Forschung* (pp. 223–241). Wiesbaden: VS Verlag für Sozialwissenschaften.

Lascoumes, P., & Le Gales, P. (2007). Introduction: Understanding public policy through its instruments—From the nature of instruments to the sociology of public policy instrumentation. *Governance-An International Journal of Policy and Administration, 20*(1), 1–21.

Lasswell, H. (1956). *The decision process: Seven categories of functional analysis*. College Park: University of Maryland Press.

Lasswell, H. (1971). *A pre-view of policy sciences*. New York: American Elsevier Publishing Company.

Laumann, E., & Knoke, D. (1987). *The organizational state. Social in national policy domains*. Madison: University of Wisconsin Press.

Laumann, E., Marsden, P., & Prensky, D. (1983). The boundary specification problem in network analysis. In R. Burt & M. Minor (Eds.), *Applied network analysis: A methodological introduction*. Beverly Hills: Sage.

Leibenstein, H. (1968). Entrepreneurship and development. *The American Economic Review, 58* (2), 72–83.

Leifeld, P., & Schneider, V. (2012). Information exchange in policy networks. *American Journal of Political Science, 56*(3), 731–744.

Lijphart, A., & Crepaz, M. (1991). Corporatism and consensus democracy in eighteen countries: Conceptual and empirical linkages. *British Journal of Political Science, 21*(2), 235–246.

Lindblom, C. (1959). The science of "Muddling Through". *Public Administration Review, 19*(2), 79–88.

Lindblom, C. (1977). *Politics and markets*. New York: Basic Books.

Lubell, M., & Fulton, A. (2007). Local diffusion networks as pathways to sustainable agriculture. *California Agriculture, 61*(3), 131–137.

Lubell, M., Robins, G., & Wang, P. (2014). Network structure and institutional complexity in an ecology of water management games. *Ecology and Society, 19*(4), 23.

Marin, B., & Mayntz, R. (1991). *Policy network: Empirical evidence and theoretical considerations*. Frankfurt am Main: Campus Verlag.

Marsden, P. (2005). Recent developments in network measurement. In P. Carrington, J. Scott, & S. Wasserman (Eds.), *Models and methods in social network analysis* (pp. 8–30). New York: Cambridge University Press.

Marsh, D. (1998). *Comparing policy networks*. Open University Press.

Marsh, D., & Rhodes, R. (1992). *Policy networks in British government*. Oxford, GB: Clarendon Press.

Mayntz, R., & Scharpf, F. (1995). *Gesellschaftliche Selbstregulierung und politische Steuerung* (Vol. 23, Schriften des Max-Planck-Instituts für Gesellschaftsforschung Köln). Frankfurt am Main: Campus Verlag.

Metz, F., & Ingold, K. (2014). *Policy instrument selection under Uncertainty: The case of micropollution regulation*. Paper presented at the Conference Paper presented at the Swiss Political Science Association Annual Congress, Berne, January 31, 2014.

Mintrom, M. (2000). *Policy entrepreneurs and school choice*. Washington, DC: Georgetown University Press.

Mintrom, M., & Norman, P. (2009). Policy entrepreneurship and policy change. *Policy Studies Journal, 37*(4), 649–667.

Nohrstedt, D. (2008). The politics of crisis policymaking: Chernobyl and Swedish nuclear energy policy. *Policy Studies Journal, 36*(2), 257–278.

North, D. (1990). *Institutions, institutional change and economic performance*. Cambridge, New York: Cambridge University Press.

O'Toole, L. (1997). Treating networks seriously: Practical and research-based agendas in public administration. *Public Administration Review, 57*(1), 45–52.

Olson, M. (1965). *The logic of collective action*. Cambridge, MA: Harvard University Press.

Ostrom, E. (1990). *Governing the commons. The evolution of institutions for collective actors*. Cambridge, New York: Cambridge University Press.

Pappi, F., & Henning, C. (1999). The organization of influence on the EC's common agriculturalpolicy: A network approach. *European Journal of Political Research, 36*(2), 257–281.

Pappi, F. U., & Henning, C. (1998). Policy networks: More than a metaphor? *Journal of Theoretical Politics, 10*(4), 553–575.

Pitts, F. (1979). The medieval river trade network of Russia revisited. *Social Networks, 1*, 285–292.

Provan, K., & Kenis, P. (2008). Modes of network governance: Structure, management, and effectiveness. *Journal of Public Administration Research and Theory, 18*(2), 229–252.

Prüfer, L., Vazansky, L., & Wystup, D. (2003). Antworskalen im ALLBUS und ISSP. Eine Sammlung. *ZUMA-Methodenbericht* (Vol. 11).

Qualtrics. (2013). *Qualtrics, survey platform, first release 2005* (Version 2013 ed.). Provo, Utah, USA.

R Development Core Team (2014). *R: A language and environment for statistical computing* (Version 3.1.2 ed.). Vienna, Austria: R Foundation for Statistical Computing.

Rhodes, R. (1988). *Beyond Westminster and Whitehall*. London: Unwin Hyman.

Rhodes, R., & Marsh, D. (1992). New directions in the study of policy networks. *European Journal of Political Research, 21*(1–2), 181–205.

Ripley, R., & Franklin, G. (1984). *Congress, the bureaucracy and public poilcy* (2nd ed.). Dorsey: Homewood.

Robins, G., Lewis, J. M., & Wang, P. (2012). Statistical network analysis for analyzing policy networks. *Policy Studies Journal, 40*(3), 375–401.

Robinson, S. (2006). A decade of treating networks seriously. *Policy Studies Journal, 34*(4), 589–598.

Sabatier, P. (2007). *Theories of the policy process*. Boulder: Westview Press.

Sabatier, P., & Jenkins-Smith, H. (1993). *Policy change and learning: An advocacy coalition approach*. Boulder: Westview Press.

Sabatier, P., & Weible, C. (2014). *Theories of the policy process*. Boulder: Westview Press.

Sager, F. (2009). Governance and coercion. *Political Studies, 57*(3), 537–558.

Sandström, A., & Carlsson, L. (2008). The performance of policy networks: The relation between network structure and network performance. *The Policy Studies Journal, 36*(4), 497–524.

Saris, W., Revilla, M., Krosnick, J., & Shaeffer, E. (2010). Comparing questions with agree/disagree response options to question with item-specific response options. *Survey Research Methods, 4*(1), 61–69.

Schaeffer, N. C., & Presser, S. (2003). The science of asking questions. *Annual Review of Sociology, 29*(1), 65–88.

Schneider, M., & Teske, P. (1992). Toward A theory of the political entrepreneur: Evidence from local government. *The American Political Science Review, 86*(3), 737–747.

Schneider, M., Teske, P., & Mintrom, M. (2011). *Public entrepreneurs: Agents for change in American government*. Princeton: Princeton University Press.

Schneider, V. (2014). Akteurskonstellationen und Netzwerke in der Politikentwicklung. In K. Schubert & N. Bandelow (Eds.), *Lehrbuch der Politikfeldanalyse* (pp. 259–288). München: Oldenbourg.

Schubert, K., & Bandelow, N. (2014). *Lehrbuch der Politikfeldanalyse 2.0* (3rd ed.). München: Oldenbourg Wissenschaftsverlag.

Schumpeter, J. (1939). *Business cycles: A theoretical, historical and statistical analysis of the capitalist process*. New York, Toronto, London: McGraw-Hill Book Company.

Schumpeter, J., & Opie, R. (1934). *The theory of economic development: An inquiry into profits, capital, credit, interest, and the business cycle*. Cambridge, MA: Harvard University Press.

Sciarini, P. (1996). Elaboration of the Swiss agricultural policy for the GATT negotiations: A network analysis. *Schweizerische Zeitschrift für Soziologie, 22*(1), 85–115.

Sciarini, P., Fischer, A., & Nicolet, S. (2004). How Europe hits home: Evidence from the Swiss case. *Journal of European Public Policy, 11*(3), 353–378.

Scott, J. (2000). *Social network analysis: A handbook*. London, Thousands Oaks: Sage.

Serdült, U. (2002). Soziale Netzwerkanalyse: eine Methode zur Untersuchung von Beziehungen zwischen sozialen Akteuren. *Österreichische Zeitschrift für Politikwissenschaft, 31*(2), 127–141.

Thatcher, M. (1998). The development of policy network analyses: From modest origins to overarching frameworks. *Journal of Theoretical Politics, 10*(4), 389–416.

Thurber, J. (1996). Political power and policy subsystems in American politics. In G. Peters & B. Rockman (Eds.), *Agenda for excellence: Administering the state* (pp. 76–104). Chatham, NJ: Chatham House.

Van Selm, M., & Jankowski, N. (2006). Conducting online surveys. *Quality & Quantity, 40*(3), 435–456.

Van Waarden, F. (1992). Dimensions and types of policy networks. *European Journal of Political Research, 21*(Special Issue), 29–52.

Ward, M., Stovel, K., & Sacks, A. (2011). Network analysis and political science. *Annual Review of Political Science, 14*, 245–264.

Wasserman, S., & Faust, K. (2009). *Social network analysis: Methods and applications* (18th printing ed.). Cambridge: Cambridge University Press.

Weaver, K., & Rockman, B. (1993). *Do institutions matter?* Brookings Institution: Washington D.C.

Weible, C., Pattison, A., & Sabatier, P. (2010). Harnessing expert-based information for learning and the sustainable management of complex socio-ecological systems. *Environmental Science & Policy, 13*(6), 522–534.

Weible, C., & Sabatier, P. (2005). Comparing policy networks: Marine protected areas in California. *Policy Studies Journal, 33*(2), 181–201.

Weible, C., & Sabatier, P. (2006). A guide to the advocacy coalition framework: Tips for researchers. In F. Fischer (Ed.), *Handbook of public policy analysis: Theory, politics, and methods*. New York: CRC Press.

Weible, C., & Sabatier, P. (2007). The advocacy coalition framework: Innovations and clarifications. In P. Sabatier (Ed.), *Theories of the policy process*. Boulder: Westview Press.

Wildavsky, A. (1964). *The politics of the budgetary process*. Boston, MA: Little, Brown.

Zafonte, M., & Sabatier, P. (1998). Shared beliefs and imposed interdependencies as determinants of ally networks in overlapping subsystems. *Journal of Theoretical Politics, 10*(4), 473–505.

Chapter 4
A Network Approach to Policy Design

4.1 Linking Network Structure to Policy Design

A substantial number of theories have emerged, (mainly) since the 1970s, that put forward numerous factors for explaining policy design (Varone 1998, p. 41–63). Rational models are among the earliest approaches to decision-making. They assume that 'best' outcomes can be achieved when decision-makers are well-informed, and therefore able to identify and select the policy alternative that solves the problem at the *least cost* (Weiss 1977; Carley 1980; after Howlett and Ramesh 2003). Many critiques to the rational model emerged (e.g., bounded rationality, prospect theory), among which the best-known alternative was developed by Charles Lindblom in his famous article titled '*The science of muddling through*' (1953). His theory of *incrementalism* posits that, generally, only small changes from the status quo are made. Hence, the policy option that brings an incremental change is most likely to be chosen. This idea has been slightly adapted by Doern and Wilson (1974), who argue that politicians move from *the least coercive instruments to the most coercive* ones. Another research group around Trebilcock et al. (1982) developed the idea of *public choice,* according to which policy design is the result of political rationality aiming at *maximizing votes*. A different body of literature around Baxter-Moore (1987) and Woodside (1986) adopts a neo-marxist-inspired approach and stipulates that governments serve the powerful. From this perspective, the *power of target groups* influences the choice of policy designs. Their hypothesis is that governments select more coercive sanctions when seeking compliance from less powerful and influential groups, than when dealing with powerful actors. Schneider and Ingram (1993) add the aspect of decision-makers' *perception of the target group*—a phenomenon termed *social construction of target population*. They hypothesize that for the powerful positively viewed groups, policy designs will emphasize capacity-building and voluntary

© Springer International Publishing AG 2017

F. Metz, *From Network Structure to Policy Design in Water Protection*,
Springer Water, DOI 10.1007/978-3-319-55693-2_4

measures, whereas policies for powerless and negatively viewed populations are expected to be more coercive, often involving sanctions (p. 339). Bennett and Howlett (1992) and also Sabatier (1987) forwarded the idea of *policy learning*. These authors assume that policy design is a result of gaining knowledge of the efficiency of policy instruments. Bennett (1991) focuses on existing interdependencies between states and argues that governments design public policies by coordinating and harmonizing their national policies with supra- and international institutions as a result of international regimes, or treaties. Howlett (1991) developed the widely acknowledged notion of *national policy styles* by building on the conventional idea that polity (structures) and politics (processes) determine policy designs (Lowi 1964, 1972). This body of literature postulates that every state has its own style of policymaking, defined by the constitution and informal traditions that influences the selection of policy designs. In the same line of thought, Freeman (1985) as well as Wilks and Wright (1987) stipulate that the selection of policy designs does not only differ between states, but also between different *policy subsystems*—the idea being that the nature of the policy subsystem drives policy design. Other scholars argue that the type of the *social problem* itself is a major driver of what determines the choice of suitable policy designs (Peters and Hoornbeek 2005; Metz and Ingold 2014b). Another very prominent body of literature stipulates that *institutions matter*. Atkinson and Nigol (1989) and other neo-institutionalists advance that policy makers choose policy designs in accordance with the values promoted by institutions. The scholars put forward that institutions influence individual preferences and collective choices. For Atkinson and Nigol (1989) the institutional variable implies three further thoughts. Firstly, the design of a policy is never a real choice, but always depends on already-existing policies. In addition to the idea of *path-dependency*, the authors point to policy makers' *values and preferences* as an explanation for policy design choices. Finally, the characteristics of institutions, meaning *internal objectives, resources, and procedures*, are crucial for the definition of public choices. Likewise, Linder and Peters' (1989) idea concerns different categories of variables influencing the selection of policy designs on three different levels: the national policy style on the macro-level; the nature of the policy problem and politico-administrative institutions on the meso-level; and personal values, preferences, and experiences influencing the perception of decision-makers on the microlevel. Finally, there is a growing body of literature on policy networks that has received considerable attention in the last 30 years (Laumann and Pappi 1976; Laumann and Knoke 1987; Knoke et al. 1996; Weible and Sabatier 2005; Lubell and Fulton 2007a; Ingold 2011; Fischer 2013). The network approach postulates that public choices are determined by the actors involved in policymaking and, above all, their web of relations (Bressers and O'Toole 2005; Marsh 1998). The increased interest in policy networks can be explained by the growing recognition that existing conceptions were no longer suited to explain policy formulation. For example, existing notions of actors' values and preferences (Atkinson and Nigol 1989) neglect the influence of social relations on values and preferences (Granovetter 1985). Moreover, the popularity of the network approach is due to various actors' involvement in policy choices today, and contemporary governments' inability to move unilaterally without incorporating other social actors (Sabatier and Jenkins-Smith 1993). Traditional conceptions (Schneider and Ingram 1993; Woodside 1986), on the contrary, relate the state to target groups, but neglect all other types of interactions among social actors. Hence, the policy network approach offers a promising explanation for policy design today.

In summary, the explanations for why certain policy designs are chosen over others can be located on three levels. On the macro-level, institutional drivers were put forward in the literature—the basic idea being that policy design is not a result of individual decisions, but is largely impacted by the political system in which actors are embedded. Microlevel approaches, on the contrary, explain policy design by drawing attention to actors' values, preferences, experiences, knowledge, power, and alike. Meso-level concepts propose an alternative explanation by focusing on policymaking processes. They acknowledge that neither institutional variables nor actors' attributes alone can explain policy design, but both may influence meso-level drivers.

The present work considers the explanatory power of policy network structures promising but understudied and therefore focuses on a network approach to policy design. The network approach is a meso-level explanation, which encompasses macro- and microlevel factors. The political system, in which actors are embedded (macro), as well as actors' attributes (micro), can influence the structure of policy networks. Because actors are constitutive elements of policy networks, microlevel foundations are incorporated into the study of policy networks (meso).

This chapter addresses the questions of why certain policy designs are chosen over others. While the previous Chaps. 2 and 3 presented the results for policy designs and network structures separately, Chap. 4 establishes a link between the two by analyzing the covariance of policy network structures and policy design. The ultimate goal here is to assess the extent to which different structural network properties might impact the comprehensiveness of policy designs.

One should note here that a classical hypothesis testing is beyond the scope of this exploratory study. Nevertheless, results are discussed in light of the formulated hypotheses in order to pretest them, and to gain preliminary insights that can then inform future research. Due to this restriction, one must take into consideration that the results presented below apply foremost to the studied cases and that they cannot be generalized beyond a population of similar cases (Haverland 2000), i.e., wealthy, western democratic nations, which are environmentally conscious and technologically advanced with regard to water protection. In order to avoid an over-interpretation of the underlying data, the discussion of hypotheses is confined to presenting results, while broader implications deserve the attention of future research, as will be explained in Chap. 5.

4.1.1 The Policy Network Approach as an Explanation for Policy Design and Research Gaps

Today's policies are made in a bargaining and coordination process involving multiple actors. Policy networks reflect those interactions in policymaking processes and are therefore composed of nodes (actors) and ties (some sort of social relation among actors). The linkages reveal a certain structural pattern, thereby uncovering the logic of interaction between members of the policy network. Network scholars therefore posit that the structure of a policy network reflects dependencies between actors in policymaking processes, and thus, has a major influence on policy design (Knoke 1990; Sciarini 1996; after Börzel 1998).

Even though many policy researchers agree that the structure of actors' ties is a key to understanding policymaking processes or outputs, there are still divergences with regard to the definition of the term *policy network*. Two main concepts have emerged (Börzel 1998): The Anglo-Saxon stream conceptualizes policy networks as a generic concept, which captures all kinds of public–private relations—hierarchical as well as non-hierarchical ones (Marsh and Rhodes 1992; Van Waarden 1992; Adam and Kriesi 2007; Knoke 1990). For the German Max-Planck-School representatives include Renate Mayntz, Fritz Scharpf, Patrick Kenis, Volker Schneider, Edgar Grande and network governance scholars (e.g., Provan and Kenis 2008), on the other hand, policy networks describe one specific form of public–private interaction, namely a non-hierarchical form of governance (Marin and Mayntz 1991). Recent policy network scholars have reconciled the two concepts and note that policy networks reflect both hierarchy as well as informal elements of policymaking. 'Institutions and networks are not substitutes; they simultaneously and interdependently influence behavior in any policy system' (Lubell et al. 2012). Hence, *policy networks are defined here as organized entities without predefined hierarchical structures consisting of a set of formal and informal interdependent relations linking a plurality of private and public actors, who are engaged in a policymaking process to address a joint problem* (*adapted from Sandström and Carlsson* 2008).

The added value of the network approach as an explanatory variable

The structure of policy networks can uncover both formal and informal structures, so therefore the network approach offers a particularly encompassing explanation for policy design. In addition to its meso-level foundation, the network approach also reflects macro- and microlevel factors (see Fig. 4.1) (Lubell et al. 2012; Daugbjerg and Marsh 1998). Macro-political institutional structures (a macro-level concept), defined by the constitution, lay down standardized procedures for decision-making processes and attribute formal decision-making power to state actors. In this regard, institutional arrangements shape network structures by influencing the logic of interaction of network members. Yet institutions shape network structures only to some degree, policy networks go beyond formal institutional structures (Börzel 1998). In fact, policy is made in a bargaining and coordination process involving multiple actors with and without formal decision-making power. This is especially true today, where scholars have observed a smooth transition between state and society, going along with the privatization of governmental tasks (Kenis and Schneider 1991; Laumann and Knoke 1987; Knoke et al. 1996). Hence, the linkages in the policy network uncover formal national as well as informal subsystem styles of policymaking, and capture all types of actors participating in policymaking (Dimitrios et al. 2011). Since policy networks reflect channels of influence beyond formal structures, they offer a more encompassing explanation for policy design than do institutions alone.

The network approach also shows that analyzing social structures provides additional explanatory power compared to microlevel concepts (Wellmann and Berkowitz 1988 after Börzel 1998, p. 259). The latter explain policy design by the choices individual actors make based on their preferences (Sandström and Carlsson 2008). Granovetter's work has shown, however, that choices or actions are grounded socially (1985, 1992), which means that interactions among network members socially influence actors' attitudes, perceptions, behavior, and policy preferences (Richey and Ikeda 2006; Knoke 1990; Zuckerman 2005). Hence, not only actors' preferences need to be understood to explain policy choices, but also actors' ties and their embeddedness in the overall network. Compared to purely microlevel explanations, the advantage of the network approach is its ability to go beyond the mere aggregation of policy actors' attributes by taking into consideration actors' interdependencies (Sandström and Carlsson 2008).

In summary, the policy network approach combines macro-level and microlevel explanations for policy choice, thereby creating its own meso-level explanation for policy choice. Network scholars conciliate elements of individualism with a structural approach and demonstrate that networks create their own governing structures (Lubell et al. 2012; Granovetter 1985, 1992). The argument is that interactions in policymaking processes, through which networks evolve, lead to building trust and to promoting the likelihood of overcoming diverging interests (Coleman 1990; Burt 2000). The structure of those interactions, in turn, impacts (facilitates or inhibits) the realization of collective gains. Hence, policy networks do not only mirror national institutional settings, but they also form governing-entities themselves with their own structural features. In fact, the structure of policy networks might not always correspond to formal hierarchies. The analytical strength of the network approach is its exhibition of both formal hierarchies as well as the informal process of bargaining and negotiation. In this way, the network approach conciliates the dialectical relationship between institutional structure and individual action through network interaction (Sandström and Carlsson 2008). Lubell et al. (2012, p. 352) conclude that 'policy networks as a "meso-"level concept mediate causal relationship between macro-level political institutions (both formal and informal) and microlevel individual behavior of political actors [...].' Therefore, the network approach can be used to assess the influence of macro-level and microlevel variables on the structure of policy networks, as well as the way in which the combination of macro-, microlevel and network structure, in turn, influences policy outputs (Lubell et al. 2012).

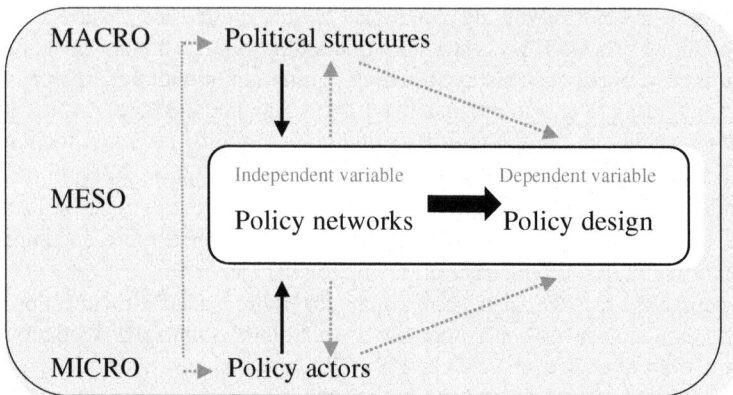

Fig. 4.1 The policy network approach as explanation for policy design: between macro- and microlevel concepts. *Note* This figure confines the policy system figure of Lubell et al. (2012) to the network approach. From a network approach, political structures and policy actors influence the structure of policy networks. Policy networks serve here as the main explanatory factor for policy design

Figure 4.1 illustrates the network approach as an explanation for policy design. Specific to the network approach is that macro- and microlevel variables are considered influencing the structure of policy networks, which in turn influences policy design (black arrows in Fig. 4.1).[1] The reverse relationships, i.e., the impact of networks on macro- and microlevel variables, as well as the direct influence of macro- and microlevel variables on policy design might, of course, exist in the empirical reality as well, but are not subject of this study (gray arrows in Fig. 4.1).

4.1.2 Hypotheses: Linking Network Structures to Policy Design

Treating networks seriously entails understanding whether specific types of network structures promote certain kinds of policy designs. To relate network configurations to policy design, I build on the previously developed concepts in this work: firstly, policy network characteristics (Sect. 3.1.3) and, secondly, the index of policy

---

[1]The present work considers the impact of macro-level political institutions on network structures as well as the impact of micro-level actor attributes on network structures in order to assess the relationship between network structure and policy design. This book analyzes neither the impact of a country's political structure on policy design, nor the impact of actor-level variables on policy design and, therefore, Fig. 4.1 shows these linkages in gray. The black arrows in Fig. 4.1 exclusively reflect the analytical perspective of this book, which does not mean that the macro- and micro-level variables might not impact policy design in the empirical reality.

comprehensiveness (Sect. 2.1.2). The research aim is to explore a potential linkage between policy network structures and the quality of policy outputs, measured in the form of policy designs that address policy problems more or less comprehensively.

Past literature repeatedly highlighted the relevance of the network approach in explaining policy design (Crona and Bodin 2006; Sciarini 1996; Marsh and Rhodes 1992; Fischer 2012; Sandström and Carlsson 2008; Ingold 2011; Lubell and Fulton 2007a; Knoke et al. 1996; Marin and Mayntz 1991; Howlett 2002), but failed so far in establishing systematic hypotheses and empirical tests beyond simple attestations that networks impact outputs (Klijn et al. 2010). To overcome this gap in research, the present study combines three streams of literatures: firstly, network literature (Burt 2000; Coleman 1990; Granovetter 1973); secondly, literature about the impact of policy networks on policy change or compromise (e.g., Bressers and O'Toole 1998; Ingold 2011; Fischer 2015); and thirdly, environmental governance literature (e.g., Koontz and Thomas 2006; Newig and Fritsch 2009; Bodin et al. 2006). The combination of the three literatures is necessary because network scholars have advanced the knowledge on structural network characteristics and their effects on social capital, but do not study policy networks or policy design. Policy scholars went one step further by studying the impact of policy networks on policy learning, policy change, or policy compromise,[2] but do not analyze whether network structures promote or inhibit comprehensive problem solving. The present study goes beyond the existing policy literature in order to explore theoretically under which circumstances the network variables that have served to explain policy change or compromise may also promote comprehensive policy designs.

Although the here-developed ideas are explorative, they are supported to some extent by scholars of environmental governance, who argue that *strong networks* lead to *more effective policy outputs* (Newig and Fritsch 2009; Koontz and Thomas 2006; Klijn et al. 2010). Environmental governance scholars evaluate a policy against its ability to solve or reduce an environmental problem and, hence, provide a normative claim about what a 'good' policy output is. This normative claim is a test of the network approach's impact on alleviating policy problems. Without demonstrating the relevance of the network approach to explaining policy designs ability solving real-world policy problems, the characterization of policy network structures remains more of a descriptive endeavor. Only when research can

---

[2]Policy scholars tend to frame policy outputs as compromises (for example in Fischer 2014) and analyze the impact of network structures on reaching compromise. However, not every policy output is a compromise; most outputs reflect the dominance of certain ideas over others. If only a subset of policy outputs represents a compromise, it may not—depending on the research goal—be helpful to conceptualize policy outputs as compromise. Moreover, the definition of 'public policy' adopted in this work involves the idea of problem solving (see Sect. 2.1). A compromise, however, is not equivalent to problem solving. While some compromise may be quite faulty in addressing a problem, others may be able to establish balance between competing interests, and at the same time, achieve a high level of problem solving. Due to these conceptual difficulties related to the term 'compromise,' the present study refrains from conceiving policy design as such and instead adopts a problem-solving perspective.

demonstrate that certain network structures promote or inhibit problem solving, critics will stop questioning the relevance of the network approach.[3]

One reason that the literature has failed so far to establish a systematic link is due to a lack of conceptualizing the performance of policy designs, be it their effectiveness (Newig and Fritsch 2009; Sandström and Carlsson 2008) or problem-solving capacity (Klijn et al. 2010). To overcome this gap and evaluate policy designs against their prospective ability to comprehensively address a policy problem, Sect. 2.1.2 establishes a new index of policy comprehensiveness. Section 3.1.3 then highlights those structural properties that policy process research discerned as crucial features of policy networks, i.e., the combination of interconnected actors exhibiting belief cohesion and uniting into coalitions, brokerage, and entrepreneurship. In the following section, these concepts are addressed in several hypotheses in order to explore under which conditions network properties, or their combination, may affect the emergence of comprehensive policy designs.

The first notion of *policy beliefs* is grounded in the ACF literature and stands for deeply rooted values, which guide the action of policy actors (Sabatier and Jenkins-Smith 1993; Ingold 2011; Henry 2011). Network scholars aggregated this idea to the overall policymaking process and labeled policy networks as *belief-cohesive*, when there is an overall consensus between actors about policy beliefs (Marsh 1998; Bressers and O'Toole 1998; Henry 2011). Belief cohesion indicates that network members share common worldviews and that low barriers for reaching common decisions exist, which also strengthens actors' ability to deliver policy outputs (Kickert et al. 1997; Ingold and Fischer 2014). If policy actors share common values, they are likely to agree about the general need to address a policy problem and on the direction of policy measures. A high level of goal consensus can hence also improve networks participants' ability to perform (Provan and Kenis 2008, p. 239). Overall, belief cohesion can be considered an indicator for low

---

[3]There are some distinctions between the policy network and the governance network literature that are necessary to point out here: Rooted in public administration research, the literatures on environmental governance (Bodin and Prell 2011; Koontz and Thomas 2006; Newman and Dale 2005; Newig and Fritsch 2009) and network governance (Provan and Kenis 2008; Provan and Milward 1995; Van Meerkerk et al. 2015; Klijn et al. 2010; Kickert et al. 1997; Klijn 1996) largely conceptualize networks as a collaborative form of governance, in opposition to markets, or hierarchies. In the present study, on the contrary, policy networks are defined such that they include all types of interactions among policy actors, hierarchical and non-hierarchical. A further distinction between the different streams of literature is that governance networks include interactions between actors during the implementation of a policy, such as service provision, or natural resource management. The present work, however, focuses on a different stage of the policy cycle, namely policymaking, i.e., stages preliminary to implementation. Policy networks, as defined in this work, capture all interactions between policy actors during the policymaking process. Despite these differences in focus, I employ environmental governance and network governance literatures in order to formulate some of my hypotheses, because these literatures advance meaningful ideas about the linkage between network structure and effective policy outcomes lacking in the policy network literature. I believe that literatures are compatible to the extent that both governance literature and policy network literature are investigating challenges that networks face when adopting policy decisions.

barriers in reaching common decisions (Börzel 1998, see also Sect. 2.2.3; Marsh and Rhodes 1992).

While there exists agreement in the policy literature that low belief cohesion impedes reaching some sort of common decision, the role of belief cohesion in reaching comprehensive policy designs is less understood. The present study therefore seeks to explore whether belief cohesion facilitates or inhibits the emergence of comprehensive policy designs.

Policy scholars have argued that if a policy network faces high constraints due to conflicts of belief dissimilarities, the policy output will most likely be invisible and target only specific subgroups (Howlett and Ramesh 2003; Salamon 2000). Invisible, partial solutions are not very likely to comprehensively solve a policy problem, and hence, the literature suggests that low belief similarities impede comprehensive policy designs. By contrast, with high belief similarities and low constraints within the policy network, the policy is more likely to target all groups contributing to a problem and thus be more visible to the general public. In other words, the more cohesive the policy network is in terms of belief similarity, the more likely that network is in comprehensively solving a collective problem. Translated into the language of this study, the following hypothesis (H) can be formulated:

H1   A high level of belief cohesion between network members tends to have a positive effect on the degree of comprehensiveness of policy design

The literature on policy networks introduces *interconnectedness* as a second element characterizing policy networks. By definition, interconnectedness refers to the amount of present ties in the network (Bressers and O'Toole 1998). Several streams of literature, such as that of environmental governance (Innes and Booher 2003; Robins et al. 2011; Klijn et al. 2010; Koontz and Thomas 2006; Bodin and Prell 2011) and policy networks (Ingold and Fischer 2014; Lubell and Fulton 2007b), put forward the idea that ties between actors, and most importantly, collaboration, are decisive for the production of collective outputs. Grounded in the social capital literature is the idea that well-connected networks lead to the creation of trust and common norms and increase the network's capacity to perform (Coleman 1990). Moreover, the ACF literature suggests that well-interconnected policy networks enable collective outputs that present a change in policy direction (Weible and Sabatier 2007). The driving mechanism is that with more ties, actors are more likely to share resources, such as information (Leifeld and Schneider 2012), knowledge, or ideas about potential solutions, which can be beneficial to policy actors in several ways: (a) Exchanges can induce learning about policy measures (Weible et al. 2010; Crona and Parker 2012). The gained knowledge may reduce actors' uncertainties about costs and risks of different policy options and strengthen their ability to choose the most effective and efficient policy design (Metz and Ingold 2014a). Furthermore, ties can (b) increase actors' organizing capacity and (c) improve actors' overall ability to problem solving (Fischer 2014). Coordination also (d) increases actors' ability to address complex problems, and by pooling resources, provides them with a higher steering capacity (Provan and Kenis

2008, p. 229). Since exchanges produce multiple benefits for policy actors, the insight arose that network ties positively impact overall network performance (Sandström and Carlsson 2008), effective governance modes (Provan and Kenis 2008), environmental practices (Lubell and Fulton 2007a), or adaptive governance responses (Crona and Parker 2012).

Building on these findings, the present study seeks to assess the impact of interconnectedness on the quality of policy outputs. As will be explained below, previous research about collaborative governance (Innes and Booher 2003; Robins et al. 2011; Klijn et al. 2010; Koontz and Thomas 2006; Bodin and Prell 2011) and network performance (Sandström and Carlsson 2008) suggests that interconnectedness might be very decisive for adopting comprehensive policy designs.

Since no single actor has enough steering capacity (such as time, knowledge, money, and power) to determine the result of policymaking processes, interdependencies exist. Interconnectedness is a way of dealing with such interdependencies, and by sharing resources, improving the overall organizing and problem-solving capacities of network members (Provan and Kenis 2008). Such improved capacities may also increase the chances of network members to find and adopt comprehensive policy designs for complex policy problems.

Additionally, network ties have been found to increase the level of trust between policy actors (Lubell 2005). Some authors argue that trust, in turn, enhances the common understanding of belonging together and working toward the common good (Bodin et al. 2006; Ostrom 1990). Others mention that trust promotes norms that place the common good at center stage (Jones et al. 2009). Similarly, trust may support network members' effort to agree on solutions that stress the common good, rather than individual interests, and enable a higher level of environmental protection in the form of comprehensive policy designs (Newig and Fritsch 2009).

Finally, network ties have been found to prompt information exchange (Leifeld and Schneider 2012), which also increases actors' access to multiple types of knowledge and social memory (Bodin et al. 2006). Interconnectedness has been said to promote social learning and enable informed decisions that reveal win-win situations (Newig and Fritsch 2009). Consequently, interconnectedness may encourage reflection on policy designs, which are acceptable to the majority of actors, *and* which comprehensively address an underlying policy issue. In conclusion, Newig and Fritsch (2009, p. 198) put forward that, 'collaborative forms of governance are expected to lead to more effective improvements in environmental quality,' or in the words of Koontz and Thomas (2006, p. 113), 'collaboration leads to better decisions.' From these theoretical thoughts, the following hypothesis can be deduced:

H2 A high level of interconnectedness between network members tends to have a
    positive effect on the degree of comprehensiveness of policy design

Network scholars bring both elements together (Bressers and O'Toole 1998), i.e., interconnectedness and belief cohesion, and argue that social relations enable actors to create common norms, and, thus, overcome divergences or prevent blockades (Lubell et al. 2012; Coleman 1990; Berardo and Scholz 2010). If blockades are prevented, discussing and achieving a comprehensive policy design seems more likely than in situations of blockades.

The network governance literature (Bodin et al. 2006), along with the literature on policy networks (Adam and Kriesi 2007), discusses the ambiguous role of *network homogeneity* and the potentially positive aspects of *network heterogeneity*. Network homogeneity points to a situation in which many similar types of policy actors with congruent values participate in the policymaking process. On the one hand, it has been argued that homogeneity promotes a higher level of trust, which may facilitate conflict resolution (Jones et al. 2009). In the same line of thought is the argument that a consensus about policy goals also increases overall policy performance (Provan and Kenis 2008). On the other hand, a very high level of homogenization involves uniform experiences, ideas, or knowledge about the underlying policy problem and potential solutions (Bodin et al. 2006; Newman and Dale 2005). Heterogeneity, by contrast, indicates that diverse types of actors, with different knowledge, organizational, and monetary capacities, are part of the network (Adam and Kriesi 2007). Hence, low belief cohesion may still be a positive sign indicating an open network, which is receptive toward new potentially constructive ideas on how to best solve an underlying policy problem (Howlett 2002). An open network might also incorporate a large number of resources to thoroughly address a policy problem (Adam and Kriesi 2007) and stimulate particularly innovative (Provan and Kenis 2008) or even comprehensive policy designs. From this perspective, belief dissimilarities in the network are not necessarily negative for network performance. What is decisive in such cases is the level of interconnectedness.

In this regard, the social capital literature's famous idea (Coleman 1990; Putnam et al. 1993) concerning *bonding ties* suggests that interconnectedness might be pivotal for adopting comprehensive policy decisions. More precisely, bonding ties closely interlink network members and create cohesive structures of redundant network linkages, which are said to promote social capital and facilitate network performance. The policy network literature has taken up the concept of bonding ties and argues that a higher level of interconnectedness in policy networks points to a process where concrete policy measures are being worked out and negotiated between opposing sides (Fischer 2014). With a higher level of exchanges, bonding ties, trust, and common norms are created (Berardo and Scholz 2010) through which divergences may be overcome between heterogeneous network members (Adam and Kriesi 2007).

In conclusion, policy design has the potential to be particularly comprehensive with the combination of low belief cohesion and high interconnectedness, because a higher level of heterogeneity (and potentially conflict) can also signify a higher diversity of resources and ideas about how to comprehensively address an underlying policy issue (Provan and Kenis 2008; Bodin et al. 2006). Once divergences can be overcome through collaboration, various resources can be shared and used for adopting particularly legitimate and accepted (Skogstad 2003; Sandström et al. 2014) and potentially also comprehensive policy designs. Provided that divergences can be overcome through collaboration, then it should be true that:

H3   With low belief cohesion, a comprehensive policy design is possible if there is
     a high level of interconnectedness between network members

While the previous hypotheses look at the overall policy network structure, the
ACF introduces a more fine-grained structural network property through the concept of *coalitions* (Sabatier and Jenkins-Smith 1993). Policy scholars have provided
empirical evidence that coalition structures are decisive in understanding policy
change and compromise (Ingold 2011; Nohrstedt 2008; Fischer 2015). Following
the ACF, actors form coalitions based on shared policy beliefs (Sabatier and
Jenkins-Smith 1993), with the ultimate aim of pooling resources and influencing the
policy design to the own advantage (Weible and Sabatier 2006; Ingold 2011). As a
result, advocacy coalitions compete over policy influence. If this competition leads
to the formation of several equally powerful coalitions, coalitions face the risk of
blocking each other and of weak solutions, which present only a minor change to
the status quo, being adopted (Fischer 2014).

Applying these insights to the aim of the present study suggests that coalitions
that block each other not only hamper compromise or change, but make the
adoption of a comprehensive policy design even more unlikely. Nevertheless, the
policy literature also provides for mechanisms through which oppositions between
coalitions can be overcome. The label *collaborative subsystems* have been introduced to describe situations in which opposing coalitions form, but still collaborate
across coalitions (Weible et al. 2010; Ingold and Gschwend 2014).

In the social capital literature, linkages that cross structural holes—such as
existing between advocacy coalitions—are labeled *bridging ties* (Burt 2003). This
concept is rooted in Burt's famous work (2000), which combined Granovetter's
concept of structural holes (1973) with Coleman's (1990) proposition that interconnectedness promotes network performance. Granovetter demonstrates that
missing linkages in social networks lead to the so-called structural holes, which
indicate that diverse types of resources exist in the network. According to Burt
(2003), missing and present linkages matter for network performance. While
structural holes provide a source for new ideas or resources, bridging ties across
those holes makes these resources accessible to other network members (Burt 2000,
p. 398).

The network governance literature argues that both bridging and bonding ties are
complementary and necessary in order to deal with complex policy problems
(Newman and Dale 2005). This literature has introduced the term *modularity*,
which denotes the tendency of network members to form multiple groups (Bodin
et al. 2006). Scholars argue that modularity, on the one hand, increases the ability of
different groups to develop diverse knowledge systems and creates dense network
structures and trust; on the other hand, modularity creates opposition and locks
actors into fixed political positions, thereby inhibiting their common ability to act
and address underlying policy problems (Bodin et al. 2006). Such hurdles can be
overcome through bridging ties, which transgress group boundaries, provide access
to diverse resources, and as such, enlarge the scope of visions concerning potential
policy solutions (Newman and Dale 2005). Hence, the coexistence between

bridging and bonding ties enables the diversity needed to design comprehensive policies and to maintain congruence necessary to form a common decision.

Applied to policy networks, these insights suggest that collaborative subsystems combine the advantages of bonding ties (occurring within advocacy coalitions) and bridging ties (occurring across advocacy coalitions). With the relevant knowledge and resources from diverse coalitions about how to best solve an underlying policy problem, it may be possible to develop policy designs that represent a particularly comprehensive policy design. Bridging ties between opposing coalitions may facilitate congruence necessary to agree on comprehensive policy designs. From this basis, the following hypothesis can be deduced.

H4a   With two or more opposing coalitions, a comprehensive policy design is possible if actors collaborate across coalitions

Collaboration across coalitions can also be facilitated by brokerage (see Sect. 3.1.3). According to the ACF, policy brokers exhibit neutral beliefs and, as mediators, form an essential link between conflicting coalitions (Ingold and Varone 2012; Christopoulos and Ingold 2015). While ACF scholars tend to stress the role of brokers as seekers for compromise and stability, network governance literature places an emphasis on brokers' ability to bridge structural holes (Burt 2003; Bodin et al. 2006). In the latter literature, brokers are portrayed as actors who embody bridging links and therefore have access to diverse group-specific knowledge, which enables them to learn from diverse groups and to synthesize a large pool of information.[4]

According to Burt (2003), brokers have the capacity to innovate due to their early access to critical information, which enables them to generate new understandings and opportunities often overlooked by others. Empirical studies confirm that brokers can have a positive impact on the entire network by building transitive social relationships between policy actors with diverse problem perceptions and interests (Henry et al. 2010) and by spreading new and diversified types of resources (Crona and Parker 2012; Crona and Bodin 2006). Brokers' advantageous structural position furthermore enables them to understand which actors to connect or to not connect, this way, coordinating the actions of the entire network (Bodin et al. 2006). Through their boundary-spanning activities, brokers are considered able to recognize goal intertwinement or win-win situations and hence may promote the attainment of comprehensive policy solutions (Van Meerkerk et al. 2015; Klijn et al. 2010).

Compared to the previously portrayed situation where actors from opposing coalitions drive collaboration in a diffuse way, here, cross-coalition collaboration

---

[4]Note that some governance scholars use the term *broker* (Bodin et al. 2006), while others employ the label *network managers* (Klijn et al. 2010). Network governance and policy network scholars seem to adopt similar conceptions of brokers, as both literatures operationalize brokerage with betweenness centrality and describe similar effects of brokers on the network (Bodin et al. 2006; Christopoulos and Ingold 2011). Additionally, further research is needed to assess the exact differences and similarities of the *brokerage* concept in different literatures.

may be weak, and instead pooled through the broker. Similarly to what has been argued before, brokers may combine the necessary conditions to enable both problem solving and agreement. While coalition formation may be a source of diverse resources, and hence, lay the ground for a particularly comprehensive policy design, brokers' ability to bridge structural holes between opposing coalitions may provide access to those resources and also further the networks' ability to come to agreement. Through brokerage, there may be great potential for designing particularly comprehensive policies due to the pooled knowledge, ideas, and resources from opposing coalitions. In conclusion, networks that are divided between several coalitions may still adopt a comprehensive policy design if there is simultaneously a sufficient level of interconnectedness mediated through policy brokers. These theoretical thoughts point toward the following hypothesis:

H4b With two or more opposing coalitions, a comprehensive policy design is possible through brokerage connecting opposing coalitions

Similar to policy brokers, entrepreneurs hold strategic network positions (Kingdon and Thurber 2011; Baumgartner and Jones 1993; Mintrom and Norman 2009). There are a couple of differences between the two types of exceptional actors which are worth pointing out: Firstly, brokers have neutral belief systems and mediate compromise between opposing coalitions (Christopoulos and Ingold 2015). Entrepreneurs, on the contrary, are said to push their preferred policy solution (Kuhnert 2001; Christopoulos 2008). Secondly, while brokers' strategic network position consists of being connected to actors from all coalitions, entrepreneurs' strategic position consists of being connected to powerful and well-connected other network members. Thirdly, brokers are typically positioned between coalitions, while entrepreneurs are likely to be partisans and members of the dominant coalition, because by definition, dominant coalitions consist of powerful, well-connected network members and entrepreneurs tend to be connected to powerful, well-connected others. A further distinction between the two types of exceptional actors is that brokers have the ability to mediate; entrepreneurs, on the contrary, are not neutral (Mintrom and Norman 2009; Schneider et al. 2011). In summary, entrepreneurs are self-interested actors who are able to promote their preferred policy designs through their connections to powerful others, as well as their membership in dominant coalitions.

By contrast, entrepreneurs lack the ability to mediate and are therefore less likely than brokers to help overcome divergences where low belief cohesion exists in the network. With low belief cohesion, entrepreneurs might face constraints that they are not able to surmount. Consequently, entrepreneurs are more likely to succeed in situations where a majority of network members share a good level of belief cohesion and are not opposed to addressing a policy issue. In such situations, entrepreneurs might be successful at promoting even very comprehensive policy designs, if a comprehensive design is in line with their own preferences.

More concretely, in policy fields where resources are short or where other policy issues are more urgent, policy actors may not become involved in the policymaking process, although they are generally in favor of addressing the policy issue at stake.

Where policymaking processes suffer from such inertia (i.e., low level of inter-connectedness and good level of belief cohesion), entrepreneurs may be particularly successful at promoting their own policy preferences. Through their contacts to powerful, well-connected other network members, entrepreneurs can diffuse arguments favorable to their own preferred policy design and mobilize coalition members to their own advantage. Where a policymaking process suffers from inertia, entrepreneurs can even be particularly beneficial to the achievement of comprehensive policy designs, given that a comprehensive design is in line with their own preferences.

In summary, where policy networks are only loosely interconnected within coalitions and share a good level of belief cohesion, policy entrepreneurs can mobilize network activity through their connections and push for a comprehensive policy response, if a comprehensive design is in line with their own preferences. The following hypothesis can be deduced:

H5   With low interconnectedness within coalitions, policy entrepreneurs can push for a comprehensive policy design if it is in line with their preferences

Lastly, policy subsystems can also be characterized by one dominant coalition (Weible et al. 2010; Ingold and Gschwend 2014). A dominant coalition includes a majority of network members and powerful actors with formal authority for making policy decisions. Typically, actors within the dominant coalition are well-connected internally and display similar beliefs, but they are disintegrated from actors belonging to minority coalitions. Policy process literature maintains that major policy change is possible if a major change is in line with the preferences of the dominant coalition (Fischer 2015). According to Fischer (2014), dominant coalitions have the required resources, such as authority, knowledge, and money, to impose their preferred policy outputs.

While dominant coalitions typically exist in adversarial (or collaborative) subsystems with two or more coalitions, unitary subsystems are composed of one coalition only (Weible et al. 2010; Ingold and Gschwend 2014). Despite this different setting, both dominant coalitions and unitary subsystems should have the required resources to impose their will. With the exception of isolated actors, unitary subsystems include all network members, and therefore are likely to exude enough power to impose their preferred policy design.

Applying this reasoning for a policy design perspective suggests that dominant coalitions, and unitary subsystems composed of one coalition, bear the potential to introduce a comprehensive policy design if a comprehensive design is in line with their preferences. If the dominant coalition promotes policy action, it should be likely that a comprehensive policy solution is agreed upon. Likewise, if members of unitary subsystems are adept at problem solving, a comprehensive policy design becomes more likely. Hence, in situations of one dominant coalition and unitary subsystems, there should be the potential for comprehensive policy action, which points to the following hypotheses:

H6a   With one dominant coalition, policy design can be comprehensive if it is in
      line with the preferences of the dominant coalition
H6b   In unitary subsystems, policy design can be comprehensive if it is in line with
      the preferences of members of a unitary subsystem

In summary, a high level of belief and preference similarity in the network, i.e.,
belief cohesion, is expected to decrease barriers for reaching a common decision.
Likewise, scholars have concluded 'that a high level of interconnectedness within a
network facilitates performance because of enhanced communication, the creation
of common norms, and the possibility to restrain opportunistic behavior'
(Sandström and Carlsson 2008, p. 508). This study therefore considers interconnected policy networks to be more likely to address policy problems comprehensively. Networks with a low level of belief cohesion are assumed to be able to
overcome divergences through a good level of interconnectedness and, hence,
address policy problems comprehensively. Moreover, strong brokerage and
entrepreneurship, as well as a dominant coalition supportive of an encompassing
policy solution, are considered conducive to the emergence of comprehensive
policy designs.

4.2   Structured, Focused Comparison

The study relies on a specific type of co-variational analysis termed *structured,
focused comparison* to relate network properties to the comprehensiveness of policy
designs (George and Bennett 2005, p. 67–72). This method was chosen because it
particularly well suits analyzing the relationship between a few defined variables of
theoretical interest in small-N research designs. The purpose of the method is to
overcome limitations of small-N research as much as possible by (a) concentrating
on one specific research focus and (b) accumulating insights through systematic
comparison. The accumulation of insights is achieved by the combination of three
prerequisites explained in more detail below:

The first prerequisite that this method involves is a *structured* analysis of cases,
which means that the same research questions are asked for all cases under study.
Likewise, data collection is standardized so that data on the same variables exists
across all units. To ensure that the data is comparable across cases, data gathering
must be carefully developed so as to reflect the research objective and theoretical
focus of the study (George and Bennett 2005, p. 69). Only if the same data exists
across cases, can systematic comparisons and an accumulation of findings across
cases then be possible to contribute to an orderly development of knowledge of the
phenomenon in question (George and Bennett 2005, p. 60).

The second prerequisite of the method is that research needs to be *focused*,
meaning that the researcher is supposed to focus on a certain defined aspect of
interest characterizing the cases. With a small number of cases, researchers quickly
run into the problem of indeterminate research designs by considering all possible

explanatory variables. In indeterminate research designs, a small number of observations obstruct too many inferences, which makes testing hypotheses impossible. In order to cope with this insufficiency, the method suggests focusing on a very specific, delimited research objective (George and Bennett 2005, p. 70). Hence, the 'focusing method' is not suitable to fully explain a social phenomenon, but it serves to examine the impact of one or a few variables of interest on a social phenomenon. To justify the focus on a partial explanation, the research objective should be carefully grounded in theory (George and Bennett 2005, p. 71) and rely on variables of theoretical interest for purposes of explanation. Through theoretical foundation, the study is of value for generating general knowledge about relationships among the studied variables (George and Bennett 2005, p. 69).

Thirdly, the method relies on systematic *comparisons* that allow for an orderly accumulation of insights with regard to a theory-oriented, specific research focus. The systematic comparison builds on the assumption of unit homogeneity and constant effects. Both will be explained below in more detail with examples from the present case.

The underlying study applies the method of structured, focused comparison by centering around one research objective, namely assessing the explanatory value of network structures for comprehensive policy designs.

The *focus* on one specific explanation for policy designs enables the accumulation of insights despite the small number of cases, and hence for a preliminary hypotheses testing.[5]

In the context of this study, *structured* implies that the same information is compiled for the studied variables through systematic data collection across the four cases under investigation. Likewise, the same hypotheses are systematically tested for the four cases. Sections 2.2 and 3.2 lay down the procedures of systematic data compilation used in this study.

The *comparison* between four cases rests on the assumption of *unit homogeneity*. The latter means that if two (or more) cases have the same value for the independent variable X, they should also be similar with regard to their scores for the dependent variable Y (King et al. 1994, p. 116). If X and Y correlate across the studied cases, results point to a relationship between the two variables. On the contrary, if X and Y are imperfectly correlated, i.e., cases with the same value of X do not display similar values of Y, are taken as a sign that there exists no relationship between the two variables. For example, if policy actors were similarly well interconnected in two policy networks, and simultaneously policy design

---

[5]A fully specified model including all potential explanations for policy design would be needed to fully understand the explanatory value of the network variable on policy design. In the framework of this research project, it was not possible to gather data on more than four policy networks, and hence, the study is restricted to four cases only. With such small-N, it was not possible to increase the number of independent variables without running into the problem of an indeterminate research design, where hypotheses are not testable. Nevertheless, by applying *focused comparison*, insights about the impact of network structures on policy design can be accumulated, and therefore, preliminary hypothesis testing can be possible.

displayed the same level of comprehensiveness, a relationship between interconnectedness and policy design would be assumed. On the contrary, if the studied policy networks displayed different levels of interconnectedness, but policy designs were similarly comprehensive, no relationship would be inferred.

Aside from unit homogeneity, *constant effects* are assumed as in co-variational analysis, which means that similar variations in values of the explanatory variable should lead to the same variation of the dependent variable. By contrast, the absence of such co-variation is considered an disconfirming evidence for a relationship between X and Y (Gerring 2004; Blatter and Haverland 2012). For example, if one network was more interconnected than another, and likewise, policy design was more comprehensive in the former than in the latter case, a positive impact of interconnectedness on policy design would be inferred. On the contrary, if the empirical observation showed that in weakly interconnected networks, policy designs were more comprehensive, a negative impact of interconnectedness on policy design would be inferred. No relationship would be deduced if no analogy between interconnectedness and the degree of comprehensiveness of policy design could be established. One should note that establishing an analogy is not identical to linearity. For example, the constant increase in interconnectedness going along with an increase, after a tipping point, with a decrease in comprehensive policy designs, would be a sign for a nonlinear effect of interconnectedness on policy designs.

4.3   Network Structures of the Rhine River Riparian States and Their Micropollutants Policies

The present chapter discusses empirical results with regard to their implication for each of the formulated hypotheses. In order to establish the link between the network structure and policy design theoretically, Sect. 4.1.2 formulates hypotheses concerning a possible effect of different network properties on the comprehensiveness of policy designs.

4.3.1   Linking Belief Cohesion to Policy Design

A first structural network characteristic, i.e., belief similarity, is rooted in the ACF literature and indicates that network members share common worldviews (Sabatier and Jenkins-Smith 1993). The aggregation of actors' belief similarities to the overall policy network is termed belief cohesion and refers to a general consensus between network members concerning policy content (Bressers and O'Toole 1998; Marsh 1998). From a theoretical perspective, a high level of belief cohesion signals that network members agree about the general need and direction of policy design, while a low degree of belief cohesion implies high constraints, particularly regarding the adoption of comprehensive policy designs. In the literature, belief cohesion has been associated with the ability of network members to deliver policy outputs (Ingold and Fischer 2014), and to perform (Provan and Kenis 2008), which should also facilitate the adoption of comprehensive policy designs—defined here as policies designed to be able to thoroughly address and alleviate a problem of public policy. Building on these definitions, the first hypothesis put forward that a high level of belief cohesion between network members should positively affect the degree of comprehensiveness of policy designs.[6]

Table 4.1 provides an overview about the empirical findings of this study that serve to discuss the first hypothesis. The table displays the comprehensiveness of the four studied policy designs, as well as the level of belief cohesion of the four policy networks; both sets of variables will subsequently be explained in more detail.

With regard to the dependent variable, the Swiss policy design consists of a technological upgrade of selected sewage treatment plants in order to better filter micropollutants from wastewater. With a score of 0.75 out of 1.00, the Swiss policy design was evaluated as highly comprehensive for the reduction of micropollutants in surface waters according to the policy comprehensiveness index (see Sect. 2.3.3.2 for an explanation).

The German policy design aims at controlling concentration limits of 162 selected chemical substances in surface waters through the adoption of environmental quality norms. This policy design achieves a medium level of comprehensiveness with a score of 0.48, where 1.00 would be the maximum (see Table 4.1). The French Micropollutants Plan includes a comprehensive, but not legally binding policy mix, and therefore, it scores with 0.34 only medium to low in the index (see Table 4.1). The Dutch policy debate on policy measures for the reduction of pharmaceutical micropollutants remained without output in the time frame of this study and can therefore only attain a low level of comprehensiveness and a score of 0.2 following the policy comprehensiveness index (see Table 4.1). As a result, Switzerland ranks highest, followed by Germany, France, and the Netherlands on the policy comprehensiveness index. This ranking is displayed on the right of Table 4.2 and reflects policy designs' ability to comprehensively reduce micropollutants in surface waters.

With regard to the independent variable, results for belief cohesion reveal that the Swiss policy network is highly belief-cohesive (5.92% density of belief similarity ties), followed by Germany with a good level of overall belief similarity (2.98%) among policy actors (see Table 4.1). The French and Dutch networks

---

[6]While the ACF employs policy beliefs as an argument for policy change, this book assesses if beliefs also serve as an explanation for comprehensive policy designs. This theoretical argument is considered valuable as change is not necessarily equivalent to problem solving. At the same time, a move towards more comprehensive policy designs represents a specific type of change. Hence, from a problem solving *and* a policy change perspective it is of particular relevance to understand under which network conditions comprehensive policy designs may be facilitated.

Table 4.1 Results for belief cohesion and policy performance

	CH	G	F	NL
Dependent variable				
Policy comprehensiveness index score	0.75 (high)	0.48 (medium)	0.34 (medium/low)	0.2 (low)
Independent variable (operationalization)				
Belief cohesion (mean density of belief similarity)	5.92% (high)	2.98% (good)	1.00% (low)	2.63% (low)

Table 4.2 Rankings for the policy comprehensiveness index and belief cohesion

Independent variable (IV): belief cohesion	Dependent variable (DV): policy comprehensiveness index
CH, G, NL, F	CH, G, F, NL

under investigation, on the contrary, attained only a low degree of belief cohesion with 1.00 and 2.63% density of belief similarity ties. The barriers for reaching a comprehensive policy design appear highest in France and the Netherlands, followed by Germany and Switzerland (see Table 4.2).

Comparing the ranking for belief cohesion with the ranking for the policy comprehensiveness index suggests some co-variance for the Swiss and German case. The Swiss network scores highest on belief cohesion followed by Germany; the same ranking applies for the comprehensiveness of policy designs. However, the order of the French and Dutch cases is reversed: While the Dutch network is more belief-cohesive, its policy design seems to be less comprehensive than in the French case.

From a comparative perspective, results suggest that the networks with a medium level of belief cohesion or higher also resulted in more comprehensive policy designs, while networks with a lower than medium level of belief cohesion resulted in policy designs characterized by a low degree of comprehensiveness. These findings support the first hypothesis of this study, according to which belief cohesion among network members has a positive effect on the degree of comprehensiveness of policy designs.

The present findings are in line with results from similar studies. Sandström and Carlsson (2008), for example, find that networks that display a low level of heterogeneity in terms of actor diversity work more efficiently. The authors' interpretation is that 'networks containing a less diverse set of actors are not as exposed to differences and competing interests, and therefore need less time to come to terms with collective action' (Sandström and Carlsson 2008, p. 515). In a similar vein, Gerber et al. (2013) found that political homophily of local governments, defined as the tendency to connect with others who are politically similar, increases the chances of collaboration, because transaction costs of bargaining over collective goods decreases. The authors conclude, 'politically similar local governments also have higher benefits of collaboration because they can learn from each other's policy experiments' (Gerber et al. 2013). Both studies indicate that belief similarity represents a strong predictor of collaboration and also positively impacts results of policymaking processes. The present study goes one step further by assessing the impact of belief cohesion on the quality of policy outcomes. In this regard, the present findings suggest a positive implication of belief cohesion on comprehensive policy designs, and hence, support previous studies' claim according to which belief cohesion matters to policy design.[7]

4.3.2 Linking Interconnectedness to Policy Design

Besides belief similarity, policy networks are characterized by their degree of interconnectedness, which refers to the amount of present ties connecting network members (Bressers and O'Toole 1998). The social capital literature puts forward that well-connected networks improve the creation of trust and common norms, and hence the network's capacity to perform (Coleman 1990). Building on the notion of trust, policy scholars' research revealed that network ties also increase the level of trust between policy actors (Lubell 2005), as well as overall network performance (Sandström and Carlsson 2008). Going one step further, environmental governance scholars argue that trust promotes norms that place the common good center stage (Bodin and Crona 2009; Ostrom 1990; Jones et al. 2009), which has been found to positively impact environmental practices (Lubell and Fulton 2007b), as well as to facilitate a higher level of environmental protection (Newig and Fritsch 2009). Aside from promoting collective outputs, research could demonstrate that network ties exhibit multiple benefits for policy actors, such as information exchanges about potential policy solutions (Leifeld and Schneider 2012) and learning (Weible and Sabatier 2007). Such resources enable policy actors to make informed policy decisions (Newig and Fritsch 2009) and also promote the adoption of effective governance modes (Crona and Parker 2012). Previous research suggests that interconnectedness may have important implications for the adoption of comprehensive policy designs. Applied to the language of this work, the second hypothesis

---

[7]Inclusiveness represents one of the six indicators of the policy comprehensiveness index; its results do not lead to the same ranking between the four countries as do the overall index results. Therefore, the co-variance between network structures and inclusiveness were analyzed in addition to the co-variance between network structures and overall index performance. For inclusiveness, i.e., the ability of a policy design to target all sectors contributing to a policy issue, France ranks highest, because policies target household, industrial, and agricultural emissions of micropollutants. The ranking is followed by the Swiss and Dutch policy designs, and lastly, by the German one, which targets more specific emitters of micropollutants while not considering others. However, linking the ranking for belief cohesion (CH, G, NL, F) with the ranking for inclusiveness (F, CH/NL, G) does not enable any significant conclusions about the impact of belief cohesion on the inclusiveness of a policy design. As this is also the case for subsequent hypotheses, inclusiveness is not discussed separately to the overall index results.

concludes that interconnectedness should have a positive effect on the comprehensiveness of policy designs.

Results shown in Table 4.3 do not support this theoretical claim at first sight. There seems to be no clear colinearity between the ranking of policy designs' comprehensiveness and the ranking of networks' interconnectedness, even when taking different ways of calculating interconnectedness into account. While the Dutch network scores highest on interconnectedness, its policy design displays only a low degree of comprehensiveness. The Swiss case scores second on network interconnectedness, but highest according to the policy comprehensiveness index. The German and French networks have the lowest levels of interconnectedness, but occupy a middle position following the ranking of the policy comprehensiveness index.

Upon further investigation, however, empirical results support the hypothesis: When excluding the Dutch network from analysis, there exists linearity between the ranking of policy designs' comprehensiveness (CH, G, F) and the ranking of networks' interconnectedness (triangles per tie: CH, G, F). Interconnectedness is highest in Switzerland, medium in Germany, and low in France, when considering triangles per tie.[8] The Dutch policy network appears to be an outlier with an exceptionally high density of collaboration ties between policy actors. The outlier position is supported by the fact that the analyzed Dutch policymaking process lasted for 12 years, while the other three processes took between 3 and 7 years. The Dutch policymaking process lasted much longer, because on the one hand, key players were among the first to recognize that micropollutants in surface waters were of public concern. On the other hand, there remains a need for a concrete policy design, as thus far certain Dutch policy actors successfully questioned the risks arising from micropollutants, and doubted the urgency for regulating micropollutants (interview with NEFARMA on 28.4.2014).[9] A consequence of this long process is that Dutch policy actors received more opportunities to interact, which might explain a comparably high density of collaboration ties. The comparably high level of interconnectedness did not translate into a comprehensive policy design in the Dutch case. These findings may suggest that interconnectedness positively impacts the comprehensiveness of policy design, but only to a point. Too much interconnectedness, on the contrary, might indicate that the network faces high constraints that can hardly be compensated by the increased network activity.

---

[8]Triangles per tie are a count of the total number of undirected triangles divided by the number of ties in the collaboration network (see Sect. 3.2.3.2). Triangles indicate that ego's alters collaborate with each other; many ties being involved in triangles indicates extensive redundancy of collaboration ties.

[9]The delay of the Dutch policymaking process may also be attributed to other factors, such as extensive administrative reforms and budget cuts affecting water management, as well as a political crisis in 2012, henceforth the issue of micropollutants may have lost urgency on the political agenda.

Table 4.3 Results and rankings for interconnectedness and policy performance

IV (operationalization)	CH	G	F	NL	Ranking	
					IV	DV Policy compr. index score
Interconnectedness (ties per actor)	3.98	2.97	3.5	4.34	NL, CH, F, G	CH, G, F, NL
Interconnectedness (triangles per tie)	1.54	1.47	1.44	2.16	NL, CH, G, F	

In summary, results support hypothesis 2, according to which interconnectedness has positive implications to comprehensive policy design, but only to a tipping point after which interconnectedness negatively affects comprehensive policy design. Future research is needed in order to assess these preliminary findings in more depth, and to analyze whether the mentioned tipping point can be quantified or is solely contextual.

The present findings relate to the collaborative governance literature, which stresses the importance of network interactions for policy outcomes. For example, Sandström and Carlsson (2008) demonstrate that densely interconnected network structures are also more efficient in producing policy outputs. Schneider et al.'s (2003) work demonstrates that networks with more frequent interactions also create greater faith in the ability of policies to improve environmental conditions. The present results for Switzerland, Germany, and France further support the observation of collaborative governance scholars regarding the positive impact of interconnectedness on policy outputs. However, the underlying findings for the Netherlands further indicate that a very high degree of collaboration may impede collective outputs. In conclusion, future research should explore the circumstances under which collaboration has either positive or negative implications for the adoption of a comprehensive policy design.

4.3.3 The Combined Effect of Belief Cohesion and Interconnectedness on Policy Design

Belief cohesion and interconnectedness do not necessarily exist independently of one another, but are instead interrelated such that belief dissimilarities might be overcome through increased collaboration. The theoretical reasoning behind this claim is that low belief cohesion can point to a network open for diverse actors and their manifold, potentially constructive ideas about how to solve a policy problem (Kriesi et al. 2006; Bodin et al. 2006; Newman and Dale 2005; Howlett 2002). Such situations of network heterogeneity can be particularly conducive to comprehensive problem-solving, because actors hold diverse resources, such as knowledge about how to comprehensively address an underlying problem (Bodin et al. 2006; Provan and Kenis 2008). In order to exploit network members' resources for a comprehensive policy design, it is decisive that diverse actors engage in social interactions, which promote the creation of bonding ties (Putnam et al. 1993), mutual learning (Sabatier and Jenkins-Smith 1993), and the creation of common norms (Lubell et al. 2012). From this perspective, interconnectedness indicates that diverse actors negotiate and are working to overcome divergences (Fischer 2014), which may result in a particularly comprehensive policy design. Building on this reasoning, the third hypothesis states that with low belief cohesion, a comprehensive policy design is possible if there exists a high level of interconnectedness between network members.[10]

The empirical results presented in Table 4.4 indicate that not all of the four studied cases serve to discuss this hypothesis. The German and French cases are not suited, as the level of interconnectedness is comparably low with 1.47 and 1.44 triangles per tie in both networks, and hence, existing divergences were most likely not overcome through collaboration, if they were overcome at all. The remaining Swiss and Dutch cases serve to discuss hypothesis 3 due to their sufficient level of interconnectedness.

The Swiss case is characterized by a high degree of belief cohesion (5.92% density of belief similarity ties), good interconnectedness (1.54 triangles per tie), and a highly comprehensive policy design (score of 0.75 on the policy comprehensiveness index). Since the levied data represents a snapshot of policy actors' interactions, it is not possible here to assess whether an initially low level of cohesion has been overcome through collaboration over time. Nevertheless, what can be observed here is that the simultaneous occurrence of a high level of interconnectedness and belief cohesion coincides with a comprehensive policy design. Hence, results suggest that when both conditions coexist, they can positively affect comprehensive policy designs.[11]

Lastly, the Dutch case is characterized by high interconnectedness (2.16 triangles per tie), which in theory should point to a situation in which common norms can be created and divergences can be overcome so that the adoption of comprehensive policy designs is facilitated. However, the empirical results show that the Dutch policy network is poorly belief-cohesive (2.63% density of belief similarity ties), and therefore indicate that divergences were not overcome through

---

[10]The reversed causality, i.e., high belief cohesion fostering interconnectedness, is considered in two further hypotheses: Hypothesis 5 claims that with a high level of belief cohesion, but low interconnectedness, a comprehensive policy design is possible when entrepreneurs foster interconnectedness.

Hypothesis 1 states that high belief cohesion should have a positive effect on a comprehensive policy design independently of the level of interconnectedness, which includes cases of low interconnectedness.

[11]One cannot exclude that there was a high level of belief similarities in the first place, which facilitated interconnectedness, because research has shown that actors with a high level of belief similarity are more likely to collaborate (Gerber et al. 2013). In such a case, high interconnectedness might have served to overcome remaining divergences about the concrete policy design.

Table 4.4 Results and rankings for belief cohesion, interconnectedness, and policy performance

IV (operationalization)	CH	G	F	NL	Ranking	
					IV	DV Policy compr. index score
Interconnectedness (triangles per tie)	1.54	1.47	1.44	2.16	NL, CH, G, F	CH, G, F, NL
Belief cohesion (mean density of belief similarity)	5.92% (high)	2.98% (good)	1.00% (low)	2.63% (low)	CH, G, NL, F	

interconnectedness. Interconnectedness seems to not have had the expected positive effect on belief cohesion. Simultaneously, policy design is not particularly comprehensive (score of 0.2 on the policy comprehensiveness index). These findings indicate that collaboration alone is not enough to overcome deeply rooted belief dissimilarities such that those common norms are created, which would promote a more comprehensive policy design. As a positive effect on a comprehensive policy design is hypothesized, logically the beliefs that are exchanged through an increased interconnectedness are assumed to be favorable to a comprehensive policy design as well. However, it is possible that interconnectedness fosters norms that go against policy comprehensiveness, and that destructive ideas neglecting the need to address a policy problem are exchanged. The empirical findings for the Dutch network, where the opposing coalition holds more power than the one representing water quality, indicate that critical stances may have been exchanged. Actors who oppose the regulation of micropollutants might have taken over the policy debate through their high level of integration in the collaboration network. In this way, a comprehensive policy might have been impeded through interconnectedness instead of being fostered.

This interpretation is no counter-evidence to the hypothesis, because results in fact support the claim that interconnectedness and belief cohesion have a combined impact on policy design. The hypothesis neglects to include the fact that high interconnectedness does not necessarily need to have a positive effect on the comprehensiveness of policy design; high interconnectedness can have a negative effect when beliefs are fostered and exchanged that advocate for a more limited policy design or a complete non-intervention, as with the Dutch case. Nevertheless, a situation with consensus about fundamental values, as well as with an adequate level of interconnectedness, such as in the Swiss collaboration network, will likely facilitate comprehensive policy designs.

The empirical results also demonstrate that when belief cohesion is low, as in the French and Dutch network, policy design tends to be less comprehensive independent of the level of interconnectedness. This observation portrays that belief similarity is an important precondition that strongly facilitates comprehensive policy designs. Such an observation supports ACF scholars' work on policy beliefs and further confirms the value in analyzing policy beliefs in order to better understand policymaking processes (Zafonte and Sabatier 1998; Ingold and Fischer 2014). Existing literature has demonstrated that actors with a high level of belief similarities are more likely to collaborate (Gerber et al. 2013; Henry et al. 2010; Ingold and Fischer 2014). The results presented here further suggest that the coexistence of both conditions, i.e., belief similarities and interconnectedness, can facilitate the adoption of comprehensive policy designs. Less well understood in the literature is whether an initially low level of belief cohesion can be overcome through network interactions. Future research is needed that takes the sequential aspect of the mentioned effects into account in order to fully comprehend the interaction between belief cohesion and interconnectedness, and that assesses to what extent divergences can be overcome through collaboration.

4.3.4   Linking Inter-coalition Collaboration to Policy Design

With the fourth hypothesis, the theoretical part of this work introduces the idea of coalitions rooted in the ACF literature, according to which actors form coalitions based on shared policy beliefs during policymaking processes in order to pool resources, coordinate their action, and influence policy design (Ingold 2008; Weible and Sabatier 2006; Sabatier and Jenkins-Smith 1993). When opposing coalitions strongly compete and block one another, they are likely to also hamper the adoption of a comprehensive policy design. By contrast, blockades between advocacy coalitions may be prevented in collaborative subsystems, where actors from opposing coalitions interact, despite their diverse belief systems. In this regard, Burt's (2003) research suggests that the existence of structural holes—such as those existing between coalitions—may be particularly conducive to comprehensive policy designs, because the creation of several groups indicates that the network exhibits diverse resources necessary for comprehensive problem-solving. As such, the network governance literature additionally emphasized the value of ties bridging group boundaries in order to access diverse group-specific resources (Bodin et al. 2006; Newman and Dale 2005). Collaborative subsystems in policy networks combine the advantages of bonding ties within coalitions with bridging ties across coalitions and hence promote the coexistence of diversity and exchange, which together may also facilitate the development of comprehensive policy designs. Building on this reasoning, hypothesis 4a reads as follows: With two or more opposing coalitions, a comprehensive policy design is possible if actors collaborate across coalitions.

Before presenting the results for collaboration across coalitions, the precise coalition structure of the four networks under investigation, exposed in Sect. 3.3.5.4, should be revisited. In the four studied networks, Swiss and German actors ally into coalitions; Dutch actors to a lesser extent, while French actors exhibit a weak tendency to coalition formation. More precisely, coalition formation is considered weak in the French case, due to a group of non-affiliated actors as well as one coalition, which is internally rather incohesive. In the Dutch network, the two existing groups of actors technically match the definition of coalitions. However, the water quality coalition, in particular, is internally disconnected and weak, which indicates that coalition members do not strongly coordinate their action.

Combining the information about coalition formation and policy design reveals that policy designs are more comprehensive in the Swiss and German cases, where coalition formation is more pronounced, than for the French and Dutch. One should note that this observation supports the theory claiming that group formation is a positive indication that the network integrates diverse types of resources necessary for designing policy (Bodin et al. 2006; Newman and Dale 2005). Moreover, from an ACF perspective (Sabatier and Jenkins-Smith 1993), coalition formation can be interpreted as evidence for the creation of alliances indicating that actors pool resources and coordinate their actions in order to work toward a common strategic goal. By contrast, if actors neglect the gravity of a policy issue, they are likely to refrain from adopting a stance and forming coalitions, which then also reduces the pace of a policymaking process, and potentially the comprehensiveness of policy design as well. In conclusion, underlying results suggest that coalition formation is a positive sign of a network's ability to design comprehensive policies, while the absence of coalition formation represents a somewhat negative indication for problem solving.

Table 4.5 displays the results for collaboration across coalitions, or groups of actors. Since French actors exhibit a lower tendency to coalition formation, the discussion of hypothesis 4a about collaboration across coalitions is mainly restricted to the Swiss, German, and Dutch data here.

In the Swiss and German cases, policy networks display comparably low densities across coalitions (CH: 0.41%, G: 0.87%), but policy designs are rather comprehensive. This observation is noteworthy from a theoretical perspective because it highlights that the lack of collaboration across coalitions does not coincide with an incomprehensive policy design. Results for the Swiss and German cases indicate that the creation of two opposing coalitions that are rather disconnected, as in adversarial subsystems, does not impede comprehensive policy designs.

These findings do not disapprove of the claim that collaboration can be conducive to comprehensive policy designs. Rather, results highlight that a lack of collaboration between members of opposing coalitions can be compensated by other network properties, which will be discussed in the next chapter.

In order to assess the implications of strong collaboration across coalitions on policy design, it is necessary here to refer to similar studies in the field (see for example Ingold 2008, 2007; Mintrom and Vergari 1996). Fischer (2014), for example, analyzed coalition structures conducive to policy change and status quo in the context of the Swiss consensus democracy. He finds that strong conflict paired with strong collaboration across opposing coalitions only leads to status quo policy solutions, and not to major policy change. His interpretation is that strong conflict, together with strong collaboration across coalitions, represents a sign of negotiations, which are more likely to result in policy solutions resembling the lowest

Table 4.5 Results and rankings for interconnectedness across coalitions and policy performance

IV (operationalization)	CH	G	F	NL	Ranking	
					IV	DV Policy compr. index score
Across-coalition collaboration (density of collaboration ties across coalitions)	0.41%	0.87%	14.5%	34.7%	NL, F, G, CH	CH, G, F, NL

common denominator. Within the Swiss consensus democracy, which requires a high level of consent from all the involved parties (Fischer 2014), situations of strong conflict and collaboration across coalitions may be conducive to seeking compromise solutions that are less optimal from a problem-solving perspective. Future research is needed to understand whether collaboration across coalitions, together with conflict, generally hampers a comprehensive policy, or if this is specific to consensus democracies, while majoritarian democracies exhibit different mechanisms.[12]

Among the cases studied here, not only Switzerland, but also the Netherlands, represents a consensus democracy. It is noteworthy that the findings for the Dutch network presented here are in line with Fischer's interpretation and reveal that Dutch actors display comparably high densities of collaboration ties across coalitions (NL: 34.7%). However, the policy design is, thus far, rather deficient. Through their interactions, Dutch actors who oppose a comprehensive policy design, as well as those who demand more comprehensive policy action, are provided numerous opportunities to exchange and diffuse their contradictory arguments. This disaccord between network members, together with the previously mentioned perception of low urgency of the matter, seems to have thus far contributed to blocking and retarding the entire policymaking process. From this perspective, collaborative subsystems—where actors with diverging beliefs exchange intensively—can impede a comprehensive policy design in consensus democracies. Nevertheless, one must consider that the Dutch policymaking process has not terminated, and the issue of micropollutants may gain urgency, for example, after upcoming EU reforms. In such a situation, intensive collaboration could be used to access actors' diverse resources, and to design a comprehensive Dutch micropollutants policy, which conciliates both compromise and problem solving.

---

[12]Note that political compromises do not, by definition, exhibit a low level of comprehensiveness. In fact, policy designs can conciliate both problem solving and compromise, and therefore, collaboration across coalitions does not necessarily have to lead to 'weak' compromises (i.e., policy designs exhibiting a low level of comprehensiveness). It is a matter of future research to understand the conditions under which collaboration across coalitions promotes both compromise and problem-solving, as well as those conditions where collaboration across coalitions hampers comprehensive policy designs.

In summary, results from this study suggest that a low level of collaboration across coalitions does not per se exhibit negative implications to the comprehensiveness of policy designs. Moreover, the results for the Dutch case contradict the theoretical assumption of a positive effect of collaboration across coalitions on policy design, although one should be wary of this conclusion since the policy-making process is not yet terminated. More research is needed to assess the exact conditions under which a high level of collaboration across coalitions facilitates or hampers comprehensive policy designs. In conclusion, hypothesis 4a does not point to an adequate mechanism for the studied cases and cannot be affirmed here.

4.3.5   Linking Brokers to Policy Design

In the previously portrayed network situation, actors from opposing coalitions collaborate directly, which leads to a diffuse inter-coalition collaboration. Next, a network structure is considered with weak inter-coalition collaboration and strong brokerage resulting in a pooled and controlled inter-coalition collaboration through the broker. Brokers bridge structural holes between coalitions and provide members of opposing coalitions with access to different resources, such as knowledge, authority, or funds (Henry et al. 2010; Crona and Parker 2012). Through their strategic network position between coalitions (Christopoulos and Ingold 2011), brokers have access to diverse group-specific resources, enabling them to synthesize a large pool of information and understand opportunities overlooked by others concerning goal intertwinement and win-win situations (Klijn et al. 2010; Bodin et al. 2006). Consequently, when adversarial subsystems coincide with strong brokerage connecting coalitions, great potential may arise for overcoming divergences and designing comprehensive policies. Since brokers bridge structural holes between coalitions and provide access to group-specific resources necessary for a comprehensive policy design, they can combine the necessary conditions for enabling both problem solving and agreement. Building on these theoretical ideas, hypothesis 4b stated that with two or more opposing coalitions, a comprehensive policy design is possible through brokerage connecting opposing coalitions.

The empirical findings about brokerage have been displayed previously in very detail in Table 3.14. For illustrative purposes, results are synthesized here in Table 4.6.

Observation affirms the importance of brokers for the overall network structure because of their ability to connect and mediate between coalitions. Section 3.3.2.5 explained that brokerage is the strongest in the Swiss and German networks and weaker in the French and Dutch networks. Likewise, policy comprehensiveness is the highest in the Swiss and German case and lower in the French and Dutch case. High betweenness centrality scores point to two strong brokers in the Swiss and German networks (see Table 4.6). Moreover, high densities of collaboration ties between brokers and both coalitions indicate that brokers channel interconnectedness in a controlled way. Strong brokerage seems to compensate a lack of direct

Table 4.6 Results for brokerage and policy performance

	IV nBetweenness centrality score of broker (mean)	DV Policy comprehensiveness index score
CH	BAFU-W 49.54 UREKS 9.06 (1.88)	High (score of 0.75)
G	BMU 27.26 UBA 16.13 (2.74)	Medium (score of 0.48)
F	DEB 33.85 (2.99)	Medium/low (score of 0.34)
NL	IenM 18.86 (2.41)	Low (score of 0.2)

collaboration between members of opposing coalitions, shown by low densities of collaboration ties across coalitions in the Swiss and German cases (CH: 0.41%; G: 0.87%). Moreover, strong brokerage occurs simultaneously with comparably good scores for inter-coalition belief similarity. Even though the present data does not enable an assessment of the precise causality, the co-occurrence of both conditions could point to positive implications of brokerage on belief cohesion across coalitions. Such an interpretation aligns with ACF's scholars' view on brokers as mediators, who are able to foster agreement across coalitions through their bridging position (Christopoulos and Ingold 2015; Ingold and Varone 2012). In this regard, Ingold (2011), for example, finds that even when coalitions block each other, brokers have the ability to mediate compromise.

The fact that strong brokerage co-occurs with comparably comprehensive policy designs in both the Swiss and German cases, additionally indicates that brokerage is not only conducive to compromise, but also to comprehensive policy designs. Hence, brokerage may facilitate the adoption of policy designs that combine problem solving together with compromise seeking. This interpretation connects to the network governance literature, which emphasizes brokers' ability to bridge structural holes, and to synthesize group-specific resources in order to promote collective action (Crona and Bodin 2006; Bodin et al. 2006; Van Meerkerk et al. 2015; Kickert et al. 1997). The access to diverse resources may also enable brokers to recognize opportunities for both comprehensive policy design and compromise. In line with this interpretation are, for example, Klijn et al. (2010) findings according to which successful policy outcomes are more likely to occur when network managers activate and connect crucial network members. Despite the different terminology, the study provides support to the present results thanks to its description of the positive impact of actors with a strategic network position on the quality of policy outcomes.

Further findings of the present study point to an absence of real brokerage in the French network, because the broker (DEB) is (a) not neutral, and (b) with its 90% density of collaboration ties particularly well connected to the members of the main coalition. Hence, the French actor DEB occupies a less strategic position and is not a typical mediator. Moreover, brokerage also appears as weak in the Dutch network, as indicated by the comparably low betweenness centrality score for the broker IenM (18.86). In both cases, there exists a high level of direct interactions across coalitions (F: 14.5%, NL: 34.7% density of collaboration ties), which cir-cumvent the broker. As a consequence, the broker does not seem to be in the position of channeling interconnectedness and mediating compromise across coalitions.

In summary, strong brokerage occurs simultaneously with comprehensive policy designs in the studied cases, while the absence of strong brokerage occurs simul-taneously with a low degree of comprehensiveness of policy designs. Inter-coalition collaboration seems to be channeled in a controlled way through brokers, which might improve the creation of common norms favorable to a comprehensive policy design, more so than when actors with diverging beliefs directly collaborate in a diffuse way. Where actors who oppose policy intervention have numerous oppor-tunities to exchange directly with supporters of a comprehensive design, as in the Dutch case, beliefs might clash. Consequently, a low level of belief cohesion across coalitions may materialize, which hampers the design of comprehensive policies. Brokers, by contrast, may be able to promote specifically those ideas and exchanges between actors that are conducive to comprehensive problem solving. Building on this reasoning, the present findings are supportive of hypothesis 4b.

4.3.6 Linking Entrepreneurship to Policy Design

The previously described brokers are actors who can improve inter-coalition interconnectedness and belief cohesion through their position as neutral mediators between coalitions. Entrepreneurs, on the contrary, are self-interested (Christopoulos and Ingold 2015; Ingold and Varone 2012). Their strategic net-work position consists of being connected to other powerful and well-connected network members, and therefore, one can expect entrepreneurs to be members of dominant coalitions; only in the case of two equally strong opposing coalitions may they be part of both. As a consequence of their embeddedness in one coalition, entrepreneurs are on the one hand, unable to mediate between opposing coalitions, but on the other hand, able to improve interconnectedness within their coalition. They have vested interest in connecting coalition members to one another, because through their activity, entrepreneurs can diffuse arguments favorable to their own preferred policy design. If entrepreneurs are in favor of and push for comprehensive problem solving, they may successfully facilitate a comprehensive policy design due to their ability to mobilize network activity. However, given their inability to mediate, entrepreneurs can be expected to best succeed when network members exhibit a certain level of belief cohesion. The underlying conclusion, formulated in the fifth hypothesis, is that with low interconnectedness within coalitions, policy entrepreneurs can push for a comprehensive policy design if it is in line with their preferences.

The empirical results necessary to pretest the hypothesis are shown in Table 3.15 (see Sect. 3.3.5.6). For illustrative purposes, they are synthesized here in Table 4.7. Results show that whenever entrepreneurs exist in a coalition, interconnectedness

Table 4.7 Results for entrepreneurship and policy performance

	IV		DV
	Number of entrepreneurs in dominant coalition	Density of collaboration ties within dominant coalition	Policy comprehensiveness index score
CH	4 Entrepreneurs in water quality coalition	18%	High (score of 0.75)
G	4 Entrepreneurs in both coalitions	18.4% in both coalitions	Medium (score of 0.48)
F	5 Entrepreneurs	25.8%	Medium/low (score of 0.34)
NL	4 Entrepreneurs in opposing coalition	51.1%	Low (score of 0.2)

within that coalition is comparably high as well (see Table 4.7). Entrepreneurs seem to be an important source of tie-creation for the four investigated collaboration networks.

Combining the results from entrepreneurs' network activity, with those for the studied policy designs, reveals that in the Swiss and German cases, entrepreneurs co-occur with a comparably good level of policy designs' comprehensiveness. In the French and Dutch cases, by contrast, entrepreneurs co-occur with incomprehensive policy designs. Hence, entrepreneurs seem to be conducive to comprehensive policy designs more in some cases than in others. The question here is: Under which conditions do entrepreneurs contribute to a comprehensive policy design, and when do they not?

One possible interpretation can be deduced from theory, according to which entrepreneurs lack the ability to mediate between actors when actors' beliefs fundamentally clash, because entrepreneurs are not neutral and do not hold a network position allowing them to mediate. As a consequence of such an inability, entrepreneurs cannot help in overcoming divergences when coalition members display a low degree of belief cohesion. In this regard, the data analyzed here shows that the French coalition displays high belief dissimilarities (1.60% density of belief similarity ties), while belief cohesion is stronger within members of the Swiss (6.40%) and German dominant coalition (4.32%). These findings reveal that entrepreneurs faced higher constraints in the French network than in those of Switzerland and Germany, and therefore might have been more successful in pushing for a comprehensive policy design in the latter cases.

In the Dutch case, on the contrary, the majority of entrepreneurs are members of the coalition that opposes a micropollutants policy, which is an example of a situation where a comprehensive policy design is not in line with the preferences of all entrepreneurs.

One can conclude that entrepreneurs seem to facilitate a comprehensive policy design under two conditions: (a) comprehensive problem solving must be in line with the preferences of entrepreneurs, which is not the case for the Dutch entrepreneurs here; and (b) entrepreneurs should not face too high of constraints within their coalition through a low degree of belief cohesion, such as in the French network. Consequently, present findings support hypothesis 5 under the condition that entrepreneurs are members of a belief-cohesive coalition favorable to a comprehensive policy design.

Previous research mainly focused on the impact of entrepreneurs on policy innovations (Arnold 2013) and policy change (Schneider et al. 2011; Mintrom and Norman 2009). The present findings are an extension of the existing literature, as they assess the conditions under which entrepreneurs can be particularly successful in pushing for a comprehensive policy design.

Although there remains a lack of policy-oriented research concerning the impact of entrepreneurs on policy performance, there do exist studies in other academic fields investigating similar phenomena. For example, in their article, Meier and O'Toole (2001) find that the performance of school districts, measured as the percentage of students in each school district who pass required tests, improves when the head of a school district activates network interactions. The authors find a positive effect of managerial action on performance even when controlling for other effects. Similar to policy entrepreneurs, heads of school districts occupy a strategic network position and have a particular stake in the issue. Meier and O'Tooles' results are relevant to the policy literature since they underline the effect of entrepreneurship on the quality of policy outcomes.

4.3.7   Linking One (Dominant) Coalition to Policy Design

The previous hypotheses covered interconnectedness between coalitions, either directly between members of coalitions (H4a), mediated through the broker (H4b), or within coalitions fostered by entrepreneurs (H5). In contrast, the last hypotheses concern the power and preferences of coalitions in collaborative or adversarial subsystems (H6a) and unitary subsystems (H6b). Common to dominant coalitions in collaborative or adversarial subsystems, on the one hand, and unitary subsystems, on the other, is that in both situations, a coalition accumulates power and is internally well-connected, as well as belief-cohesive (Weible et al. 2010; Ingold and Gschwend 2014). The literature on coalition structures in policy networks maintains that dominant coalitions have the required resources and authority to impose their preferred policy design (Fischer 2014), and the same should hold in unitary subsystems with only one coalition. Hence, the sixth hypotheses put forward that comprehensive problem solving is possible if it is in line with the preferences of the dominant coalition (H6a) or with the majority of members of unitary subsystems (H6b).

The results necessary to discuss hypotheses 6 (a) and (b) have been examined in detail in Sect. 3.3.5.4 and are synthesized here in Table 4.8. Findings reveal a particularly powerful dominant coalition that exhibits beliefs favorable to an improvement of water protection in the Swiss network. Here, the existence of a dominant water quality coalition coincides with a comprehensive policy design. In the German case, there are two competing coalitions that are almost equal in size (19 vs. 17 actors) and densities (18.4% densities of collaboration ties within both coalitions). The water quality coalition is only slightly dominant due to its higher power scores (7.10 vs. 5.59 mean degree; 7.63 vs. 5.88 mean reputational power) compared to the opposing coalition. These results suggest that both coalitions have strongly competed to impose their preferred policy design throughout the whole policymaking process. In fact, until the very last moment, the list of substances on the German environmental quality norms list (EQN-list) was subject of strong debate (see Sect. 2.3.4.1). The competition between the two German coalitions may also have inhibited a high level of policy performance. Nevertheless, the German policy design displays a medium level of comprehensiveness, which may relate to the greater power of the German water quality coalition that advocates for high water protection standards.

The French network comprises only one coalition together with a few remaining, non-affiliated policy actors. On the one hand, results indicate that coalition members, on average, exhibit comparably high support for a pro-environmental policy orientation (see Table 4.8). On the other hand, French coalition members display a comparably low degree of belief cohesion (1.60% mean density of belief similarity ties, see Table 3.13). In conjunction, these results indicate that there exists dissent regarding micropollutants reduction policies. Hence, even if the water quality coalition does not face opposition from another coalition, the internal disagreement among coalition members can partially relate to the comparably low performance of the French policy design. Of particular interest, here is the observation that comprehensive policy designs may not only depend on the support of majorities, but also on a general consent. Even small political minorities can interrupt such consent, and therefore impact the quality of policy designs.

In the Dutch network, opposition toward a comprehensive policy design is even more pronounced than in the French case. The powerful opposing coalition is divided, and in some parts, strongly opposed to a comprehensive policy design for the reduction of micropollutants. Along with this opposition goes that the Dutch policy design lacks comprehensiveness.

All in all, findings for the Swiss and German data indicate that where dominant coalitions support high environmental standards, a comprehensive policy design is facilitated. By contrast, results for the Dutch case suggest that a dominant coalition objecting to comprehensive policy design can negatively implicate problem solving. Likewise, French data suggests that a coalition in which actors are internally divided may impede policy performance. In conclusion, results from this study provide support for the mechanisms portrayed in hypotheses 6a and 6b.

The underlying results relate to the observation made by ACF scholars according to which the power of coalitions matters for policy change (Nohrstedt 2010; Smith 2000; Heaney 2006; Ingold 2007). Fischer's (2014) study, for example, demonstrates that dominant coalitions have the required resources to enable major policy

Table 4.8 Results for coalition structure, mean beliefs, and policy performance

	IV Coalition structure	Mean beliefs for pro-environmental policy	DV Policy comprehensiveness index score
CH	Dominant water quality coalition, minority opposing coalition	3.1 (4 denotes strong support)	High (score of 0.75)
G	Two competing coalitions, where water quality coalition is slightly more powerful than opposing coalition	3.15	Medium (score of 0.48)
F	One water quality coalition, remaining actors	3.3	Medium/low (score of 0.34)
NL	Powerful opposing coalition, weak water quality coalition	2.8	Low (score of 0.2)

changes. The present findings are an extension of the existing literature about coalitions and policy change and indicate that dominant coalitions not only facilitate policy change, but may also promote comprehensive policy designs if it is in line with their preferences.

4.4   Explanatory Strength of the Network Approach to Policy Design

The aim of this research is to assess the relevance of the network approach by examining whether specific structural network properties are conducive to comprehensive problem solving and policy designs. The underlying question then is which combination and level of network configurations, i.e., interconnectedness, belief cohesion, brokerage, entrepreneurship, and coalition structure, promote or inhibit networks' ability to design comprehensive policies. The four cases studied here are a first test at illustrating the relationship between network configurations and comprehensive policy designs.

The following sections begin by briefly summarizing results and then go on to outline the implications of the results in light of the formulated hypotheses (see Table 4.9 for an overview).

Summary of the four cases

Empirical findings suggest that the Swiss network adopted a comprehensive policy design through the combination of a well-connected and belief-cohesive dominant coalition, which promotes a comprehensive policy design and faces opposition only from a minority coalition; the existence of entrepreneurs in the dominant coalition improving interconnectedness and advocating for a comprehensive policy design; and low interconnectedness between opposing coalitions paired with strong brokerage promoting interconnectedness and belief cohesion across coalitions in a

Table 4.9 Discussion of hypotheses—an overview

Hypotheses positive implications of the following network properties on comprehensive policy design:		Results support (✓)/do not support (✗) hypotheses
		Confinement, further insights:
H1	Belief cohesion	✓	
H2	Interconnectedness	✓	Unless very high level of interconnectedness
H3	Belief cohesion and interconnectedness	✓	Belief cohesion is an important precondition for interconnectedness in collaboration networks
H4a	Collaboration across coalitions	✗	
H4b	Brokerage across coalitions	✓	
H5	Entrepreneurship favorable to a comprehensive policy design	✓	If good level of belief cohesion in dominant coalition
H6	Dominant coalition, or unitary subsystem, favorable to a comprehensive policy design	✓	

channeled way. The results indicate that such a network structure has the potential to remediate policy problems comprehensively. Following the policy comprehensiveness index, the Swiss policy design attained a high score because it has the ability to considerably reduce micropollutants in surface waters through the improvement of wastewater treatment technology.

The German policy design attains a medium-level score according to the policy comprehensiveness index, which is less comprehensive than the Swiss one, because EQNs control concentration limits of micropollutants without reducing emissions. The German network structure reflects this less-comprehensive policy design through its lower degree of overall interconnectedness and belief cohesion compared to the Swiss case, as well as the opposition of two, almost equally large coalitions. The opposing coalitions and entrepreneurs have been strongly competing to impose their preferred policy design throughout the entire policymaking process. Until the very last moment, the list of substances on the EQN-list was object of strong debates (see Sect. 2.3.4.1). Nevertheless, the network had enough steering capacity for a satisfactory policy design, because of strong brokerage promoting belief cohesion and interconnectedness across coalitions, and a slightly more powerful pro-water quality coalition, which successfully advocated an encompassing list of substances for the control of micropollutants in surface waters.

For the time frame of this study, the French and Dutch policy designs score lower on the policy comprehensiveness index than do the Swiss and German policy designs. Likewise, findings indicate that the studied French and Dutch policy networks lack certain structural network properties, which one can associate with the promotion of comprehensive policy designs. Characteristic features of both network structures include their low degree of belief cohesion among policy actors, indicating diverging beliefs on micropollutants regulation, paired with weak brokerage, and thus, a lack of mediation between opposing blocks of actors. Moreover, not all actors affiliate in coalitions in order to achieve common strategic goals, which can thus contribute to their comparably weak policy performance. One can observe actors' moderate tendency to coalition formation because the scores for intra-coalition belief cohesion and interconnectedness are lower than for inter-coalition belief cohesion and interconnectedness, at least for one cluster in both networks.

Despite these similarities, French and Dutch network structures display certain differences, which may help to explain why the Dutch policy design is less comprehensive than the French one. The Dutch network exhibits two distinct coalitions, where one coalition is well-connected and strong, while actors in the other coalition are disconnected and display a weak tendency to coalition formation. The French network, on the contrary, reveals only one coalition and a conglomeration of non-affiliated actors. Differing from both cases is also that the French network displays a low level of interconnectedness, while Dutch actors are particularly well-interconnected. The Dutch network structure is nevertheless unfavorable to a comprehensive policy design, because so far the coalition opposing the regulation of micropollutants is more powerful than the water quality coalition. The latter is not only less powerful, but also poorly interconnected and insufficiently belief-cohesive. In the French case, entrepreneurs are part of the water quality coalition, while in the Dutch network, entrepreneurs are more numerous in the opposing coalition. These network differences are also reflected in the comprehensiveness of policy designs, which is higher in the French case than for the Dutch.

All in all, the findings indicate that the analysis of structural network properties can contribute to elucidating differences in the comprehensiveness of policy design. Evidently, the four cases in this study represent only a first test of the underlying hypotheses, and results cannot be generalized. Nonetheless, the empirical results for the cases studied point to some critical insights, which are identified in the following section in order to inform future research.

Summary of discussion of hypotheses

The first hypothesis put forward that a high level of belief cohesion between network members should have a positive effect on the degree of comprehensiveness of policy designs. As illustrated in Table 4.9, results support hypothesis 1 and indicate that belief similarities between network members facilitate comprehensive policy designs, while divergences between policy actors represent unfavorable conditions for the adoption of comprehensive policy designs. In short, when a disagreement occurs regarding the existence of a policy problem or basic values, it may be difficult to find common ground in the form of a comprehensive policy design.

Intensifying collaboration, as suggested by hypothesis 2, is not always conducive to a comprehensive policy design. Results for the Dutch case indicate that the intensity of collaboration can reach a tipping point after which increased activity points to the presence of high constrains and divides between policy actors. In such situations of conflict, the creation of working groups or collaborative platforms for dialogue and exchange might not assist in persuading those policy actors who firmly oppose a comprehensive policy design. Intensifying collaboration among policy actors may not only strengthen beliefs favorable to a comprehensive policy solution, but may also diffuse arguments that question the relevance of a public policy issue. The integration of actors who oppose an encompassing policy design can consequently hamper comprehensiveness and lead to a rather vague policy design, or even block the entire policymaking process.

By contrast, below a certain tipping point, collaboration signals that actors are working on finding common ground and on designing a common policy output. Hence, results support hypothesis 2 concerning the effect of interconnectedness on comprehensive policy designs, under the condition that interconnectedness does not reach an exceptionally high level. Whether this level can be quantified, or is context-dependent, is subject to future research. In conclusion, the second insight from this study is that collaboration may have positive—but under certain conditions also negative—implications for the quality of policy design.

Results also support hypothesis 3, according to which a high level of belief cohesion in combination with high interconnectedness in policy networks has a positive effect on policy networks. Findings allow more differentiated conclusions whereby interconnectedness alone holds less explanatory power than belief cohesion and interconnectedness taken together. Results showed that even if interconnectedness is good, when paired with low belief cohesion, it may have negative implications on the comprehensiveness of policy designs. A possible interpretation of this observation is that when actors collaborate intensively who also disagree about fundamental values, disagreements may foster even more through the exchange. Moreover, results indicated that with low belief cohesion, policy designs tend to be less comprehensive, independent of the level of interconnectedness. Present results suggest that belief cohesion represents an important precondition in policy networks, which facilitates comprehensive policy designs. The analysis of tie-formation in policy networks consequently leads to stronger insights about policy designs, when paired with the analysis of actors' beliefs.

Another insight from this study regarding the hypotheses on coalitions (4a–6b) is that the mere formation of a coalition structure is a positive signal for political activity taking place and pace. As exemplified by the Swiss and German networks, the tendency of actors to form coalitions indicates that actors work toward a concrete strategic goal, coordinate, and pool resources. The French and Dutch networks, on the contrary, are characterized by a weaker tendency of actors to ally in coalitions, which might indicate that actors attribute a low priority to the issue of micropollutants and therefore invest fewer resources in coordinating with others. Coalition formation is, thus, a positive signal for political activity taking place in a policy field.

Another theoretical claim that can be adopted was formulated in hypothesis 4b, concerning the effect of brokers on comprehensive policy designs. The empirical results of this study suggest that brokers play a crucial role in the adoption of comprehensive policy designs. Brokers enjoy a high level of trust and credibility in the policy network. Therefore, they are able to work on overcoming divergences among policy actors through mediation. Moreover, results highlight that more collaboration across coalitions is not necessarily better (H4a), but a channeled type of interconnectedness through the broker (H4b) seems to be more conducive to comprehensive policy designs than untamed connections among actors of opposing coalitions.

In addition, findings also corroborate hypothesis 5, which theorized that policy entrepreneurs can push for a comprehensive policy design if it is in line with their preferences. Results of this study show that entrepreneurs can further activate interconnectedness and advocate their preferred policy design within the more powerful coalition, and hence, successfully encourage comprehensive policy designs if in line with their preferences. An important precondition for the success of entrepreneurs is, however, that coalition members share a certain level of belief cohesion.

Lastly, empirical results approve hypotheses 6a and 6b according to which the creation of one powerful coalition, either in the form of a dominant coalition or unitary subsystem with only one coalition, favorable to a comprehensive policy design, can have a strong impact on policy comprehensiveness. Hence, when groups of actors ally, who have the required resources and authority, they can push for a comprehensive policy design if it is in line with their preferences.

Hypothesis 4a concerning collaboration across coalitions portrays mechanisms that cannot be sustained by observation here. The expectation was that members from separate coalitions profit from each other's resources through their interactions, which could then be conducive to a comprehensive policy design. However, results for the Swiss and German cases show that a lack of collaboration across coalitions does not coincide with an incomprehensive policy design. These findings are no counter-evidence to a potential positive impact of cross-coalition collaboration on comprehensive policy designs. Results rather highlight that a lack of cross-coalition collaboration can be compensated through brokerage. As mentioned previously, brokerage across coalitions seems to be more successful in mitigating discrepancies and fostering common norms favorable to a comprehensive policy design, than a direct actor-to-actor collaboration across coalitions (if belief cohesion is weak across coalitions).

Additionally, results highlight that belief cohesion across coalitions is more decisive for a comprehensive policy design than collaboration across coalitions, because actors holding diverging beliefs are given numerous opportunities to intensify disagreements through their exchange. Such a situation may block the entire policymaking process, which in turn can have a negative effect on a comprehensive policy design.

All in all, results suggest that relating structural network configurations to policy design can considerably improve one's understanding of why certain policy designs are chosen over others in a policymaking process. The network approach provides a particularly powerful explanation for the comprehensiveness of policy designs, as policy performance can be considered a reflection of social cleavages in the policymaking process. The degree of comprehensiveness of a policy output depends on the existing network features and social divides in the network. Findings of this study indicate that a comprehensive policy design is more likely with the combination of the following structural network characteristics: a belief-cohesive and well-connected, but not too excessively interconnected overall network structure, in which actors tend to ally in coalitions and where strong brokerage mediates across coalitions (rather than diffuse inter-coalition collaboration); and the formation of one well-interconnected and belief-cohesive dominant coalition favorable to a comprehensive policy design, which comprises particularly active entrepreneurs promoting comprehensive policy designs. In the absence of those network configurations, a comprehensive policy design is less likely.

The insights are particularly valuable to the environmental governance literature that has discussed the question of network structures conducive to environmental protection (Koontz and Thomas 2006). Bodin et al. (2006), for example, conclude that a 'beneficial structure [...] appears to be a network containing separate groups with internal trust and with some degree of trust among them, linked together by motivated brokers who are interested in using their structural position to initiate and maintain adaptive co-management.' These theoretical expectations about network structure are corroborated by the findings of the present study. Findings affirm that coalition formation, paired with belief cohesion within and across coalitions, and strong brokerage is conducive to comprehensive policy designs. Additionally, the present study expands insights on relevant structural network characteristics by borrowing concepts from policy process research, and in particular from ACF literature (Weible et al. 2010; Sabatier and Jenkins-Smith 1993; Ingold and Varone 2012). Further insights, for example, concern the importance of entrepreneurship, coalition formation, and coalition structures, i.e., power and preferences of advocacy coalitions, for policy design.

To conclude, policy networks can be considered maps of the policymaking process, which unfold what happens in the 'black box' between input and output of a policymaking process. The analysis of policy networks unravels the roles, positions, or relations actors involved in the policymaking process exhibit, and therefore, policy networks contribute to explaining variance in the comprehensiveness of policy design. The underlying research question of this study was: '*Do structural characteristics of policy networks help us understand some part of the variance of policy designs' prospective problem-solving ability?*' Findings of this study provide first indications that the relational structure between policy actors is an important element for comprehending how to achieve comprehensive policy designs and promote problem solving. Chapter 5 discusses the broader conclusions of these findings.

References

Adam, S., & Kriesi, H.-P. (2007). The network approach. In P. Sabatier (Ed.), *Theories of the policy process* (pp. 129–154). Boulder: Westview Press.

Arnold, G. (2013). Street-level policy entrepreneurship. *Public Management Review, 17*(3), 307–327.

Atkinson, M., & Nigol, R. (1989). Selecting policy instruments. Neo-institutional and rational choice interpretations of automobile insurance in Ontario. *Canadian Journal of Political Science, 22*(1), 107–135.

Baumgartner, F., & Jones, B. (1993). *Agendas and instability in American politics.* Chicago: University of Chicago Press.

Baxter-Moore, N. (1987). Policy implementation and the role of the state. A revisited approach to the study of policy instruments. In R. Jackson, D. Jackson, & N. Baxter-Moore (Eds.), *Contemporary Canadian Politics.* Scarborough: Prentice-Hall.

Bennett, C. (1991). Review article. What is policy convergence and what causes it? *British Journal of Political Science, 21,* 215–233.

Bennett, C., & Howlett, M. (1992). The lessons of learning: Reconciling theories of policy learning and policy change. *Policy Science, 25*(3), 275–294.

Berardo, R., & Scholz, J. (2010). Self-organizing policy networks: Risk, partner selection, and cooperation in Estuaries. *American Journal of Political Science, 54*(3), 632–649.

Blatter, J., & Haverland, M. (2012). *Designing case studies. explanatory approaches in Small-N research* (Research Methods Series). Basingstoke: Palgrave Macmillan.

Bodin, Ö., Crona, B., & Ernstson, H. (2006). Social networks in natural resource management: What is there to learn from a structural perspective? *Ecology and Society, 11*(2).

Bodin, Ö., & Crona, B. I. (2009). The role of social networks in natural resource governance: What relational patterns make a difference? *Global Environmental Change, 19,* 366–374.

Bodin, Ö., & Prell, C. (Eds.). (2011). *Social networks and natural resource management. Uncovering the social fabric of environmental governance.* Cambridge: Cambridge University Press.

Börzel, T. (1998). Organizing Babylon. On the different conceptions of policy networks. *Public Administration, 76,* 253–273.

Bressers, H., & O'Toole, L. (1998). The selection of policy instruments: A network-based perspective. *Journal of Public Policy, 18*(3), 213–239.

Bressers, H., & O'Toole, L. (2005). Instrument selection and implementation in a networked context. In P. Eliadis, M. Hill, & M. Howlett (Eds.), *Designing government: From instruments to governance* (pp. 132–153). Montreal, Kingston: McGill-Queen's University Press.

Burt, R. (2000). The network structure of social capital. In B. Staw & R. Sutton (Eds.), *Research in organizational behavior* (pp. 345–423). Greenwich, CT: JAI Press.

Burt, R. (2003). The social capital of structural holes. In M. Guillen, R. Collins, P. England, & M. Meyer (Eds.), *The new economic sociology: Developments in an emerging field* (pp. 148–189). New York: Russell Sage Foundation.

Carley, M. (1980). *Rational techniques in policy analysis.* London: Heinemann Educational Books.

Christopoulos, D. (2008). Political entrepreneurs: Network structure and power. Published online: http://www.researchgate.net/publication/265495932. Accessed on July 9, 2015.

Christopoulos, D., & Ingold, K. (2011). Distinguishing between political brokerage and political entrepreneurship. *Procedia—Social and Behavioral Sciences, 10,* 36–42.

Christopoulos, D., & Ingold, K. (2015). Exceptional or just well connected? Political entrepreneurs and brokers in policy making. *European Political Science Review, 7,* 475–498.

Coleman, J. (1990). *Foundations of social theory.* Cambridge, MA: Harvard University Press.

Crona, B., & Bodin, Ö. (2006). What you know is who you know? Communication patterns among resource users as a prerequisite for co-management. *Ecology and Society, 11*(2).

Crona, B., & Parker, J. (2012). Learning in support of governance: Theories, methods, and a framework to assess how bridging organizations Contribute to adaptive resource governance. *Ecology and Society, 17*(1).

Dahl, R., & Lindblom, C. (1953). *Politics, economics and welfare*. Chicago: The University of Chicago Press.

Daugbjerg, C., & Marsh, D. (1998). Explaining policy outcomes: integrating the policy network approach with macro-level and micro-level analysis. In D. Marsh (Ed.), *Comparing policy networks* (pp. 53–71). Philadelphia: Open University Press.

Doern, B., & Wilson, S. (1974). *Issues in Canadian public policy*. Toronto: McMillan.

Fischer, M. (2012). *Entscheidungsstrukturen in der Schweizer Politik zu Beginn des 21. Jahrhunderts*. Glarus, Chur: Rüegger.

Fischer, M. (2013). *Policy network structures, institutional context, and policy change*. Paper presented at the COMPASSS Working Paper 73.

Fischer, M. (2014). Coalition structures and policy change in a consensus democracy. *Policy Studies Journal, 42*(3), 344–366.

Fischer, M. (2015). Institutions and coalitions in policy processes: A cross-sectoral comparison. *Journal of Public Policy, 35*(2), 1–24.

Freeman, G. (1985). National styles and policy sectors. Explaining structured variation. *Journal of European Public Policy, 5*(4), 467–496.

George, A., & Bennett, A. (2005). *Case studies and theory development in the social sciences*. Cambridge, MA: MIT Press.

Gerber, E., Henry, A. D., & Lubell, M. (2013). Political homophily and collaboration in regional planning networks. *American Journal of Political Science, 57*(3), 598–610.

Gerring, J. (2004). What is a case study and what is it good for? *The American Political Science Review, 98*(2), 341–354.

Granovetter, M. (1973). The strength of weak ties. *American Journal of Sociology, 78*(6), 1360–1380.

Granovetter, M. (1985). Economic action and social structure: The problem of embeddedness. *American Journal of Sociology, 91*, 481–510.

Granovetter, M. (1992). Economic institutions as social construction: A framework of analysis. *Acta Sociologica, 35*, 3–11.

Haverland, M. (2000). National adaptation to European integration: The importance of institutional veto points. *Journal of Public Policy, 20*(01), 83–103.

Heaney, M. (2006). Brokering health policy: Coalitions, parties, and interest group influence. *Journal of Health Politics, Policy and Law, 31*(5), 887–944.

Henry, A. D. (2011). Ideology, power, and the structure of policy networks. *Policy Studies Journal, 39*(3), 361–383.

Henry, A. D., Lubell, M., & McCoy, M. (2010). Belief systems and social capital as drivers of policy network structure: The case of California regional planning. *Journal of Public Administration Research and Theory, 21*(3), 419–444

Howlett, M. (1991). Policy instruments, policy styles, and policy implementation. *Policy Studies Journal, 19*(2), 1–21.

Howlett, M. (2002). Do networks matter? Linking policy network structure to policy outcomes: Evidence from four Canadian policy sectors 1990–2000. *Canadian Journal of Political Science, 35*(2), 235–267.

Howlett, M., & Ramesh, M. (2003). *Studying public policy: Policy cycles and policy subsystems*. Oxford: Oxford University Press.

Ingold, K. (2007). The influence of actors' Coalition on policy choice: The case of the Swiss climate policy. In T. Friemel (Ed.), *Applications of social network analysis*. UVK: Constance.

Ingold, K. (2008). *Analyse des mécanismes de décision: Le cas de la politique climatique suisse*. Zürich and Chur: Rüeggger Verlag.

Ingold, K. (2011). Network structures within policy processes: Coalitions, power, and brokerage in Swiss climate policy. *Policy Studies Journal, 39*(3), 435–459.

Ingold, K., & Fischer, M. (2014). Drivers of collaboration to mitigate climate change: An illustration of Swiss climate policy over 15 years. *Global Environmental Change, 24*, 88–98.

Ingold, K., & Gschwend, M. (2014). Science in policy-making: Neutral experts or strategic policy-makers? *West European Politics, 37*(5), 993–1018.

Ingold, K., & Varone, F. (2012). Treating policy brokers seriously: Evidence from the climate policy. *Journal of Public Administration Research and Theory, 22*(2), 319–346.

Innes, J., & Booher, D. (2003). *Collaborative policymaking: Governance through dialogue. Deliberative policy analysis*. Cambridge: Cambridge University Press.

Jones, N., Sophoulis, C., Iosifides, T., Botetzagias, I., & Evangelinos, K. (2009). The influence of social capital on environmental policy instruments. *Environmental Politics, 18*(4), 595–611.

Kenis, P., & Schneider, V. (1991). Policy networks and policy analysis: Scrutinizing a new analytical toolbox. In B. Marin & R. Mayntz (Eds.), *Policy networks—Empirical evidence and theoretical considerations*. Frankfurt am Main: Campus Verlag.

Kickert, W., Klijn, E.-H., & Koppenjan, J. (1997). *Managing complex networks*. London: Sage.

King, G., Keohane, R., & Verba, S. (1994). *Designing social inquiry: Scientific inference in qualitative research*. Chichester: Princeton University Press.

Kingdon, J., & Thurber, J. (2011). *Agendas, alternatives, and public policies*. New York: Longman.

Klijn, E.-H. (1996). Analyzing and managing policy processes in complex networks: A theoretical examination of the concept policy network and its problems. *Administration & Society, 28*(1), 90–119.

Klijn, E.-H., Steijn, B., & Edelenbos, J. (2010). The impact of network management strategies on the outcomes in governance networks. *Public Administration, 88*(4), 1063–1082.

Knoke, D. (1990). *Political networks. The structural perspective*. New York: Cambridge University Press.

Knoke, D., Pappi, F. U., Broadbent, J., & Tsujinaka, Y. (1996). *Comparing policy networks. Labor politics in the U.S., Germany, and Japan*. Cambridge UK, New York: Cambridge University Press.

Koontz, T., & Thomas, C. (2006). What do we know and need to know about the environmental outcomes of collaborative management? *Public Administration Review, 66*, 111–121.

Kriesi, H., Adam, S., & Jochum, M. (2006). Comparative analysis of policy networks in Western Europe. *Journal of European Public Policy, 13*(3), 341–361.

Kuhnert, S. (2001). An evolutionary theory of collective action: Schumpeterian entrepreneurship for the common good. *Constitutional Political Economy, 12*(1), 13–29.

Laumann, E., & Knoke, D. (1987). *The organizational state. Social in national policy domains*. Madison: University of Wisconsin Press.

Laumann, E., & Pappi, F. U. (1976). *Networks of collective action. A perspective on community influence systems*. New York: Academic Press.

Leifeld, P., & Schneider, V. (2012). Information exchange in policy networks. *American Journal of Political Science, 56*(3), 731–744.

Linder, S., & Peters, G. (1989). Instruments of government: Perceptions and contexts. *Journal of Public Policy, 9*(1), 35–58.

Lowi, T. (1964). American Business, Public Policy, Case-Studies, and Political Theory. *World Politics, 16*(04), 677–715.

Lowi, T. (1972). Four systems of policy, politics and choice. *Public Administration Review, 32*(4), 298–310.

Lubell, M. (2005). Do watershed partnerships enhance beliefs conducive to collective action? In P. Sabatier, W. Focht, M. Lubell, Z. Trachtenberg, A. Vedlitz, & M. Matlock (Eds.), *Swimming upstream: Collaborative approaches to watershed management*. Cambridge, MA: MIT Press.

Lubell, M., & Fulton, A. (2007a). Local diffusion networks as pathways to sustainable agriculture. *California Agriculture, 61*(3), 131–137.

Lubell, M., & Fulton, A. (2007b). Local policy networks and agricultural watershed management. *Journal of Public Administration Research and Theory, 18*(4), 673–696.

Lubell, M., Scholz, J., Berardo, R., & Robins, G. (2012). Testing policy theory with statistical models of networks. *Policy Studies Journal, 40*(3), 351–374.

Marin, B., & Mayntz, R. (1991). *Policy network: Empirical evidence and theoretical considerations.* Frankfurt am Main: Campus Verlag.

Marsh, D. (1998). *Comparing policy networks*: Open University Press.

Marsh, D., & Rhodes, R. (1992). *Policy networks in British government.* Oxford, GB: Clarendon Press.

Meier, K., & O'Toole, L. (2001). Managerial strategies and behavior in networks: A model with evidence from U.S. public education. *Journal of Public Administration Research and Theory, 11*(3), 271–294.

Metz, F., & Ingold, K. (2014a). *Policy instrument selection under uncertainty: The case of micropollution regulation.* Paper presented at the Conference Paper presented at the Swiss Political Science Association Annual Congress, Berne, January 31, 2014

Metz, F., & Ingold, K. (2014b). Sustainable wastewater management: Is it possible to regulate micropollution in the future by learning from the past? A policy analysis. *Sustainability, 6*(4), 1992–2012.

Mintrom, M., & Norman, P. (2009). Policy entrepreneurship and policy change. *Policy Studies Journal, 37*(4), 649–667.

Mintrom, M., & Vergari, S. (1996). Advocacy coalitions, policy entrepreneurs, and policy change. *Policy Studies Journal, 24*(3), 420–434.

Newig, J., & Fritsch, O. (2009). Environmental governance: Participatory, multi-level—And effective? *Environmental Policy and Governance, 19*(3), 197–214.

Newman, L., & Dale, A. (2005). Network structure, diversity, and proactive resilience building: A respone to Tompkins and Adger. *Ecology and Society, 10*(1).

Nohrstedt, D. (2008). The politics of Crisis policymaking: Chernobyl and Swedish nuclear energy policy. *Policy Studies Journal, 36*(2), 257–278.

Nohrstedt, D. (2010). Do advocacy coalitions matter? Crisis and CHANGE in swedish nuclear energy policy. *Journal of Public Administration Research and Theory, 20,* 309–333.

Ostrom, E. (1990). *Governing the commons. The evolution of institutions for collective actors.* Cambridge, New York: Cambridge University Press.

Peters, G., & Hoornbeek, J. (2005). The problem of policy problems. In P. Eliadis, M. Hill, & M. Howlett (Eds.), *Designing government.* Montreal, Kingston: McGill-Queen's University Press.

Provan, K., & Kenis, P. (2008). Modes of network governance: Structure, management, and effectiveness. *Journal of Public Administration Research and Theory, 18*(2), 229–252.

Provan, K., & Milward, B. (1995). A preliminary theory of interorganizational network effectiveness: A comparative study of four community mental health systems. *Administrative Science Quarterly, 40*(1), 1–33.

Putnam, R., Leonardi, R., & Nanetti, R. (1993). *Making democracy work: Civic traditions in modern Italy.* Princeton: Princeton University Press.

Richey, S., & Ikeda, K. I. (2006). The influence of political discussion on policy preference: A comparison of the United States and Japan. *Japanese Journal of Political Science, 7*(3), 273–288.

Robins, G., Bates, L., & Pattison, P. (2011). Network governance and environmental management: Conflict and cooperation. *Public Administration, 89*(4), 1293–1313.

Sabatier, P. (1987). Knowledge, policy-oriented learning and policy change. *An Advocacy Coalition Framework. Science Communication, 8*(4), 649–692.

Sabatier, P., & Jenkins-Smith, H. (1993). *Policy change and learning: An advocacy coalition approach.* Boulder: Westview Press.

Salamon, L. (2000). The new governance and the tools of public action: An introduction. *Fordham Urban Law Journal, 28*(5), 1611–1674.

Sandström, A., & Carlsson, L. (2008). The performance of policy networks: The relation between network structure and network performance. *The Policy Studies Journal, 36*(4), 497–524.

Sandström, A., Crona, B., & Bodin, Ö. (2014). Legitimacy in Co-management: The impact of preexisting structures, social networks and governance strategies. *Environmental Policy and Governance, 24*(1), 60–76.

Schneider, A., & Ingram, H. (1993). Social construction of target populations: Implications for politics and policy. *The American Political Science Review, 87*(2), 334–347.

Schneider, M., Scholz, J., Lubell, M., Mindruta, D., & Edwardsen, M. (2003). Building consensual institutions: Networks and the national estuary program. *American Journal of Political Science, 47*(1), 143–158.

Schneider, M., Teske, P., & Mintrom, M. (2011). *Public entrepreneurs: Agents for change in American government*. Princeton: Princeton University Press.

Sciarini, P. (1996). Elaboration of the Swiss agricultural policy for the GATT negotiations: A network analysis. *Schweizerische Zeitschrift für Soziologie, 22*(1), 85–115.

Skogstad, G. (2003). Legitimacy and/or policy effectiveness? Network governance and GMO regulation in the European Union. *Journal of European Public Policy, 10*(3), 321–338.

Smith, A. (2000). Policy networks and advocacy Coalitions: Explaining policy change and stability in UK industrial pollution policy? *Environment and Planning C, Government and Policy, 18*, 95–114.

Trebilcock, M., Hartle, D., Prichard, R., & Dewees, R. (1982). *The choice of governing instrument. A study prepared for the economy council of Canada*. Ottawa: Canadian Government Publishing Centre.

Van Meerkerk, I., Edelenbos, J., & Klijn, E.-H. (2015). Connective management and governance network performance: The mediating role of throughput legitimacy. Findings from survey research on complex water projects in the Netherlands. *Environment and Planning C: Government and Policy, 33*(3), 746–764.

Van Waarden, F. (1992). Dimensions and types of policy networks. *European Journal of Political Research, 21*(Special Issue), 29–52.

Varone, F. (1998). *Le choix des instruments des politiques publiques. Une analyse comparée des politiques d'efficience énergétique du Canada, du Danemark, des Etats-Unis, de la Suède et de la Suisse*. Bern: Paul Haupt Verlag.

Weible, C., Pattison, A., & Sabatier, P. (2010). Harnessing expert-based information for learning and the sustainable management of complex socio-ecological systems. *Environmental Science & Policy, 13*(6), 522–534.

Weible, C., & Sabatier, P. (2005). Comparing policy networks: Marine protected areas in California. *Policy Studies Journal, 33*(2), 181–201.

Weible, C., & Sabatier, P. (2006). A guide to the advocacy coalition framework: Tips for researchers. In F. Fischer (Ed.), *Handbook of public policy analysis: Theory, politics, and methods*. New York: CRC Press.

Weible, C., & Sabatier, P. (2007). The advocacy coalition framework: Innovations and clarifications. In P. Sabatier (Ed.), *Theories of the policy process*. Boulder: Westview Press.

Weiss, C. (1977). Research for polilcy's sake: The enlightenment function of social research. *Policy Analysis, 3*(4), 531–545.

Wellmann, B., & Berkowitz, S. (1988). *Social structure: A network approach*. Cambridge: Cambridge University Press.

Wilks, S., & Wright, M. (1987). *Comparative government-industry relations. Western-Europe, the United States and Japan*. Oxford: Claredon Press.

Woodside, K. (1986). Policy instruments and the study of public policy. *Canadian Journal of Political Science, 19*(4), 775–793.

Zafonte, M., & Sabatier, P. (1998). Shared beliefs and imposed interdependencies as determinants of ally networks in overlapping subsystems. *Journal of Theoretical Politics, 10*(4), 473–505.

Zuckerman, A. (2005). *The social logic of politics: Personal networks as contexts for political behavior*. Philadelphia: Temple University Press.

Chapter 5
Conclusion

5.1 Main Findings

The ultimate goal of this book is to assess the relevance of the network approach by examining whether structural characteristics of policy networks may help us in understanding some part of the variance of policy designs' prospective problem-solving ability. With this goal in mind, this study also works toward understanding which combination of structural network properties promotes or inhibits networks' abilities to design comprehensive policies.

To analyze the impact of policy networks on the comprehensiveness of policy designs, this study compares the micropollutants policies of the four Rhine riparian states: Switzerland, Germany, France, and the Netherlands. The comparison is performed by evaluating policy designs against their prospective ability to effectively reduce micropollutants in waters, by means of the policy comprehensiveness index introduced here. In order to explain the resulting variations of policy designs' comprehensiveness, the present study pays particular attention to the relational aspect of policymaking, because actors' ability to formulate comprehensive policy designs is considered dependent upon the Web of relations in which they are embedded. Relations between policy actors are analyzed by borrowing insights from the literatures on policy processes, social capital, and network governance about crucial structural properties of policy networks, including belief cohesion, interconnectedness, coalition structure, brokerage, and entrepreneurship. This study operationalizes network structure with the help of a formal social network analysis so that the interactions between policy actors, who were involved in policymaking processes on micropollutants, are analyzed.

Since gathering data about policy networks is particularly resource-intensive, this study is limited to the analysis of four cases. Consequently, results should be considered a first test of the formulated hypotheses, which may not be generalized, but still provide important indications about the influence of policy networks on the comprehensiveness of policy design.

© Springer International Publishing AG 2017

F. Metz, *From Network Structure to Policy Design in Water Protection*,
Springer Water, DOI 10.1007/978-3-319-55693-2_5

A first indication resulting from this study concerns the importance of belief cohesion (hypothesis 1), defined as belief similarity between network members, which seems to considerably facilitate the adoption of a comprehensive policy design. Belief cohesion highlights that network members share common ground, which, in turn, seems to facilitate comprehensive problem solving. These findings further support policy process literature emphasizing that policy beliefs represent a strong explanation for policy outputs (Zafonte and Sabatier 1998; Ingold and Fischer 2014; Henry et al. 2010). Observations further suggest that policy beliefs not only matter for explaining policy change, but also provide important insights into a network's ability to design comprehensive policies.

A second implication of the underlying results is that interconnectedness (hypothesis 2) between network members seems to positively impact the comprehensiveness of policy designs to a tipping point. A very high level of interconnectedness, on the contrary, points to strong negotiations and high constraints, which may hardly be compensated by increased network activity. In their article, Koontz and Thomas claim that because research has not analyzed the effects of collaboration on environmental outcomes so far, 'the excitement over collaborative processes has not been matched by evidence that collaboration has actually improved the environment' (Koontz and Thomas 2006, p. 111). The present findings delve into one piece of this postulated research agenda by highlighting that collaboration may have positive—but under certain conditions also negative— implications for the quality of policy design.

This study also suggests that there are important interaction effects between belief cohesion and interconnectedness (hypothesis 3). In the Swiss case analyzed here, a high level of belief cohesion and interconnectedness occurs alongside with a comprehensive policy design. While the causal processes behind this observation are less clear, findings by other network scholars suggest that belief cohesion increases the chances of collaboration (Henry et al. 2010; Ingold and Fischer 2014), which in turn may have positive implications for policy performance. Less well understood is whether divergences may be overcome through collaboration in order to achieve a comprehensive policy design. Dutch data indicates that arguments opposed to policy intervention may be diffused through intense collaboration, which in turn may hamper policy performance. In order to assess such a mechanism more thoroughly, the combined analysis of belief cohesion and interconnectedness leads to better insights about policy design than does the analysis of interconnectedness, or policy beliefs, alone.

A further finding of this study suggests that collaboration across coalitions (hypothesis 4a) may not necessarily be conducive to the adoption of comprehensive policy designs, if belief cohesion is weak across coalitions. Following Fischer's (2014) interpretation, cross-coalition collaboration may promote compromising in the form of policy solutions resembling the lowest common denominator. From a problem-solving perspective, however, collaboration across coalitions may hamper optimal policy designs, if there simultaneously exists high conflict across coalitions. While these conclusions apply to consensus democracies that rely on a high level of general consent, the same network properties, i.e., collaboration paired with conflict across coalitions, may have other implications for policy design in majoritarian democracies.

A further observation here is that coalition formation represents a positive sign that a network is able to design comprehensive policies, while the absence of coalition formation represents a negative indication for problem solving. Coalition formation demonstrates that network members coordinate their actions, which also reveals that actors attribute significance and urgency to a policy issue. Furthermore, group formation suggests that the network includes diverse actor groups that also have the potential to contribute constructively to problem solving (Crona and Bodin 2006; Newman and Dale 2005).

Additionally, the discussion of hypotheses 3 and 4a highlights that coalition formation can provide important insights into the interaction effects between belief cohesion and interconnectedness in policy networks. Concerning hypothesis 3, findings suggest that collaboration (if not excessively high) has positive implications for comprehensive policy design. Regarding hypothesis 4a, by contrast, results indicate that collaboration across coalitions (when paired with conflict) may hamper policy performance. In order to better understand the similarities and differences between the underlying network mechanisms, future research is needed that considers interaction effects between the mentioned network properties.

Moreover, as opposed to a direct actor-to-actor type of collaboration across coalitions as portrayed in hypothesis 4a, brokerage across coalitions seems to successfully foster common norms favorable to a comprehensive policy design. Results suggest that there is great potential to design comprehensive policies where brokers mediate between opposing coalitions (hypothesis 4b). Brokers hold strategic network positions that may enable them to promote policy designs that are conducive to problem solving.

Additionally, findings indicate that not only brokers, but also entrepreneurs, are able to use their strategic network position to push for a comprehensive policy design (hypothesis 5). If entrepreneurs favor a comprehensive policy design and do not face too high of constraints within their coalitions through a low degree of belief cohesion, they are then able to mobilize their network contacts and promote a comprehensive policy design.

In conclusion, present results underline the relevance of agency, i.e., brokerage and entrepreneurship, for the quality of policy outcomes. Agency appears as an important structural property of policy networks that impacts the ability of policy networks to deliver comprehensive policy designs. In conclusion, agency provides useful insights into the study of policy networks and policy design.

Finally, results indicate that the formation of dominant coalitions, which are favorable to comprehensive problem solving, facilitates the adoption of comprehensive policy designs. Conversely, dominant coalitions, or a majority of actors in unitary subsystems, can hamper problem solving by objecting to a comprehensive policy design.

These findings also provide a first indication about the combination of structural network properties promoting or inhibiting networks' ability to design comprehensive policies. A belief-cohesive network, in which policy actors collaborate and ally within coalitions, and where strong brokers mediate across coalitions, seems to promote comprehensive policy designs. Further conditions that are conducive to problem solving are belief-cohesive and well-interconnected dominant coalitions favorable to comprehensive policy designs, together with particularly active entrepreneurs pushing for a comprehensive design.

On the contrary, belief dissimilarities between network members seem to hamper comprehensive policy designs, as divergences complicate achieving common ground and all the more inhibit finding a comprehensive policy design. Moreover, constraints may be emphasized when belief dissimilarities are combined with an exceptionally high level of interconnectedness in the network, or across coalitions. In such situations, actors opposing a comprehensive policy design are given numerous opportunities to diffuse their arguments and block the entire policy-making process, which also hinders problem solving. Moreover, the lack of strong brokerage mediating between opponents may be associated with incomprehensive policy designs. A low tendency of actors to ally into coalitions furthermore highlights that actors do not attribute high importance to an issue and, hence, that policy action is neither taking place nor pace. Finally, chances for problem solving decrease where entrepreneurs, dominant coalitions, or a majority of network members in unitary subsystems oppose a comprehensive policy design.

The cases under investigation serve to pretest the plausibility of the theoretical argument about the impact of policy networks on policy design. It was possible to demonstrate that networks make a difference in the cases studied. In conclusion, findings provide first indications that the analysis of structural network properties contributes to explaining variance of the comprehensiveness of policy design. With this finding, the present research contributes to the broader theoretical discourse concerning the question as to whether networks matter. It is beyond the scope of this study to assess the explanatory power of the network variable in relation to other explanations for policy design. Future research is needed to investigate whether and to what extent policy networks matter given other effects.

5.2 Limitations of This Study and Learnings for Future Research

The theory of this work seeks to explore the explanatory power of one key causal mechanism, namely the link between network structures and comprehensive policy designs. The empirical portion of this work is restricted to studying four cases, because policy networks rely on a large number of observations, and gathering data from policy actors is particularly resource demanding. Due to these limitations, findings serve as a plausibility probe of the hypothesized linkages and generate first indications about possible causal processes. On the one hand, the present research strategy serves to explore how future quantitative examination could be approached. On the other hand, these exploratory purposes are associated with some limitations that impact the generality of the conclusions. The following chapters make these limitations explicit and propose how future research can move forward from them.

In order to discuss how a potential future large-N research could be approached, Sect. 5.2.2 imbeds the present input to broader quantitative research by discussing alternative factors, aside from policy networks, that may explain policy design. Policy networks are not considered the only explanation for the choice of policy designs here. Nevertheless, this study focuses on policy networks as a key explanatory variable, because the linkage between networks and policy design remains understudied to date.

5.2.1 Empirical and Methodological Limitations

The first limitation of the present study concerns its ability to be generalized. According to Blatter and Haverland, '[findings] can only be generalized to the population of cases that display the same scores on all the control and independent variables as the cases that have been studied' (Blatter and Haverland 2012, p. 69). Hence, it seems reasonable to generalize conclusions to the (rather small) population of similar cases that include wealthy, Western democratic nations, which are environmentally conscious and technologically advanced with regard to water protection (Haverland 2000).

A further limitation of the present study is linked to the policy issue of micropollutants, which represents a new, potentially less-politicized policy field than migration, labor, agriculture, or European policy, for example (see, e.g., Fischer 2012; Kriesi et al. 2006). Findings reveal a limited impact of the subsystem-specific arrangement on the structure of policy networks. However, result may be a particularity of the issue of micropollutants, which is new to political agendas, and may still not be completely incorporated into the policy subsystem of water protection policy.

Despite the restrictions related to the policy field of micropollutants, it was a deliberate choice to analyze exclusively one policy field here. The purpose of the present study is to gain explanatory leverage through the comparison of the same policy subfield across four democracies. Future research is needed to quantify the interaction effects between macro-political structures and policy networks.

Additionally, future research is necessary that investigates the linkage between network structure and policy design in several policy fields *and* political systems. To perform such a research, one could build on insights from previous studies indicating that subsystems mitigate the impact of political institutions on policy network structures more in some policy fields than in others (e.g., John and Cole 2000). While Kriesi et al.'s (2006) study considers network structures of three policy fields in seven Western European countries, they do not establish the link to policy design. By integrating several policy fields and political systems into one study, one could thoroughly assess the impact of network structures on policy design by taking into consideration the interaction effect between the political system of a country and network structure.

Furthermore, it is clear that the policy comprehensiveness index, which was introduced in this study for operationalizing policy design, merits further discussion. For the purpose of this study, the composite indicators of the policy comprehensiveness index are deduced from existing policy instrument literature and hence reflect the state of current knowledge. Such a deductive approach seems most appropriate in the absence of other information or empirical evidence about valid composite indicators and also provides a basis from which future research may begin. Further research is needed to assess whether composite indicators are exhaustive, without being redundant, as well as how to weigh indicators. In this study, the values of composite indicators rely on estimations by the analyst, because the 'raw data' is qualitative. Through estimation, this qualitative raw data is translated into quantitative units. Future research is needed to address the issue of estimation biases more thoroughly and to propose alternative ways of evaluation, for example, through expert assessment.

A general obstacle in policy network research is that relational data from policy actors is not publicly accessible and, due to the potential sensitivity of political data, may not become public in the future either. In order to deal with these data constraints, researchers are obliged to conduct surveys, which are particularly time-consuming and resource-intensive in the context of political actors. The resulting response rates may not be sufficiently high for network research in every case, where a complete dataset should ideally exist, since the elimination of a few nodes may change the entire network structure. Moreover, network surveys—like any other surveys—cannot completely exclude measurement error even if there is established scientific knowledge of how to define policy network boundaries, and how to design surveys for policy network data. The present study handles this issue by carefully following the approved rules for boundary definition (see Sect. 3.2.1.3) and survey design (see Sect. 3.2.1.4). Nevertheless, data constraints limit the utility of the network approach for future policy research. To overcome these constraints, future research is necessary on alternative ways to gather policy network data, especially longitudinal data, as well as research on how to deal with measurement errors of network data and with missing data.

The lack of longitudinal data represents a drawback for the present study, particularly for the discussion of hypotheses 3, 4a, and 5. Hypothesis 3 concerns the question of whether divergences between network members may be overcome through collaboration. Hypothesis 4a formulates a similar idea by focusing on collaboration across coalitions. Without longitudinal data, the direction of causality between collaboration and belief cohesion cannot be assessed. The question here is whether initial divergences between coalitions may be reduced through collaboration across coalitions over time, or whether divergences are reinforced through collaboration. In order to assess such questions, future research requires detailed longitudinal data.

Hypothesis 5 accounts for the role of entrepreneurs and assumes that entrepreneurs are able to activate their network contacts, thereby improving the overall interconnectedness in their coalition. To assess the exact effect of entrepreneurship on network interactions, future research is necessary to test the causality over time between the interconnectedness of a dominant coalition and entrepreneurs' activity.

Finally, the main methodological limitation here is that this study is unable to estimate the explanatory power of the network variable to policy design. Due to the aforementioned restricted accessibility of policy network data, research has often been constrained to examining a single or a few policy networks in the past (Menahem 1998; Luzi et al. 2008). Even if the present study considers a total of four policy networks, the number of cases is not sufficient to quantify the explanatory power of the key causal variable. The present study deals with this issue by (a) constructing this research around a somewhat probabilistic research question according to which policy networks may affect policy design rather than determine it and (b) considering results no more than a plausibility probe for the hypothesized linkages. With the combination of similar and diverse cases, the present analysis seeks to increase its internal validity as much as possible with small-N research designs. Nevertheless, the analysis still includes too few cases in order to make real probabilistic inferences. To assess the explanatory power of the network variable thoroughly, future large-N research on a complete model of policy design would be necessary, where policy networks represent just one among many other explanatory variables. The next chapter discusses what variables may be of relevance in such an attempt.

5.2.2   Alternative Explanations to Comprehensive Policy Design

One main conclusion of this study is that network structure may contribute to explaining variations in policy design. Future research is necessary in order to establish whether the results of the present case study can be generalized to a wider population of cases (Blatter and Haverland 2012, p. 211). In order to determine whether policy networks generally impact policy design, a large-N study is needed, which applies statistical techniques to specify the explanatory strength of the network variable compared to rival explanations. In order to assess the explanatory power of the network variable, further research is necessary that integrates the structure of policy networks, together with other valid explanations for policy design into one comprehensive model.

Alternative political science explanations that may be considered in such a model include, among others, public budgets, political majorities, the reelection calculus, public discourses or awareness, media attention, path dependency, or shocks (such as accidents putting water quality at risk) (see Sect. 4.1 for an overview). Additionally, explanations advanced by other research disciplines may serve to explain policy design. Among others, sociologists' research about cultural differences of environmental attitudes, or post-materialist values, may matter

(Ingelhardt 1997). With a generally higher level of pro-environmental values, comprehensive policy designs may be more likely to occur as well. For example, Franzen and Vogl (2013) found that in 2010 the degree of environmental concern was highest in Switzerland, followed by Germany and France. Hence, environmental attitudes of a country's inhabitants may partly account for the level of comprehensiveness of policy designs.

A complete model may also consider economists' research about the relationship between economic development and environmental protection (e.g., research on the environmental Kuznets curve), about financing schemes of environmental policies (e.g., Toke and Lauber 2007), willingness to pay for environmental protection, cost-benefit analyses (Logar et al. 2014), or payments for ecosystem services (e.g., Engel et al. 2008). Micropollutants policies provide a strong example of the importance of financing schemes for environmental policy design. While it is possible to finance the technological upgrade of wastewater treatment plants in the Swiss case through a charge that Swiss households are obliged to pay, German law does not allow for such a micropollutants charge to date (interview with the German Federation of Municipal Organizations on 6.5.2014). Hence, one option for a comprehensive policy design on micropollutants is hampered in the German case, among others, by the lack of an appropriate financing scheme.

Moreover, research of hydrologists or ecotoxicologists concerning the condition of the environment (e.g., Hollender et al. 2008) or insights by engineers about technological aspects of environmental protection (e.g., Altmann et al. 2012) may be essential. Taking micropollutants policies as an example, a country's wastewater treatment infrastructures may impact policy makers' decisions about appropriate solutions. German and French interviewees reported that compared to Switzerland, large countries exhibit numerous small wastewater treatment plants, which renders an infrastructural solution more costly (interview with the French Ministry of Ecology on 18.10.2013, and the German Federal Environmental Agency on 17.4.2014).

Finally, geostrategic aspects may play a role in water protection policy, because the location along a river may provide an explanation for the amount of policy action and thus the quality of policy design. In fact, downstream countries, such as the Netherlands, like to point to pollution originating in upstream countries as an excuse for their own inaction (interview with the Dutch Union of Water Boards on 10.4.14). Upstream countries may be, following a rational choice logic, more susceptible to be laggards when it comes to water protection policy. In the present study, however, this is not the case for the upstream country Switzerland, which represents a leader in micropollutants policies. One may also assume downstream location to be a reason for more intensive policy action, because with higher problem pressure, policy engagement rises as well as. Again, empirical evidence from the present research does not support this geostrategic assumption, since the Netherlands displays, despite its downstream location, the least comprehensive policy design among the Rhine countries for the time frame of this study. Hence, geographical location seems to provide contradictory explanations that do not apply in the cases studied here.

In particular, for the Dutch case, a number of factors may have contributed to the lack of comprehensiveness of policy design, alongside policy networks—the key explanatory variable of this study: the restructuring of the Dutch administration and the related dissolution of several ministries and agencies in 2010; the standstill of numerous legislative projects during the governmental crisis in 2012; and an economic crisis in the Netherlands that led to a policy demanding tremendous budget cuts in water management. Due to these restrictions, new investments necessary for the elimination of pharmaceuticals in wastewater were difficult to maintain.

The listing of potential rival explanations to policy networks is not exhaustive, but serves its purpose in demonstrating that future research is necessary for assessing the explanatory power of the network variable compared to competing explanations.

5.3   Contributions of This Research to Science and Practice, and Future Research

Despite its empirical limitations mentioned above, the present work contributes to several streams of literature. Research on effective environmental policy has addressed environmental performance (Scruggs 2003; Jänicke and Weidner 1995; Jänicke 1997; Crepaz 1995; Daugbjerg and Sønderskov 2012; Jahn 1998; Duit 2005) or implementation performance (Bressers 2004; Howlett 2004), but has failed so far to evaluate *policy performance*. For researchers who aim at understanding which policy designs have the ability to improve the state of the environment, however, a policy performance perspective is more suitable than the focus on environmental or implementation performance. For example, research on environmental performance assumes that environmental improvements automatically result from an introduced environmental policy. However, between the introduction of a policy and a change in the state of the environment, many causal variables may intervene that do not allow for such a conclusion. Moreover, implementation research tends to assume that a well-implemented policy (Hill and Hupe 2002) also delivers the desired environmental improvements, which may not be the case if the policy is not well designed for problem solving. In order to complement the larger causal chain between policy design, implementation, and environmental performance, the present study focuses on the performance of policy designs since it has thus far operated as the missing link.

A policy performance approach incorporates aspects of implementation and environmental performance. More specifically, it involves an ex-ante estimation about the chances of a policy design to (a) cope with an underlying problem and to (b) be implemented. Hence, a policy performance perspective may be a useful instrument for future research in order to assess policy designs' prospective ability to improve the state of the environment.

To operationalize such a policy performance approach, this work introduces an index of policy comprehensiveness. From a methodological perspective, an index approach may constitute a promising and helpful tool compared to a common use of policy typologies. With the help of such an index, a ranking of several policy designs can be established to reflect performance and to allow analysts to compare policy variations across countries or time in a standardized way.

The policy comprehensiveness index is composed of six indicators, which are deduced from insights of existing policy instrument literature. This index should be considered a first attempt and a starting point in research, which provides a basis from which future research may start to improve the current approach. For example, it may be helpful for scientists to build on the insights summarized here of previous thinkers on policy design. Moreover, the proposed index may serve to test the validity of composite indicators and further develop, weigh, or reject existing indicators, as well as to suggest new indicators.

A policy performance approach may also serve to further expand the literature invested in why certain policy designs are chosen over others. While there exists a large body of literature on potential explanations for policy choices, there remains a lack of research on how to achieve optimal policy design from a problem-solving perspective. The policy performance approach proposed here may serve to further expand the question of 'how can we explain policy choices' to the question of 'how do we achieve optimal policy designs' in future research. To answer the latter question, it is necessary to quantify a policy design's problem-solving ability, such as attempted by the policy comprehensiveness index proposed in this study.

To explain policy design, the present work seeks to contribute to the literature claiming that the network approach should be taken seriously (e.g., O'Toole 1997; Robinson 2006). While there exists a large body of literature analyzing how and why policy actors interact with one another (e.g., Kriesi and Jegen 2001; Gerber et al. 2013; Ingold and Fischer 2014), there continues to be a lack of research about the consequences resulting from network structure (Hennig et al. 2012, p. 184). In order to close this research gap, the present work employs policy networks as independent variables. Insights from policy process research are borrowed (see Sect. 3.1.3) for identifying those network properties that matter in policy networks and for systematically linking those network properties to policy design. Through the formulated hypotheses about the effects of policy networks on policy design, the present study contributes to further establishing policy network theory.

From a theoretical perspective, the present work may also be combined with research about policy change (Sabatier and Jenkins-Smith 1993). For this purpose, the policy comprehensiveness index may be employed to study a change from 'less comprehensive' to 'more comprehensive' policy designs over time.

Policy change scholars have thus far employed the network variable to assess which network properties are associated with policy change (e.g., Fischer 2013; Howlett 2002). The present research, on the contrary, makes use of the relational aspect of policymaking in order to determine how to achieve comprehensive policy designs. As it turns out, these two research goals are complementary because certain types of network structures promote policy change best described as a 'smallest common denominator' type of compromise, which lacks comprehensiveness; other network structures, meanwhile, may be conducive to the adoption of comprehensive policy designs. Understanding both types of situations may aid in understanding under which conditions compromises can be attained that are also optimal from a problem-solving perspective.

This study performs a preliminary test of the formulated hypotheses linking networks to policy design. The goal of such a discussion is to assess whether the adopted research path is worth further scholarly attention. Findings summarized above indicate that, in fact, network structures may contribute to explaining comprehensive policy design. As networks act as just one among other valid explanations to policy design, future research should move toward more complete explanatory models of policy designs, as discussed in Sect. 5.2.2.

Contribution of this research to practice

This research cannot solely be applied to science, but also to practice. Policy actors who are involved or interested in micropollutants policies find information about the Rhine countries' policymaking processes and the resulting policies for the reduction of aquatic micropollutants in Sect. 2.3 and about policy network structures and embedded water policy actors in Sect. 3.3.

Thirdly, this study may be of interest to actors involved in policymaking processes, particularly the question of how to achieve comprehensive policy designs in multi-actor settings. For policy practitioners' work, the introduced policy comprehensiveness may be of relevance in order to evaluate policy proposals against their problem-solving ability. The purpose of the index would then be to assess how a comprehensive policy could be designed in order to achieve an improvement in the state of the environment. Moreover, the findings about network structures may assist those policy practitioners interested in comprehensive policy designs to develop a promising strategy for policymaking processes. For example, one finding of this study is that belief cohesion is pivotal for achieving a comprehensive policy design. Thus, in situations with a sufficient level of belief cohesion, policy actors could adopt a strategy aiming at an ambitious policy design from a problem-solving perspective. If belief cohesion is low, a comprehensive policy design may still be possible. Decisive then is to strengthen the role of the broker, invest resources for forming a coalition—ideally a dominant one—and encourage entrepreneurs who favor a comprehensive policy design to activate their network contacts. Hence, a network approach to policy design may not only serve science, but also practice, in its search for optimal policy results.

References

Altmann, D., Schaar, H., Bartel, C., Schorkopf, D. L., Miller, I., Kreuzinger, N., et al. (2012). Impact of ozonation on ecotoxicity and endocrine activity of tertiary treated wastewater effluent. *Water Research, 46*(11), 3693–3702.

Blatter, J., & Haverland, M. (2012). *Designing case studies. Explanatory approaches in small-N research* (Research methods series). Basingstoke: Palgrave Macmillan.

Bressers, H. (2004). Implementing sustainable development: How to know what works, where, when and how. In W. Lafferty (Ed.), *Governance for sustainable development: The challenge of adapting form to function.* Cheltenham: Edward Elgar.

Crepaz, M. (1995). Explaining national variations of air pollution levels: Political institutions and their impact on environmental policy-making. *Environmental Politics, 4*(3), 391–414.

Crona, B., & Bodin, Ö. (2006). What you know is who you know? Communication patterns among resource users as a prerequisite for co-management. *Ecology and Society, 11*(2).

Daugbjerg, C., & Sønderskov, K. M. (2012). Environmental policy performance revisited: Designing effective policies for green markets. *Political Studies, 60*(2), 399–418.

Duit, A. (2005). *Understanding environmental performance of states: An institution-centered approach and some difficulties.* Paper presented at the YOG Working Paper Series 7.

Engel, S., Pagiola, S., & Wunder, S. (2008). Designing payments for environmental services in theory and practice: An overview of the issues. *Ecological Economics, 65*(4), 663–674.

Fischer, M. (2012). Entscheidungsstrukturen in der Schweizer Politik zu Beginn des 21. Jahrhunderts. Glarus, Chur: Rüegger.

Fischer, M. (2013). *Policy network structures, institutional context, and policy change.* Paper presented at the COMPASSS Working Paper 73.

Fischer, M. (2014). Coalition structures and policy change in a consensus democracy. *Policy Studies Journal, 42*(3), 344–366.

Franzen, A., & Vogl, D. (2013). Two decades of measuring environmental attitudes: A comparative analysis of 33 countries. *Global Environmental Change, 23*(5), 1001–1008.

Gerber, E., Henry, A. D., & Lubell, M. (2013). Political homophily and collaboration in regional planning networks. *American Journal of Political Science, 57*(3), 598–610.

Haverland, M. (2000). National adaptation to European integration: The importance of institutional veto points. *Journal of Public Policy, 20*(01), 83–103.

Hennig, M., Brandes, U., Pfeffer, J., & Mergel, I. (2012). *Studying Social Networks. A Guide to Empirical Research.* Frankfurt am Main: Campus Verlag.

Henry, A. D., Lubell, M., & McCoy, M. (2010). Belief systems and social capital as drivers of policy network structure: The case of California regional planning. *Journal of Public Administration Research and Theory, 21*(3), 419–444.

Hill, M., & Hupe, P. (2002). *Implementing public policy. Governance in theory and in practice.* London: Sage.

Hollender, J., Singer, H., & McArdell, C. (2008). Polar organic micropollutants in the water cycle. In P. Hlavinek, O. Bonacci, J. Marsalek, & I. Mahrikova (Eds.), *Dangerous pollutants (xenobiotics) in urban water cycle* (pp. 103–116). Dordrecht: Springer.

Howlett, M. (2002). Do networks matter? Linking policy network structure to policy outcomes: Evidence from four Canadian policy sectors 1990–2000. *Canadian Journal of Political Science, 35*(2), 235–267.

Howlett, M. (2004). Beyond good and evil in policy implementation: Instrument mixes, implementation styles, and second generation theories of policy instrument choice. *Policy and Society, 23*(2), 1–17.

Ingelhardt, R. (1997). *Modernization and postmodernization: Cultural, economic, and political change in 43 societies.* Princeton: Princeton University Press.

Ingold, K., & Fischer, M. (2014). Drivers of collaboration to mitigate climate change: An illustration of Swiss climate policy over 15 years. *Global Environmental Change, 24,* 88–98.

Jahn, D. (1998). Environmental performance and policy regimes: Explaining variations in 18 OECD-countries. *Policy Sciences, 31,* 107–111.

Jänicke, M. (1997). The political system's capacity for environmental policy. In M. Jänicke, H. Jörgens, & H. Weidner (Eds.), *National environmental policies* (pp. 1–24). Berlin: Springer.

Jänicke, M., & Weidner, H. (Eds.). (1995). *Successful environmental policy. A critical evaluation of 24 cases.* Berlin: Edition sigma.

John, P., & Cole, A. (2000). When Do Institutions, Policy Sectors, and Cities Matter? Comparing Networks of Local Policy Makers in Britain and France. *Comparative Political Studies, 33*(2), 248–268.

Koontz, T., & Thomas, C. (2006). What do we know and need to know about the environmental outcomes of collaborative management? *Public Administration Review, 66,* 111–121.

Kriesi, H., & Jegen, M. (2001). The Swiss energy policy elite: The actor constellation of a policy domain in transition. *European Journal of Political Research, 39,* 251–287.

Kriesi, H., Adam, S. & Jochum, M. (2006). Comparative Analysis of Policy Networks in Western Europe. *Journal of European Public Policy, 13*(3), 341–361.

Logar, I., Brouwer, R., Maurer, M., & Ort, C. (2014). Cost-benefit analysis of the Swiss national policy on reducing micropollutants in treated wastewater. *Environmental Science and Technology, 48*(21), 12500–12508.

Luzi, S., Abdelmoghny Hamouda, M., Sigrist, F., & Tauchnitz, E. (2008). Water policy networks in Egypt and Ethiopia. *The Journal of Environment & Development, 17*(3), 238–268.

Menahem, G. (1998). Policy paradigms, policy networks and water policy in Israel. *Journal of Public Policy, 18*(3), 283–310.

Newman, L., & Dale, A. (2005). Network structure, diversity, and proactive resilience building: A respone to Tompkins and Adger. *Ecology and Society, 10*(1).

O'Toole, L. (1997). Treating networks seriously: Practical and research-based agendas in public administration. *Public Administration Review, 57*(1), 45–52.

Robinson, S. (2006). A decade of treating networks seriously. *Policy Studies Journal, 34*(4), 589–598.

Sabatier, P., & Jenkins-Smith, H. (1993). *Policy Change and Learning: An Advocacy Coalition Approach.* Boulder: Westview Press.

Scruggs, L. (2003). *Sustaining abundance: Environmental performance in industrial democracies.* Cambridge: Cambridge University Press.

Toke, D., & Lauber, V. (2007). Anglo-Saxon and German approaches to neoliberalism and environmental policy: The case of financing renewable energy. *Geoforum, 38*(4), 677–687.

Zafonte, M., & Sabatier, P. (1998). Shared beliefs and imposed interdependencies as determinants of ally networks in overlapping subsystems. *Journal of Theoretical Politics, 10*(4), 473–505.

Annexes

Annex 1
List of Preliminary and Survey Interviews

No	Country	Date	Organization	Location
1	CH	20.12.11	Federal Office for the Environment, Department for Water (BAFU-W) (preliminary)	Berne
2	CH	28.2.13	Federal Office for the Environment, Department for Water (BAFU-W) (preliminary)	Koblenz
3	CH	3.4.13	Swiss Water Association (VSA) (preliminary)	Berne
4	CH	2.5.13	Scienceindustries (INDUS)	Zurich
5	CH	2.5.14	Swiss Water Association (VSA)	Berne
6	CH	16.5.13	Swiss Employers' Association (ECON)	Berne
7	CH	22.5.13	Cercl'eau (CERCL)	St Gallen
8	CH	23.5.13	National Council's Committee on the Environment, Spatial Planning, and Energy (UREKN)	Zurich
9	CH	24.5.13	Swiss People's Party (SVP)	Berne
10	CH	24.5.13	Council of State's Committee on the Environment, Spatial Planning, and Energy (UREKS)	Berne
11	CH	28.5.13	Swiss Gas and Water Industry Association (SVGW)	Zurich
12	CH	31.5.13	Swiss Green Party (GPS)	Berne
13	CH	3.6.13	Federal Office for the Environment, Department of Air Protection and Chemicals (BAFU-CHEM)	Phone
14	CH	17.7.13	Federal Office for the Environment, Department for Water (BAFU-W)	Berne
15	G	3.4.12	Federal Ministry for the Environment (BMU) (preliminary)	Phone

(continued)

© Springer International Publishing AG 2017
F. Metz, *From Network Structure to Policy Design in Water Protection*,
Springer Water, DOI 10.1007/978-3-319-55693-2

(continued)

No	Country	Date	Organization	Location
16	G	27.3.12	State of North Rhine-Westphalia (NRW) (preliminary)	Düsseldorf
17	G	4.4.12	German Federal Institute of Hydrology (BfG) (preliminary)	Phone
18	G	4.4.12	State of Baden-Württemberg (BAWÜ) (preliminary)	Phone
19	G	17.4.12	State of Rhineland-Palatinate (RLP) (preliminary)	Phone
20	G	26.4.12	State of Bavaria (BAY) (preliminary)	Phone
21	G	18.6.12	State of Saarland (preliminary)	Saarbrücken
22	G	18.12.13	Federal Ministry for the Environment (BMU) (preliminary)	Phone
23	G	7.3.14, 1.4.15	Federal Ministry for the Environment (BMU)	Zurich
24	G	13.3.14	State of Rhineland-Palatinate (RLP)	Mainz
25	G	13.3.14	State of Hesse (HES)	Wiesbaden
26	G	17.3.14	State of North Rhine-Westphalia (NRW)	Düsseldorf
27	G	17.4.14	Federal Environmental Agency (UBA)	Dessau
28	G	22.4.14	Federal Ministry of Economics (BMWI)	Berlin
29	G	29.4.14	German Chemical Industry Association (VCI)	Frankfurt/M
30	G	6.5.14	Federation of Municipal Organizations (BV)	Munich
31	F	18.10.13	Ministry of Ecology, Department of Water and Biodiversity (DEB) (preliminary)	Paris
32	F	7.11.13	General Directorate for Risk Prevention (DGPR)	Paris
33	F	25.11.13	Alsace Nature (ALSACENATURE)	Strasbourg
34	F	10.12.13	France Nature Environment (FNE)	Phone
35	F	11.12.13	National Agency for Water and Aquatic Environments (ONEMA)	Phone
36	F	16.12.13	National Reference Laboratory for Water Monitoring (AQUAREF/INERIS)	Phone
37	F	8.1.14	National Innovation Centre for Sustainable Development and Environment in Small Companies (CNIDEP)	Phone
38	F	9.1.14	French Professional Federation of Water Companies (FP2E)	Phone
39	F	23.1.14	Assembly of the French Departments (ADF)	Phone
40	F	20.2.14	Water Basin Delegations/Prefects of Water Basin (DELEGBASSIN)	Metz
41	F	20.2.14	Regional Authority for the Environment, Spatial Planning, and Housing (DREAL)	Metz
42	F	20.2.14	Agence de l'Eau Rhin-Meuse (AGENCE)	Metz
43	NL	11.3.14	Waterschap Vechtstromen (preliminary)	Almelo

(continued)

(continued)

No	Country	Date	Organization	Location
44	NL	10.4.14	Union of Water Boards (UvW)	The Hague
45	NL	10.4.14	Ministry for Infrastructure and Environment (IenM)	The Hague
46	NL	11.4.14	Watercycle Research Institute (KWR)	Nieuwegein
47	NL	14.4.14	Health Council of the Netherlands (GEZONDHEIDSRAAD)	The Hague
48	NL	15.4.14	Department for Water Management (RWS)	Lelystad
49	NL	23.4.14	Association for River Waterworks (RIWA)	Nieuwegein
50	NL	28.4.14	Association for Innovative Medicines in the Netherlands (NEFARMA)	The Hague
51	NL	28.4.14	Association of Dutch Drinking Water Companies (VEWIN)	The Hague
52	NL	1.5.14	National Institute for Public Health and the Environment (RIVM)	Bilthoven
53	NL	8.5.14	Foundation for Applied Water Research (STOWA)	Phone

Annex 2
Identification of Swiss Policy Actors

	List of actors per category	Reputational approach	Positional approach	Decisional approach
	Federal state			
1	Federal Department of the Environment, Transport, Energy, and Communications (UVEK)		X	X
2	Federal Office for the Environment, Department for Water (BAFU-W)	X	X	X
3	Federal Office for the Environment, Department of Air Protection and Chemicals (BAFU-CHEM)		X	
4	Federal Office for Spatial Planning (ARE)		X	
5	Federal Office for Health (BAG)		X	X
6	Federal Office for Justice (BJ)		X	X
7	Swiss Finance Administration (EFV)		X	X
8	Swiss Office of Personnel (EPA)		X	X
9	Swissmedic (SMEDIC)		X	X

(continued)

(continued)

	List of actors per category	Reputational approach	Positional approach	Decisional approach
10	Federal Office for Agriculture (BLW)		X	
11	State Secretariat for Economic Affairs (SECO)	X	X	X
12	Federal Office for Energy (BFE)		X	
13	Federal Council (BR)		X	X
	Cantonal conferences and associations			
14	Conference of Cantonal Directors of Construction, Planning, and Environmental Protection (BPUK)	X		X
15	Conference of Directors of Construction, Planning, and Environmental Protection (KVU)	X		X
16	Association of Cantonal Chemists of Switzerland (VKCS)	X		
17	Cercl'eau (CERCL)	X		X
18	Lab'Eaux (LABEAUX)	X		
	Parliament			
19	National Council's Committee on the Environment, Spatial Planning, and Energy (UREKN)		X	X
20	Council of State's Committee on the Environment, Spatial Planning, and Energy (UREKS)		X	
	Political parties			
21	Free Democratic Party. The Liberals (FDP)	X	X	X
22	Swiss Green Party (GPS)	X	X	X
23	Swiss Social Democratic Party (SP)	X	X	X
24	Swiss People's Party (SVP)	X	X	X
25	Christian Democratic People's Party (CVP)		X	
26	Green Liberal Party of Switzerland (GLP)		X	
27	Civil Democratic Party (BDP)		X	
	Environmental associations			
28	Pro Natura (PRONA)	X	X	X
29	Western Swiss Association for Water and Air Protection (ARPEA)	X		X
30	Swiss Fishery Association (FISCH)	X		
31	World Wide Fund For Nature, Switzerland (WWF)	X	X	X

(continued)

(continued)

	List of actors per category	Reputational approach	Positional approach	Decisional approach
	Swiss communal and water associations			
32	Sewage Treatment Plants in Large Cities Initiative (ERFA)	X		X
33	Western Swiss Group of Sewage Treatment Plants Operators (GRESE)	X		X
34	Swiss Water Association (VSA)	X	X	X
35	Communal Infrastructure (KI)			X
36	Swiss Cities Association (SSV)		X	X
37	Swiss Gas and Water Industry Association (SVGW)			X
38	Swiss Municipalities Association (SGV)		X	
	Economic associations			
39	Economiesuisse (ECON)	X	X	X
40	Scienceindustries (INDUS)	X		X
41	Swiss Cosmetics and Detergent Association (SKW)	X		
42	Swiss Mechanical and Electrical Engineering Industry Association (SMEM)	X		X
43	Basel Chamber of Commerce (HKBB)			X
44	Swiss Employers' Association (SAV)		X	
45	Swiss Trade Association (SGEWV)		X	
46	Centre Patronal (CP)			X
	Consumer and workers' associations			
47	Consumer Forum (KF)	X		X
48	Homeowner Association Switzerland (HEV)			X
49	Swiss Trade Union (SGB)		X	X
50	Consumer Protection Foundation (Alliance SKA-FRC-ACSI)			X
51	Swiss Farmers Association (SBV)		X	
	Science, laboratories, consultancies			
52	Swiss Federal Institute of Aquatic Science and Technology (Eawag)	X	X	X
53	Swiss Federal Institute for Technology Lausanne (EPFL)	X		X
54	Université de Lausanne (UNIL)	X		
55	Universität Basel (UBAS)			X
56	Ecotox Centre (OEKOTOX)			X

(continued)

(continued)

	List of actors per category	Reputational approach	Positional approach	Decisional approach
57	University of Applied Sciences of North-West Switzerland (FHNW)			X
58	Holinger Engineering (HOLINGER)			X
59	Hunziker-Betatech (HUNZIKER)			X
60	RWB Engineering (RWB)			X
61	BMG Engineering (BMG)			X
62	Swiss Association for Agricultural Development and Rural Areas (AGRIDEA)			X

Annex 3
Identification of German Policy Actors

	List of actors per category	Reputational approach	Positional approach	Decisional approach
	Federal state			
1	Federal Ministry for the Environment (BMU)	x	x	x
2	Federal Ministry of Economics (BMWI)	x	x	x
3	Federal Ministry of Agriculture (BML)	x	x	x
4	Federal Ministry of Health (BMG)	x	x	x
5	Federal Ministry of Justice (BMJ)	x	x	x
6	Federal Ministry of Finance (BMF)	x	x	x
7	Federal Environmental Agency (UBA)	x	x	x
	Länder			
8	Common Working Group on Water of the Federal Government and State Governments (LAWA)	x		x
9	Common Working Group on Environmental Quality Norms of the Federal Government and State Governments (BLAK-UQN)	x		x
10	State of North Rhine-Westphalia (NRW)	x	x	x
11	Rhineland-Palatinate (RLP)	x	x	x

(continued)

(continued)

	List of actors per category	Reputational approach	Positional approach	Decisional approach
12	State of Baden-Württemberg (BaWü)	x	x	x
13	State of Bavaria (BAY)	x	x	x
14	State of Hesse (HES)	x	x	x
	Environmental associations			
15	Federal Association of Citizens' Initiative on Environmental Protection (BBU)	x		x
16	Friends of the Earth Germany (BUND)	x		x
17	German League for Nature, Animal Protection, and Environment (DNR)			x
18	German Fishing Association (DFV)			x
19	Green League (GrLi)			x
20	German Association for Water Protection (VDG)			x
21	Nature Conservation Union Germany (NABU)			x
	Communal associations			
22	Federation of Municipal Organizations (BV)	x	x	x
23	Association of Municipal Companies (VKU)			x
	Water associations			
24	Federal Association of Energy and Water Industry (BDEW)	x		x
25	German Technical and Scientific Association for Gas and Water (DVGW)			x
26	German Association for Water, Wastewater, and Waste (DWA)	x		x
27	Alliance of Public Water Management (AÖW)			x
28	German Hydropower Plants Association (BDW)			x
	Economic associations			
29	German Chemical Industry Association (VCI)	x		x
30	Federation of German Industries (BDI)	x		x
31	Association of German Chambers of Commerce and Industry (DIHK)	x		x
32	German Metal Industry Association (WVM)	x		x

(continued)

(continued)

	List of actors per category	Reputational approach	Positional approach	Decisional approach
33	German Engineering Federation (VDMA)			x
34	German Automotive Industry Association (VDA)			x
35	Federal Association of Mineral Raw Materials (MIRO)			x
36	Association of German Petroleum Industry (MWV)			x
37	German Association of Construction Industry (BauInd)			x
38	Federal Association of Medicine Manufacturers (BAH)			x
39	Technical Association for Power and Heat Generation (VGB)			x
40	German Electrical and Electronic Manufacturers' Association (ZVEI)			x
41	Agrochemical Association (IVA)	x		
	Farmers' associations			
42	German Farmers' Association (DBV)			x
	Consumer and workers' associations			
43	German Confederation of Trade Unions (DGB)			x
44	Mining, Chemical, and Energy Union (IGBCE)			x
45	Union of Public Services (GÖD)			x
46	Federation of German Consumer Organizations (VZBV)		x	
	Science, laboratories, consultancies			
47	Water Chemistry Society			x
48	Association of German Engineers (VDI)			x
49	Association of Engineers for Water Management, Waste Management, and Environmental Construction (BWK)			x
50	German Federal Institute of Hydrology (BfG)	x	x	

Annex 4
Identification of French Policy Actors

	List of actors per category	Reputational approach	Positional approach	Decisional approach
	National public authorities			
1	Department of Water and Biodiversity/Ministry of Ecology (DEB)	x		x
2	General Directorate for Risk Prevention (DGPR)	x		x
3	National Agency for Water and Aquatic Environments (ONEMA)	x		x
4	Ministry of Health (MINISTSANTE)	x		x
5	Ministry of Agriculture (MINISTAGRI)	x		x
6	National Water Committee (ASN)	x		x
7	National Water Committee (CNE)	x	x	
	Public research institutes			
8	National Reference Laboratory for Water Monitoring (AQUAREF)	x		x
9	National Competence Centre for Industrial Safety and Environmental Protection (INERIS)	x	x	x
10	Geology and Mining Research Institute (BRGM)	x		x
11	French Research Institute for Exploitation of the Sea (IFREMER)	x		x
12	National Research Institute of Science and Technology for Environment and Agriculture (IRSTEA)	x	x	x
13	French National Metrology and Testing Laboratory (LNE)			
14	National Center for Scientific Research (CNRS)			x
15	French Agency for Food, Environment and Occupational Health & Safety (ANSES)	x	x	x
16	French Environment and Energy Agency (ADEME)	x	x	
	Decentralized services			
17	Prefect of Region			
18	Prefect of Department		x	x
19	Prefect of Water Basin (DELEGBASSIN)		x	

(continued)

(continued)

	List of actors per category	Reputational approach	Positional approach	Decisional approach
20	Water Police			x
21	Inspection of Classified Industrial Facilities			x
22	Regional Authority for the Environment, Spatial Planning, and Housing (DREAL)	x	x	x
23	Local Territorial Authority (DDT)		x	
	Regional and local authorities			
24	Association of French Regions (ARF)		x	x
25	Assembly of the French Departments (ADF)		x	
26	Association of French Mayors (AMF)		x	x
	Water catchment authorities			
27	Water Agency (AGENCE)	x	x	x
28	Basin Committee		x	
29	Local Water Commissions (CLE)		x	
	Environmental associations			
30	Robin des Bois (ROBINDESBOIS)	x		x
31	France Nature Environment (FNE)	x	x	x
32	World Wide Fund For Nature, France (WWF)	x	x	
33	Friends of the Earth France	x	x	
34	Alsace Nature (ALSACENATURE)	x		
	Water associations			
35	French Professional Federation of Water Companies (FP2E)	x		x
36	National Federation of Public Water Services (FNCCR)	x		x
	Economic associations			
37	National Federation of Industrial Water Users (FENARIVE)	x		x
38	French Business Confederation (MEDEF)			x
39	Chemical Industries Union (UIC)	x	x	
40	National Innovation Centre for Sustainable Development and Environment in Small Companies (CNIDEP)	x		
41	Assembly of French Chambers of Commerce and Industry (CCI)	x	x	
42	EDF Energy (EDF)	x	x	

(continued)

(continued)

	List of actors per category	Reputational approach	Positional approach	Decisional approach
43	Association of French Paper Industries (COPACEL)	x	x	
44	Union française des industries pétrolières (UFIP)	x	x	
45	Federation of Mechanical Engineering Industries (FIM)	x	x	
	Agricultural associations			
46	Farmers' Association (CHAMBREAGRI)		x	
	Consumer associations			
47	Federal Union of Consumers (UFC)	x	x	x
48	Consumers, Housing, and Well-being Association (CLCV)	x	x	

Annex 5
Identification of Dutch Policy Actors

	List of actors per category	Decisional approach	Positional approach	Reputational approach
	National public authorities			
1	Ministry of Infrastructure and Environment (IenM)	X	X	X
2	Department for Water Management (RWS)	X	X	X
3	Ministry of Health, Well-being, and Sport (VWS)	X	X	
4	National Institute for Public Health and the Environment (RIVM)	X	X	
5	Ministry of Economy (EZ)	X	X	
6	Netherlands Enterprise Agency (RVO)		X	
7	Former Ministry of Agriculture, Nature, and Food Quality (LNV)		X	
8	Former Ministry of Transport and Water Management (V&W)		X	
9	Former Ministry of Housing, Spatial Planning, and the Environment (VROM)	X		

(continued)

(continued)

	List of actors per category	Decisional approach	Positional approach	Reputational approach
10	Health Council of the Netherlands (GEZONDHEIDSRAAD)	X	X	
11	Medicines Evaluation Board (CBG)	X	X	
12	National Executive Talk Water		X	
13	Advisory Committee on Water		X	
14	Former National Institute for Coast and Sea (RIKZ)	X		
	Provinces, municipalities, and water boards			
15	Association of Dutch Provinces (IPO)		X	X
16	Union of Water Boards (UvW)	X	X	X
17	Association of Dutch Municipalities (VNG)		X	X
	Parliament and political parties			
18	Dutch Senate		X	
19	Dutch House of Representatives (TWEEDEKAMER)	X	X	
20	Parliamentary Committee for Infrastructures and Environment (VCIenM)	X	X	
21	Parliamentary Committee for Health, Well-being, and Sports		X	
22	Parliamentary Committee for Economic Affairs		X	
23	Christian Union Party (CHRISTENUNIE)	X		
24	Green Left Party (GROENLINKS)			X
25	Labor Party (PvdA)			X
26	Democrats 66 (D66)			X
	Agricultural associations			
27	Dutch Federation of Agriculture and Horticulture (LTO)		X	X
	Economic associations			
28	Confederation of Netherlands Industry and Employers (VNO-NCW)		X	
29	Association of the Dutch Chemical Industry (VNCI)		X	
30	Industrial Water User Association (VEMW)		X	

(continued)

(continued)

	List of actors per category	Decisional approach	Positional approach	Reputational approach
	Pharmaceutical and health sector			
31	Association of the Dutch Generic Medicines Industry (BOGIN)	X	X	
32	Association for Innovative Medicines in the Netherlands (NEFARMA)	X	X	
33	Royal Dutch Society for the Advancement of Pharmacy (KNMP)	X		
34	Foundation for Pharmaceutical Statistics (SFK)	X	X	
35	Dutch Working Party on Antibiotic Policy (SWAB)		X	
36	Reinier de Graaf Hospital (RDGG)	X		
	Water associations			
37	Association of Dutch Drinking Water Companies (VEWIN)	X	X	X
38	Vitens Water Supply Company (VITENS)	X		X
39	Association for River Waterworks (RIWA)	X	X	X
40	RIONED Foundation		X	
	Environmental associations			
41	Cooperative Fishery Organization		X	
42	Foundation House of the Earth (STICHTINGHA)			X
43	Society for the Preservation of Nature in the Netherlands		X	
44	World Wide Fund For Nature, the Netherlands (WWF)		X	
	Consumer associations			
45	Consumer Association		X	
	Research and consultancy			
46	Watercycle Research Institute (KWR)	X	X	
47	Deltares/Former National Institute for Integrated Freshwater and Wastewater Management (RIZA)	X	X	X
48	Foundation for Applied Water Research (STOWA)	X	X	X
49	Grontmij Consulting			X

Annex 6
Swiss Questionnaire

Laufende Änderung des Gewässerschutzgesetzes und der Gewässerschutzverordnung bezüglich des Eintrags von Spurenstoffen in die Gewässer

Akteursbefragung 2013

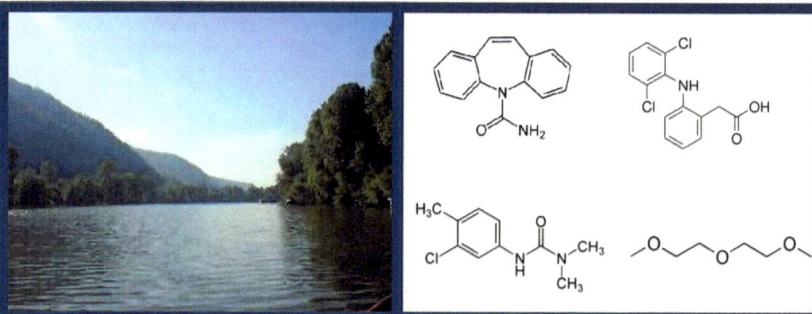

Erläuterungen zum Fragebogen

Wie im Begleitbrief erwähnt, ist dieser Fragebogen Teil eines Forschungsprojektes, das von der Universität Bern in Zusammenarbeit mit dem Wasserforschungsinstitut des ETH-Bereichs (Eawag) durchgeführt wird. Ziel ist es, den **seit 2007 laufenden Entscheidungsprozess zur Änderung des Gewässerschutzgesetzes (GSchG) und der Gewässerschutzverordnung (GSchV; SR 814.201) bezüglich des Eintrags von Spurenstoffen in die Gewässer** zu untersuchen.

Der Fragebogen setzt sich aus insgesamt 13 Fragen zusammen und gliedert sich in Teil A bis D:

> Teil A: Beteiligung Ihrer Organisation am genannten Entscheidungsprozess
> Teil B: Zusammenarbeit Ihrer Organisation mit anderen Akteuren im Entscheidungsprozess
> Teil C: Positionen Ihrer Organisation zur Regulierung von Spurenstoffen
> Teil D: Kompetenzen Ihrer Organisation im Bereich Wasser

Bitte beantworten Sie die folgenden Fragen aus der Perspektive Ihrer Organisation.

Das Forschungsprojekt wird im Rahmen einer Doktorarbeit realisiert und vom Schweizerischen Nationalfonds gefördert. Da Ihre Organisation eine zentrale Rolle im Entscheidungsprozess einnimmt, ist Ihre Teilnahme an der Befragung äusserst wesentlich für das Gelingen der Forschungsarbeit. Wir danken Ihnen daher im Vorfeld für das Beantworten des Fragebogens und Ihre wertvolle Unterstützung. Das Ausfüllen sollte **nicht mehr als 30 Minuten** in Anspruch nehmen.

Bitte senden Sie den ausgefüllten Fragebogen mit beiliegendem frankierten Antwortcouvert **bis zum 15. Mai 2013** zurück. Sobald alle Daten vorliegen, informieren wir Sie gerne über die Forschungsergebnisse.

Die von Ihnen angegebenen Informationen werden ausschliesslich zu Forschungszwecken genutzt, vertraulich behandelt und nicht an Dritte weitergegeben.

Projektleitung:

Prof. Dr. Karin Ingold
Institut für Politikwissenschaft Universität Bern, Eawag

u^b

Projektpartner und Sponsoren:

UNIVERSITÄT BERN

eawag aquatic research ooo

FN SNF
SCHWEIZERISCHER NATIONALFONDS
ZUR FÖRDERUNG DER WISSENSCHAFTLICHEN FORSCHUNG

Bei Fragen können Sie sich per Mail oder Telefon direkt an die zuständige **Kontaktperson** wenden:

Florence Metz (Doktorandin)
Universität Bern, Institut für Politikwissenschaft, Lerchenweg 36, CH-3000 Bern 9
florence.metz@ipw.unibe.ch
Tel. 031 631 48 22

Name der Person, die den Fragebogen ausfüllt: ....

Name Ihrer Dienststelle und Ihrer Organisation: ....

    Strasse, Hausnr.: .... PLZ, Ort: ....

    Telefonnr.: .... Email: ....

Möchte Ihre Organisation über die Resultate der Studie informiert werden?   Ja ❑   Nein ❑

Teil A:
Entscheidungsprozess zur Änderung des GSchG / der GSchV

1. An welchen Etappen des Entscheidungsprozesses zur Änderung des GSchG / der GSchV bezüglich des Eintrags von Spurenstoffen in die Gewässer (2007 – heute, inklusive Vorarbeiten vor 2007) war Ihre Organisation beteiligt?

   Bitte kreuzen Sie alle Etappen an, an denen Ihre Organisation beteiligt war oder voraussichtlich beteiligt sein wird.

   Unter Beteiligung wird verstanden: einen Beitrag durch aktive Mitarbeit leisten, an Arbeitsgruppen / Workshops / formellen und informellen Konsultationen teilnehmen.

Phase Datum	Etappe	Beteiligt
Änderung der Gewässerschutzverordnung (GSchV)		
Auslöser Vor 2007	Forschung (z.B. NFP 50, Projekt Fischnetz), Amtsentscheid	❑
	Projekt „**Strategie MicroPoll**" zur Optimierung der Abwasserreinigung (2007 – 2012)	
Konzeptphase 2007 – 2009	**Bericht BAFU** Umweltwissen 17 / 09 „Mikroverunreinigungen in den Gewässern. Bewertung und Reduktion der Schadstoffbelastung aus der Siedlungsentwässerung"	❑
	Studie im Auftrag des BAFU „Mikroverunreinigungen. Beurteilungskonzept für organische Spurenstoffe aus kommunalem Abwasser"	
Ausarbeitung Nov. 2009	**Entwurf** zur Änderung der GSchV	❑
Konsultation Nov. 2009 – Juli 2010	**Anhörung** „Eintrag von organischen Spurenstoffen in die Gewässer – Änderung der Gewässerschutzverordnung" (27.11.2009 – 30.4.2010)	❑
Änderung des Gewässerschutzgesetzes (GschG) und Überarbeitung der GSchV		
Auslöser Aug. 2010 – März 2011	**Motion UREK-SR** (10.3635) „Verursachergerechte Finanzierung der Elimination von Spurenstoffen im Abwasser", Beratung und Zustimmung durch beide Räte	❑
Konzeptphase April 2011 – April 2012	**Studien** zu Kosten und Finanzierung der Elimination von Spurenstoffen im Abwasser (BG Ingenieure und Berater AG, Ernst Basler + Partner, Ecoplan)	❑
	Bericht BAFU Umweltwissen 1214 „Mikroverunreinigungen aus kommunalem Abwasser. Verfahren zur weitergehenden Elimination auf Kläranlagen"	
Ausarbeitung 2011 – 2012	**Vorentwurf** zur Änderung des GSchG	❑
Konsultation Frühjahr 2012 – Frühjahr 2013	**Ämterkonsultation** zum Vorentwurf des GSchG, **Beratung im BR** über Gesetzesvorlage GSchG (25.04.2012)	❑
	Vernehmlassung „Änderung des Gewässerschutzgesetzes. Verursachergerechte Finanzierung der Elimination von Spurenstoffen im Abwasser" (25.4.12 – 31.8.12)	
	Ämterkonsultation zum Gesetzesentwurf	
Überarbeitung GSchV Frühjahr 2013	Überarbeiteter **Entwurf** zur Änderung der **GSchV**	❑
Finalisierung GSchG Mai – Juni 2013	**Finalisierung** des Gesetzesentwurfs **GSchG**	❑
	Antrag an den BR und Verabschiedung der **Botschaft** zur Änderung des GSchG	
Konsultation GSchV Okt. – Dez. 2013	**Anhörung** und **Ämterkonsultation** zu überarbeitetem Entwurf GSchV, Information über Resultate der Anhörung	❑
Finalisierung GSchV Frühjahr 2014	**Finalisierung** des Verordnungsentwurfs **GSchV**	❑
	Antrag an den BR und **Verabschiedung der GSchV**	
Parlamentarische Phase 2013 – 2014	**Beratung** in parlamentarischen Kommissionen UREK-SR / NR	❑
	Abstimmung im Erstrat und Zweitrat	
Umsetzungsphase Ab 2015	**Inkrafttreten GSchG / GSchV**	❑

2. Der Entscheidungsprozess zur Änderung des GSchG / der GSchV hat zum Ziel, den Eintrag von Spurenstoffen in die Gewässer zu verringern. Um dies zu erreichen, verfolgt die Schweiz derzeit den Ansatz, rund 100 ausgewählte Abwasserreinigungsanlagen (ARAs) technisch aufzurüsten. Dadurch sollen 80% der Spurenstoffe aus dem Abwasser eliminiert werden. Zur Finanzierung der technischen Aufrüstung der ARAs soll eine gesamtschweizerische Abwassergebühr erhoben werden. Durch die Gebühr finanziert der Bund Abgeltungen von 75% auf die Erstinvestition der Aufrüstung.

   Wie sehr unterstützt Ihre Organisation diesen Ansatz?

Meine Organisation unterstützt den Ansatz voll und ganz	Meine Organisation unterstützt den Ansatz mehrheitlich	Meine Organisation lehnt den Ansatz mehrheitlich ab	Meine Organisation lehnt den Ansatz voll und ganz ab
❑	❑	❑	❑

Weitere Anmerkungen: ....
....
....
....
....

3. *Wie gut wurden die Interessen Ihrer Organisation im Entscheidungsprozess zur Änderung des GSchG / der GSchV bezüglich des Eintrags von Spurenstoffen in die Gewässer (2007 – heute) berücksichtigt?*

Die Interessen meiner Organisation wurden voll und ganz berücksichtigt	Die Interessen meiner Organisation wurden mehrheitlich berücksichtigt	Die Interessen meiner Organisation wurden mehrheitlich nicht berücksichtigt	Die Interessen meiner Organisation wurden ganz und gar nicht berücksichtigt
❑	❑	❑	❑

Weitere Anmerkungen: ....
....
....
....
....

Teil B:
Akteure im Entscheidungsprozess

4. Wichtigkeit der Akteure

Im Entscheidungsprozess zur Änderungen des GSchG / der GSchV bezüglich des Eintrags von Spurenstoffen in die Gewässer (2007 – heute) waren zahlreiche Akteure beteiligt. Eine möglichst vollständige Liste dieser beteiligten Akteure finden Sie nachfolgend.

Bitte kreuzen Sie in der ersten Spalte all diejenigen Akteure an, die im Entscheidungsprozess Ihrer Meinung nach besonders wichtig waren.

*Bitte machen Sie in der zweiten Spalte **genau 3 Kreuzchen für die gesamte Liste**, um anzugeben, welche Ihrer Meinung nach die 3 wichtigsten Akteure im Entscheidungsprozess waren.*

Falls Sie Akteure auf der Liste vermissen, können Sie diese in den leeren Zeilen ergänzen und auch hier deren Wichtigkeit evaluieren.

Akteure	Besonders wichtig	3 wichtigste Akteure
Bund		
Bundesrat (BR)	☐	☐
UVEK – Generalsekretariat	☐	☐
BAFU, Abteilung Wasser, Sektion Oberflächengewässer Qualität	☐	☐
BAFU, Abteilung Luftreinhaltung und Chemikalien	☐	☐
Bundesamt für Energie (BFE)	☐	☐
Bundesamt für Raumplanung (ARE)	☐	☐
Bundesamt für Gesundheit (BAG)	☐	☐
Bundesamt für Justiz (BJ)	☐	☐
Eidg. Finanzverwaltung (EFV)	☐	☐
Eidg. Personalamt (EPA)	☐	☐
Bundesamt für Landwirtschaft (BLW)	☐	☐
Staatssekretariat für Wirtschaft (SECO)	☐	☐
Swissmedic	☐	☐
Konferenzen und Vereinigungen der Kantone		
Bau-, Planungs- und Umweltdirektoren-Konferenz (BPUK)	☐	☐
Konferenz der Vorsteher der Umweltschutzämter (KVU)	☐	☐
Verband der Kantonschemiker der Schweiz (VKCS)	☐	☐
Cercl'eau	☐	☐
Lab'Eaux	☐	☐
Parlament		
UREK NR	☐	☐
UREK SR	☐	☐
Parteien		
FDP. Die Liberalen (FDP)	☐	☐
Schweizerische Volkspartei (SVP)	☐	☐
Sozialdemokratische Partei der Schweiz (SP)	☐	☐
Grüne Partei der Schweiz/Grünes Bündnis (GPS)	☐	☐
Bürgerlich - Demokratische Partei (BDP)	☐	☐
Christlich demokratische Volkspartei (CVP)	☐	☐
Grünliberale Partei Schweiz (GLP)	☐	☐
Umweltverbände und –vereine		
Pro Natura	☐	☐
Association romande pour la protection des eaux et de l'air (ARPEA)	☐	☐
Schweiz. Fischereiverband	☐	☐
WWF Schweiz	☐	☐

Akteure	Besonders wichtig	3 wichtigste Akteure
Gesamtschweizerische Dachverbände der Gemeinden / Städte, Wasser- und Abwasserverbände		
ERFA Klärwerke Grossstädte CH	☐	☐
Groupement Romand des Exploitants de Stations d'Epuration (GRESE)	☐	☐
Verband Schweizer Abwasser- und Gewässerschutzfachleute (VSA)	☐	☐
Schweiz. Verein des Gas- und Wasserfaches (SVGW)	☐	☐
Kommunale Infrastruktur (KI)	☐	☐
Schweiz. Städteverband (SSV)	☐	☐
Schweiz. Gemeindeverband	☐	☐
Arbeitgeber-, Industrie-, Gewerbe- und Wirtschaftsverbände		
Economiesuisse, Verband der Schweizer Unternehmen	☐	☐
Scienceindustries (ehem. SGCI)	☐	☐
Schweiz. Kosmetik- und Waschmittelverband (SKW)	☐	☐
Swissmem	☐	☐
Handelskammer beider Basel	☐	☐
Schweiz. Arbeitgeberverband	☐	☐
Schweiz. Gewerbeverband	☐	☐
Centre Patronal	☐	☐
Arbeitnehmerverbände, Konsumentenorganisationen, Hauseigentümerverbände		
Konsumentenforum	☐	☐
Stiftung für Konsumentenschutz (Allianz SKS-FRC-ACSI)	☐	☐
Schweiz. Gewerkschaftsbund (SGB)	☐	☐
Schweiz. Bauernverband (SBV)	☐	☐
Hauseigentümerverband Schweiz (HEV)	☐	☐
Wissenschaft, Labore, Beratungsbüros		
Eawag	☐	☐
EPF Lausanne	☐	☐
Université de Lausanne	☐	☐
Universität Basel	☐	☐
Fachhochschule Nordwestschweiz	☐	☐
Oekotoxzentrum	☐	☐
Agridea	☐	☐
Holinger AG	☐	☐
Hunziker-Betatech	☐	☐
RWB	☐	☐
BMG Engineering AG	☐	☐
Weitere Akteure		
	☐	☐
	☐	☐
	☐	☐
	☐	☐

5. Zusammenarbeit mit anderen Akteuren

Im bereits erwähnten Entscheidungsprozess zur Änderungen des GSchG / der GSchV bezüglich des Eintrags von Spurenstoffen in die Gewässer (2007 – heute) haben zahlreiche Akteure zusammengearbeitet. Eine möglichst vollständige Liste dieser Akteure finden Sie nachfolgend.

Bitte kreuzen Sie all diejenigen Akteure an, mit welchen Ihre Organisation im Entscheidungsprozess <u>eng zusammengearbeitet</u> hat, unabhängig davon, ob Sie mit diesen inhaltlich übereinstimmen oder nicht.

Mit enger Zusammenarbeit ist gemeint: das Diskutieren von Erkenntnissen, das Ausarbeiten von Optionen, der Austausch über Positionen, das Bewerten von Alternativen.

Diese Frage ist wichtig, damit wir das Funktionieren von politischen Entscheidungsprozessen verstehen können. Persönliche Namen werden **nicht** veröffentlicht.

Falls Sie Akteure auf der Liste vermissen, können Sie diese in den leeren Zeilen ergänzen und auch hier Ihre Zusammenarbeit angeben.

Akteure	Enge Zusammenarbeit
Bund	
Bundesrat (BR)	❑
UVEK – Generalsekretariat	❑
BAFU, Abteilung Wasser, Sektion Oberflächengewässer Qualität	❑
BAFU, Abteilung Luftreinhaltung und Chemikalien	❑
Bundesamt für Energie (BFE)	❑
Bundesamt für Raumplanung (ARE)	❑
Bundesamt für Gesundheit (BAG)	❑
Bundesamt für Justiz (BJ)	❑
Eidg. Finanzverwaltung (EFV)	❑
Eidg. Personalamt (EPA)	❑
Bundesamt für Landwirtschaft (BLW)	❑
Staatssekretariat für Wirtschaft (SECO)	❑
Swissmedic	❑
Konferenzen und Vereinigungen der Kantone	
Bau-, Planungs- und Umweltdirektoren-Konferenz (BPUK)	❑
Konferenz der Vorsteher der Umweltschutzämter (KVU)	❑
Verband der Kantonschemiker der Schweiz (VKCS)	❑
Cercl'eau	❑
Lab'Eaux	❑
Parlament	
UREK NR	❑
UREK SR	❑
Parteien	
FDP. Die Liberalen (FDP)	❑
Schweizerische Volkspartei (SVP)	❑
Sozialdemokratische Partei der Schweiz (SP)	❑
Grüne Partei der Schweiz/Grünes Bündnis (GPS)	❑
Bürgerlich - Demokratische Partei (BDP)	❑
Christlich demokratische Volkspartei (CVP)	❑
Grünliberale Partei Schweiz (GLP)	❑
Umweltverbände und –vereine	
Pro Natura	❑
Association romande pour la protection des eaux et de l'air (ARPEA)	❑
Schweiz. Fischereiverband	❑
WWF Schweiz	❑

Akteure	Enge Zusammenarbeit
Gesamtschweizerische Dachverbände der Gemeinden / Städte, Wasser- und Abwasserverbände	
ERFA Klärwerke Grossstädte CH	❑
Groupement Romand des Exploitants de Stations d'Epuration (GRESE)	❑
Verband Schweizer Abwasser- und Gewässerschutzfachleute (VSA)	❑
Schweiz. Verein des Gas- und Wasserfaches (SVGW)	❑
Kommunale Infrastruktur (KI)	❑
Schweiz. Städteverband (SSV)	❑
Schweiz. Gemeindeverband	❑
Arbeitgeber-, Industrie-, Gewerbe- und Wirtschaftsverbände	
Economiesuisse, Verband der Schweizer Unternehmen	❑
Scienceindustries (ehem. SGCI)	❑
Schweiz. Kosmetik- und Waschmittelverband (SKW)	❑
Swissmem	❑
Handelskammer beider Basel	❑
Schweiz. Arbeitgeberverband	❑
Schweiz. Gewerbeverband	❑
Centre Patronal	❑
Arbeitnehmerverbände, Konsumentenorganisationen, Hauseigentümerverbände	
Konsumentenforum	❑
Stiftung für Konsumentenschutz (Allianz SKS-FRC-ACSI)	❑
Schweiz. Gewerkschaftsbund (SGB)	❑
Schweiz. Bauernverband (SBV)	❑
Hauseigentümerverband Schweiz (HEV)	❑
Wissenschaft, Labore, Beratungsbüros	
Eawag	❑
EPF Lausanne	❑
Université de Lausanne	❑
Universität Basel	❑
Fachhochschule Nordwestschweiz	❑
Oekotoxzentrum	❑
Agridea	❑
Holinger AG	❑
Hunziker-Betatech	❑
RWB	❑
BMG Engineering AG	❑
Weitere Akteure	
	❑
	❑
	❑
	❑

6. Diese Frage bezieht sich ebenfalls auf den bereits erwähnten Entscheidungsprozess zur Änderungen des GSchG / der GSchV bezüglich des Eintrags von Spurenstoffen in die Gewässer (2007 – heute).

   Bitte kreuzen Sie all diejenigen Akteure an, mit welchen Ihre Organisation mehrheitlich inhaltliche Übereinstimmungen bzw. inhaltliche Divergenzen hatte, unabhängig davon, ob Sie mit diesen zusammengearbeitet haben oder nicht.

   Falls Sie Akteure auf der Liste vermissen, können Sie diese in den leeren Zeilen ergänzen und auch hier das Übereinstimmungs- / Divergenz-Profil evaluieren.

Akteure	Mehrheitlich Übereinstimmungen	Mehrheitlich Divergenzen
Bund		
Bundesrat (BR)	☐	☐
UVEK – Generalsekretariat	☐	☐
BAFU, Abteilung Wasser, Sektion Oberflächengewässer Qualität	☐	☐
BAFU, Abteilung Luftreinhaltung und Chemikalien	☐	☐
Bundesamt für Energie (BFE)	☐	☐
Bundesamt für Raumplanung (ARE)	☐	☐
Bundesamt für Gesundheit (BAG)	☐	☐
Bundesamt für Justiz (BJ)	☐	☐
Eidg. Finanzverwaltung (EFV)	☐	☐
Eidg. Personalamt (EPA)	☐	☐
Bundesamt für Landwirtschaft (BLW)	☐	☐
Staatssekretariat für Wirtschaft (SECO)	☐	☐
Swissmedic	☐	☐
Konferenzen und Vereinigungen der Kantone		
Bau-, Planungs- und Umweltdirektoren-Konferenz (BPUK)	☐	☐
Konferenz der Vorsteher der Umweltschutzämter (KVU)	☐	☐
Verband der Kantonschemiker der Schweiz (VKCS)	☐	☐
Cercl'eau	☐	☐
Lab'Eaux	☐	☐
Parlament		
UREK NR	☐	☐
UREK SR	☐	☐
Parteien		
FDP. Die Liberalen (FDP)	☐	☐
Schweizerische Volkspartei (SVP)	☐	☐
Sozialdemokratische Partei der Schweiz (SP)	☐	☐
Grüne Partei der Schweiz/Grünes Bündnis (GPS)	☐	☐
Bürgerlich - Demokratische Partei (BDP)	☐	☐
Christlich demokratische Volkspartei (CVP)	☐	☐
Grünliberale Partei Schweiz (GLP)	☐	☐
Umweltverbände und –vereine		
Pro Natura	☐	☐
Association romande pour la protection des eaux et de l'air (ARPEA)	☐	☐
Schweiz. Fischereiverband	☐	☐
WWF Schweiz	☐	☐

Akteure	Mehrheitlich Übereinstimmungen	Mehrheitlich Divergenzen
Gesamtschweizerische Dachverbände der Gemeinden / Städte, Wasser- und Abwasserverbände		
ERFA Klärwerke Grossstädte CH	☐	☐
Groupement Romand des Exploitants de Stations d'Epuration (GRESE)	☐	☐
Verband Schweizer Abwasser- und Gewässerschutzfachleute (VSA)	☐	☐
Schweiz. Verein des Gas- und Wasserfaches (SVGW)	☐	☐
Kommunale Infrastruktur (KI)	☐	☐
Schweiz. Städteverband (SSV)	☐	☐
Schweiz. Gemeindeverband	☐	☐
Arbeitgeber-, Industrie-, Gewerbe- und Wirtschaftsverbände		
Economiesuisse, Verband der Schweizer Unternehmen	☐	☐
Scienceindustries (ehem. SGCI)	☐	☐
Schweiz. Kosmetik- und Waschmittelverband (SKW)	☐	☐
Swissmem	☐	☐
Handelskammer beider Basel	☐	☐
Schweiz. Arbeitgeberverband	☐	☐
Schweiz. Gewerbeverband	☐	☐
Centre Patronal	☐	☐
Arbeitnehmerverbände, Konsumentenorganisationen, Hauseigentümerverbände		
Konsumentenforum	☐	☐
Stiftung für Konsumentenschutz (Allianz SKS-FRC-ACSI)	☐	☐
Schweiz. Gewerkschaftsbund (SGB)	☐	☐
Schweiz. Bauernverband (SBV)	☐	☐
Hauseigentümerverband Schweiz (HEV)	☐	☐
Wissenschaft, Labore, Beratungsbüros		
Eawag	☐	☐
EPF Lausanne	☐	☐
Université de Lausanne	☐	☐
Universität Basel	☐	☐
Fachhochschule Nordwestschweiz	☐	☐
Oekotoxzentrum	☐	☐
Agridea	☐	☐
Holinger AG	☐	☐
Hunziker-Betatech	☐	☐
RWB	☐	☐
BMG Engineering AG	☐	☐
Weitere Akteure		
	☐	☐
	☐	☐
	☐	☐
	☐	☐

Teil C:
Positionen Ihrer Organisation

7. Nachfolgend finden Sie eine Liste mit verschiedenen Zielen bezüglich der Verringerung des Eintrags von Spurenstoffen in die Gewässer.

Bitte geben Sie den Zustimmungsgrad Ihrer Organisation zu nachfolgenden Zielen an.

	Meine Organisation			
Ziele	Stimmt dem voll und ganz zu	Stimmt dem mehrheitlich zu	Lehnt dies mehrheitlich ab	Lehnt dies voll und ganz ab
Massnahmen sollten an der Quelle (bei der Verunreinigung) ansetzen.	❑	❑	❑	❑
Massnahmen sollten end-of-pipe (bei der Abwasserreinigung) ansetzen.	❑	❑	❑	❑
Solange die Auswirkungen von Spurenstoffen nicht umfassend erforscht sind, sollten präventive Massnahmen zur Verringerung des Eintrags von Spurenstoffen ergriffen werden (Vorsorgeprinzip).	❑	❑	❑	❑
Solange die Auswirkungen von Spurenstoffen nicht umfassend erforscht sind, sollten *keine* Massnahmen zur Verringerung des Eintrags von Spurenstoffen ergriffen werden.	❑	❑	❑	❑
Massnahmen sollten Oberflächengewässer gänzlich von Spurenstoffen befreien.	❑	❑	❑	❑

8. Um den Eintrag von Spurenstoffen in die Gewässer zu verringern, können verschiedene Kompetenzbereiche rechtliche Grundlagen schaffen.

Bitte geben Sie den Zustimmungsgrad Ihrer Organisation zur Regulierung in nachfolgenden Kompetenzbereichen an.

	Meine Organisation			
Die Verringerung des Eintrags von Spurenstoffen ist Aufgabe der:	Stimmt dem voll und ganz zu	Stimmt dem mehrheitlich zu	Lehnt dies mehrheitlich ab	Lehnt dies voll und ganz ab
Gewässerschutzregulierung	❑	❑	❑	❑
Chemikalienregulierung	❑	❑	❑	❑
Landwirtschaftsregulierung	❑	❑	❑	❑
Trinkwasserregulierung	❑	❑	❑	❑
Keine Regulierungsaufgabe, sondern des individuellen Konsumentenverhaltens	❑	❑	❑	❑

9. Nachfolgend finden Sie eine Liste möglicher regulativer Instrumente, die zur Verringerung von Spurenstoffen in den Gewässern beitragen können.

Bitte geben Sie den Zustimmungsgrad Ihrer Organisation zur Regulierung von Spurenstoffen durch nachfolgende Instrumente an.

Falls Sie regulative Instrumente auf der Liste vermissen, können Sie diese in den leeren Zeilen ergänzen und auch hier Ihren Zustimmungsgrad angeben.

Regulative Instrumente	Meine Organisation			
	Stimmt dem voll und ganz zu	Stimmt dem mehrheitlich zu	Lehnt dies mehrheitlich ab	Lehnt dies voll und ganz ab
Zulassungseinschränkungen einzelner Spurenstoffe	❑	❑	❑	❑
Anwendungseinschränkungen einzelner Spurenstoffe	❑	❑	❑	❑
Entsorgungsvorschriften für Produkte, die Spurenstoffe enthalten	❑	❑	❑	❑
Festlegung der best-verfügbaren Technik zur Spurenstoffelimination (z.B. Aufrüstung der ARAs, Behandlung von Abwasserteilströmen in Betrieben)	❑	❑	❑	❑
Festlegung der besten landwirtschaftlichen Praxis zur Verringerung des Eintrags von Spurenstoffen	❑	❑	❑	❑
Festlegung von Umweltqualitätsnormen / Immissionsgrenzwerten für Spurenstoffe	❑	❑	❑	❑
Festlegung von Emissionsgrenzwerten für Spurenstoffe	❑	❑	❑	❑
Abgabe auf Produkte, die Spurenstoffe enthalten	❑	❑	❑	❑
Erhöhung der Abwassergebühr für die Finanzierung von Massnahmen zur Verringerung von Spurenstoffen	❑	❑	❑	❑
Steuerfinanzierte Subventionen (z.B. für Investitionen in Technologie zur Spurenstoffelimination, Monitoringtechnologie, optimierte Produktionsprozesse)	❑	❑	❑	❑
Kontrollverfahren (z.B. Erweiterung der Messprogramme, Registrierungspflicht von Spurenstoffen)	❑	❑	❑	❑
Freiwillige Massnahmen durch Betriebe und Zivilgesellschaft (z.B. Technologie zur Spurenstoffelimination, optimierte Produktionsprozesse, Labeling, Verzicht)	❑	❑	❑	❑
Staatliche Informationskampagnen, Beratung durch staatliche Behörden	❑	❑	❑	❑
Forschung, Bildung von Expertengruppen / Informationsplattformen	❑	❑	❑	❑
Festlegen von Reduktionszielvereinbarungen mit Privatwirtschaft (Branchenvereinbarungen) oder zwischen unterschiedlichen Regierungsebenen (öffentlich-öffentliche Partnerschaft)	❑	❑	❑	❑
Andere:	❑	❑	❑	❑
	❑	❑	❑	❑

Teil D:
Kompetenzen Ihrer Organisation und geographische Eingrenzung

10. Nachfolgend finden Sie eine Liste mit Aufgabenbereichen der Wasserregulierung.

a) Bitte kreuzen Sie in nachfolgender Tabelle an, für welche Aufgabenbereiche Ihre Organisation *formell zuständig* ist, d.h. Regulierungs-, Umsetzungs-, oder Beratungskompetenzen besitzt (Mehrfachnennungen erlaubt).
Falls Sie Aufgaben auf der Liste vermissen, können Sie diese in den leeren Zeilen ergänzen.

b) Uns interessiert auch, in welchen Aufgabenbereichen der Wasserregulierung Ihrer Meinung nach Massnahmen mit einer höheren, gleich hohen oder niedrigeren Priorität als im Bereich Spurenstoffe zu treffen sind.

Bitte schätzen Sie die *Priorität* aller nachfolgenden Aufgabenbereiche der Wasserregulierung im Vergleich zum Bereich Spurenstoffe aus der Sicht Ihrer Organisation ein.

Sollten Sie Aufgaben hinzugefügt haben, bewerten Sie bitte auch hier die Priorität.

Aufgabenbereiche der Wasserregulierung	a) Meine Organisation ist zuständig für:	b) Höhere Priorität als Spurenstoffe	Gleiche hohe Priorität wie Spurenstoffe	Niedrigere Priorität als Spurenstoffe
Spurenstoffe (Chemische Gewässerqualität)	❑			
Gewässerökologie / Renaturierung	❑	❑	❑	❑
Nährstoffe / Landwirtschaft	❑	❑	❑	❑
Industrielle Emissionen	❑	❑	❑	❑
Abwasserreinigung	❑	❑	❑	❑
Wasserführung / Restwassermengen	❑	❑	❑	❑
Grundwasser	❑	❑	❑	❑
Trinkwasser	❑	❑	❑	❑
Wasserkraft	❑	❑	❑	❑
Hochwasser	❑	❑	❑	❑
Gewässerüberwachung,-messungen	❑	❑	❑	❑
Andere:	❑	❑	❑	❑
	❑	❑	❑	❑
	❑	❑	❑	❑

11. *Bitte geben Sie an, für welches Gebiet Ihre Organisation im Bereich Gewässerschutz formell zuständig ist, d.h. Regulierungs-, Umsetzungs-, oder Beratungskompetenzen besitzt.*

Meine Organisation ist zuständig für:	Bitte präzisieren Sie
❑ Gemeinde / Stadt:	....(Name der Gemeinde/Stadt)
❑ Region bestehend aus folgenden Gemeinden / Städten:	....(Name der Region) ....(Namen der Gemeinden/Städte)
❑ Kanton:	....(Name des Kantons)
❑ Ganze Schweiz	
❑ Grenzüberschreitende Region:	....(Bitte beschreiben Sie die Region)
❑ Teil-Einzugsgebiet des Rheins:	....(Name des Einzugsgebietes)
❑ Gesamtes Einzugsgebiet des Rheins	
❑ Andere:	....

12. Bei der Ressource Wasser kann die formelle Zuständigkeit vom tatsächlichen Tätigkeitsgebiet abweichen.

Falls das Tätigkeitsgebiet Ihrer Organisation vom formellen Zuständigkeitsgebiet abweicht, zeichnen Sie bitte auf nachfolgender grosser Karte deutlich das Tätigkeitsgebiet Ihrer Organisation ein (falls es nicht abweicht, weiter zu Frage 13).

Mit Tätigkeitsgebiet ist das Gebiet gemeint, in welchem Ihre Organisation Aktivitäten unternimmt, die relevant für den Gewässerschutz sein können.

Beispiel
für das Einzeichnen eines
Tätigkeitsgebietes, das mit den
Kantons- und Landesgrenzen nicht
übereinstimmt.

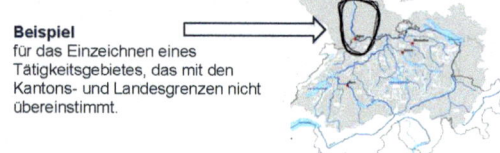

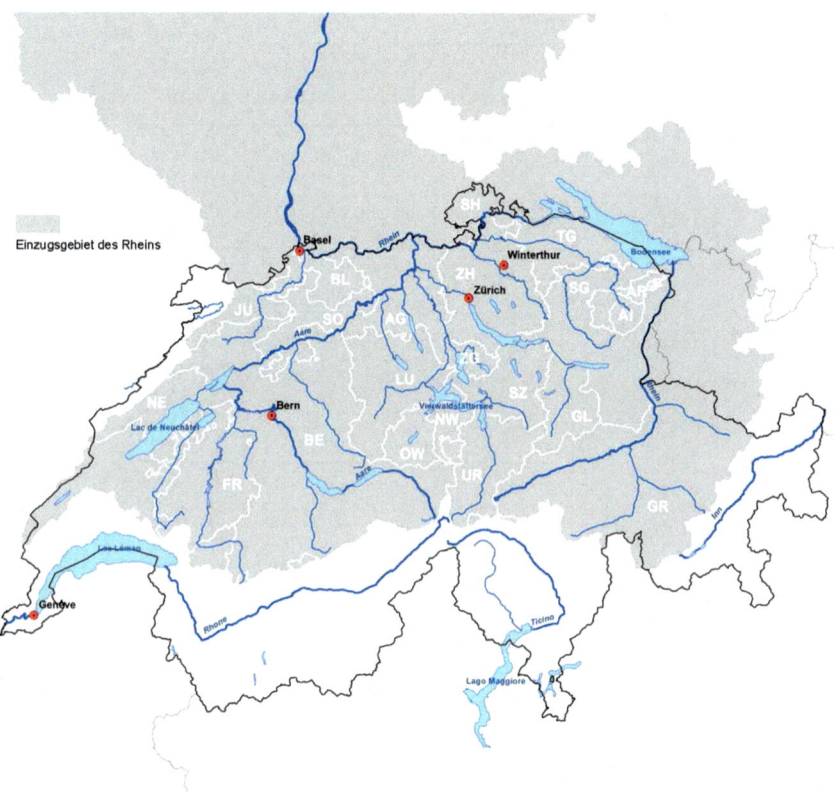

13. a) *Bitte geben Sie an, in welchen* **nationalen und internationalen** <u>Gremien oder Plattformen</u> *zum Thema Spurenstoffe Ihre Organisation im Einzugsgebiet des Rheins Mitglied ist.*

a) Meine Organisation ist Mitglied von **nationalen / internationalen**:	Bitte präzisieren Sie
❑ Wissens- oder Dialogplattformen:	....(Namen der Plattformen)
❑ Lenkungsgruppen / Begleitgruppen:	....(Namen der Lenkungsgruppen)
❑ Internationalen Gewässerschutzkommissionen:	....(Namen der Gewässerschutzkommissionen)
❑ Andere:	....

b) *Bitte geben Sie ebenfalls an, mit welchen Akteuren Ihre Organisation zum Thema Spurenstoffe im Einzugsgebiet des Rheins* **international** <u>zusammenarbeitet</u>.

b) Meine Organisation arbeitet **international** zusammen mit:	Bitte präzisieren Sie
❑ Gemeinden / Städten:	....(Namen der Gemeinden/Städte)
❑ Ländern / Regionen:	....(Namen der Länder/Regionen)
❑ Wasseragenturen / Einzugsgebietsorganisationen:	....(Namen der Agenturen)
❑ Internationalen Verbänden / nationalen Verbänden aus dem Ausland:	....(Namen der Verbände)
❑ Universitäten / Fachhochschulen:	....(Namen der Universitäten)
❑ Andere:	....

Wir danken Ihnen für Ihre wertvolle Mitarbeit!

Haben Sie noch weitere Anmerkungen oder Ideen zum Thema Spurenstoffe oder zum Fragebogen, die Sie mit uns teilen möchten?

Bitte senden Sie den ausgefüllten Fragebogen mit beiliegendem frankierten Antwortcouvert
bis zum 15. Mai 2013 an:

Florence Metz
Universität Bern, Institut für Politikwissenschaft
Lerchenweg 36
CH-3000 Bern 9

Annex 7
German Questionnaire

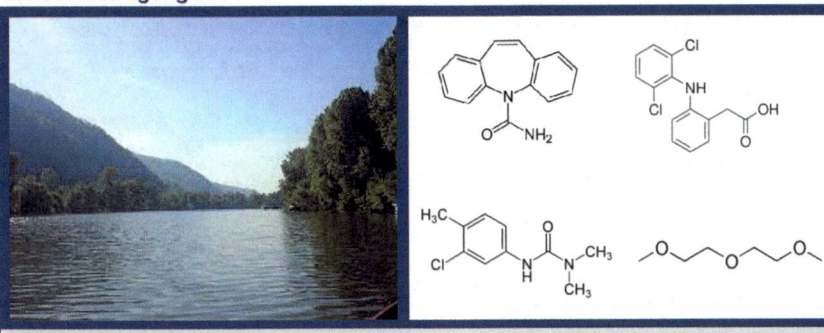

Erläuterungen zum Fragebogen

Wie im Begleitbrief erwähnt, ist dieser Fragebogen Teil eines Forschungsprojektes zum Thema Gewässerschutzpolitik und Mikroschadstoffe, das von der Universität Bern in Zusammenarbeit mit dem Wasserforschungsinstitut EAWAG durchgeführt wird. Ziel ist es, den **von 2008 bis 2011 laufenden Entscheidungsprozess zur Verabschiedung der Oberflächengewässerverordnung (OGewV)** zu untersuchen.

Der Fragebogen setzt sich aus insgesamt 12 Fragen zusammen und gliedert sich in Teil A bis D:

Teil A: Beteiligung Ihrer Organisation am genannten Entscheidungsprozess
Teil B: Zusammenarbeit Ihrer Organisation mit anderen Akteuren im Entscheidungsprozess
Teil C: Positionen Ihrer Organisation zur Regulierung von Mikroschadstoffen
Teil D: Kompetenzen Ihrer Organisation im Bereich Wasser

Bitte beantworten Sie die folgenden Fragen aus der Perspektive Ihrer Organisation.

Das Forschungsprojekt wird im Rahmen einer Doktorarbeit realisiert und vom Schweizerischen Nationalfonds gefördert. Da Ihre Organisation eine zentrale Rolle im Entscheidungsprozess einnimmt, ist Ihre Teilnahme an der Befragung äußerst wesentlich für das Gelingen der Forschungsarbeit. Wir danken Ihnen daher im Vorfeld für das Beantworten des Fragebogens und Ihre wertvolle Unterstützung. Das Ausfüllen sollte **nicht mehr als 30 Minuten** in Anspruch nehmen.

Bitte senden Sie den ausgefüllten Fragebogen mit beiliegendem frankierten Antwortcouvert **bis Ende April 2014** zurück. Die von Ihnen angegebenen Informationen werden ausschließlich zu Forschungszwecken genutzt, vertraulich behandelt und nicht an Dritte weitergegeben.

Projektleitung:

Prof. Dr. Karin Ingold
Institut für Politikwissenschaft Universität Bern, Eawag

Projektpartner und Sponsoren:

u^b

eawag
aquatic research °°°

SCHWEIZERISCHER NATIONALFONDS
ZUR FÖRDERUNG DER WISSENSCHAFTLICHEN FORSCHUNG

UNIVERSITÄT
BERN

Bei Fragen können Sie sich per Mail oder Telefon direkt an die zuständige **Kontaktperson** wenden:

Florence Metz (Doktorandin)
Universität Bern, Institut für Politikwissenschaft, Fabrikstrasse 8, CH-3012 Bern
florence.metz@ipw.unibe.ch
Tel. 0049 (0)163 60 77 689

Name(n) der Person(en), die interviewt wird (werden):..

Name Ihrer Dienststelle und Ihrer Organisation:..

Telefonnr.:...Email: ....

Möchte Ihre Organisation über die Resultate der Studie informiert werden?   Ja ❑   Nein ❑

Teil A:
Entscheidungsprozess zur Verabschiedung der Oberflächengewässerverordnung

1. An welchen Etappen des Entscheidungsprozesses zur Verabschiedung der Oberflächengewässerverordnung (2008 – 2011) war Ihre Organisation beteiligt?

   Bitte kreuzen Sie alle Etappen an, an denen Ihre Organisation beteiligt war.

   Unter Beteiligung wird verstanden: einen Beitrag durch aktive Mitarbeit leisten, an Arbeitsgruppen / Workshops / formellen und informellen Konsultationen teilnehmen.

Phase Datum	Etappe	Beteiligt
Verabschiedung der Oberflächengewässerverordnung		
Auslöser Vor 2008	Umsetzung der WRRL 2000/60/EG und der UQN-RL 2008/105/EG in Deutschland Föderalismusreform 2006	❑
Konzeptphase 2008 - 2010	Bund-Länder-Arbeitskreis UQN (BLAK-UQN) (15./16.9.2008)	❑
	138. und 139. LAWA Vollversammlung (22./23.9.2009 und 25./26.3.2010)	
	74. Umweltministerkonferenz (11.6.2010)	
Ausarbeitung 2010	Diskussionsentwurf (29.3.2010)	❑
	Referentenentwurf (1.8.2010)	
Konsultation Nov 2010	Ressortübergreifende Abstimmung	❑
	Anhörung der Verbände und der Bundesländer (November 2010)	
Finalisierung des Verordnungsentwurfs Frühjahr 2011	Vorlage für das Kabinett	❑
	Beschluss der OGewV durch das Bundeskabinett (16.3.2011) und Zuleitung an den BR	
Zustimmung Sommer 2011	Beratung in den BR Ausschüssen (17.3.2011), Drucksache 153/11	❑
	Änderungsempfehlungen der BR Ausschüsse (17.5.2011), Drucksache 153/1/11	
	Zustimmung des BR zur OGewV unter Massgabe von 25 Änderungen in seiner 883.Sitzung (27.5.2011)	
	Beschluss des Bundeskabinetts den vom BR beschlossenen Änderungen zuzustimmen (22.6.2011)	
Verabschiedung und Umsetzung Ab August 2011	Inkrafttreten der OGewV (26.7.2011) und Umsetzung	❑

❑ Meine Organisation hat am Entscheidungsprozess zur Verabschiedung der Oberflächengewässerverordnung nicht teilgenommen.

Weitere Anmerkungen: ....

....

2. *Wurden die Positionen Ihrer Organisation im Entscheidungsprozess zur Verabschiedung der Oberflächengewässerverordnung (2008 – 2011) berücksichtigt?*

Die Positionen meiner Organisation wurden			
Voll und ganz berücksichtigt	Mehrheitlich berücksichtigt	Mehrheitlich nicht berücksichtigt	Ganz und gar nicht berücksichtigt
☐	☐	☐	☐

Weitere Anmerkungen: ....
....

3. a) Die Oberflächengewässerverordnung legt Umweltqualitätsnormen (UQN) für flussgebietsspezifische Schadstoffe gemäß Wasserrahmenrichtlinie fest. *Wie bewertet Ihre Organisation das Festlegen von UQN als Instrument, um Mikroschadstoffe auf Bundesebene zu regulieren?*

Bitte geben Sie den Zustimmungsgrad Ihrer Organisation zu folgender Aussage an:

Das Festlegen von UQN ist ein gutes Instrument, um Mikroschadstoffe auf Bundesebene zu regulieren.

Meine Organisation stimmt dieser Aussage			
Voll und ganz zu	Mehrheitlich zu	Mehrheitlich nicht zu	Ganz und gar nicht zu
☐	☐	☐	☐

Weitere Anmerkungen: ....
....

3. b) Anlage 5 der Oberflächengewässerverordnung legt für insgesamt 162 Stoffe Umweltqualitätsnormen fest, darunter 13 neu aufgenommene Stoffe (neu im Vergleich zu den vorher bestehenden Regelungen auf Länderebene). Fünf weitere neue Stoffe wurden durch den Bundesrat gestrichen (Carbamazepin, Fenpropimorph, Phosphorsäuretriphenylester, Sulfamethoxazol, Uran).

Wie bewertet Ihre Organisation die bestehende Stoffliste und die festgelegten UQN der OGewV?

Aus der Sicht meiner Organisation sind Stoffliste und UQN der OGewV				
Zu weitreichend	Teilweise zu weitreichend	Gerade richtig	Teilweise nicht weitreichend genug	Nicht weitreichend genug
☐	☐	☐	☐	☐

Weitere Anmerkungen: ....
....

Teil B:
Akteure im Entscheidungsprozess

4. Wichtigkeit der Akteure

Im Entscheidungsprozess zur Verabschiedung der Oberflächengewässerverordnung (2008 – 2011) waren zahlreiche Akteure beteiligt. Eine möglichst vollständige Liste dieser beteiligten Akteure finden Sie nachfolgend.

Bitte kreuzen Sie in der ersten Spalte all diejenigen Akteure an, die im Entscheidungsprozess Ihrer Meinung nach besonders wichtig waren.

*Bitte machen Sie in der zweiten Spalte **genau 3 Kreuzchen für die gesamte Liste**, um anzugeben, welche Ihrer Meinung nach die 3 wichtigsten Akteure im Entscheidungsprozess waren.*

Falls Sie Akteure auf der Liste vermissen, können Sie diese in den leeren Zeilen ergänzen und auch hier deren Wichtigkeit evaluieren.

Akteure	Besonders wichtig	3 wichtigste Akteure
Bund		
BMU Bundesministerium für Umwelt	☐	☐
BMWI Bundesministerium für Wirtschaft und Energie	☐	☐
BMEL Bundesministerium für Ernährung und Landwirtschaft	☐	☐
BMG Bundesministerium für Gesundheit	☐	☐
BMJ Bundesministerium der Justiz und für Verbraucherschutz	☐	☐
BMF Bundesministerium der Finanzen	☐	☐
UBA Umweltbundesamt	☐	☐
Bundesländer Arbeitsgemeinschaften, BL am Rhein		
LAWA Bund/Länder-Arbeitsgemeinschaft Wasser	☐	☐
BLAK-UQN Bund/Länder-Arbeitskreis Umweltqualitätsnormen	☐	☐
Baden-Württemberg	☐	☐
Bayern	☐	☐
Hessen	☐	☐
Nordrhein-Westfalen	☐	☐
Rheinland-Pfalz	☐	☐
Andere Bundesländer	☐	☐
Dachverbände der Gemeinden, Städte, Kreise, Bezirke		
BVKS Bundesvereinigung der kommunalen Spitzenverbände	☐	☐
VKU Verband kommunaler Unternehmen	☐	☐
Wasser- und Abwasserverbände		
BDEW Bundesverband der Energie- und Wasserwirtschaft	☐	☐
DWA Dt. Vereinigung für Wasserwirtschaft	☐	☐
DVGW Dt. Verein des Gas- und Wasserfaches	☐	☐
AÖW Allianz der öffentlichen Wasserwirtschaft	☐	☐
BDW Bundesverband Dt. Wasserkraftwerke	☐	☐
Umweltverbände		
BUND für Umwelt und Naturschutz	☐	☐
NABU Naturschutzbund	☐	☐
BBU Bundesverband Bürgerinitiativen Umweltschutz	☐	☐
DFV Dt. Fischerei-Verband	☐	☐
VDG Vereinigung dt. Gewässerschutz	☐	☐
DNR Dt. Naturschutzring	☐	☐
Grüne Liga	☐	☐

Akteure	Besonders wichtig	3 wichtigste Akteure
Wirtschaftsverbände		
VCI Verband der chemischen Industrie	☐	☐
BDI Bundesverband der Dt. Industrie	☐	☐
DIHK Dt. Industrie und-Handelskammertag	☐	☐
WVMetalle WirtschaftsVereinigung Metalle	☐	☐
BAH Bundesverband der Arzneimittel-Hersteller	☐	☐
VDA Verband der Automobilindustrie	☐	☐
MIRO Bundesverband Mineralische Rohstoffe	☐	☐
MWV Mineralölwirtschaftsverband	☐	☐
Hauptverband der dt. Bauindustrie	☐	☐
VDMA Verband Dt. Maschinen- und Anlagenbau	☐	☐
VGB Technische Vereinigung der Großkraftwerksbetreiber	☐	☐
ZVEI Zentralverband Elektrotechnik- und Elektroindustrie	☐	☐
IVA Industrieverband Agrar	☐	☐
Landwirtschaftsverbände		
DBV Deutscher Bauernverband	☐	☐
Arbeitnehmerverbände, Verbraucherschutzverbände		
DGB Deutsche Gewerkschaftsbund	☐	☐
IG Chemie Industriegewerkschaft Chemie	☐	☐
GÖD Gewerkschaft öffentlicher Dienst	☐	☐
vzbv Verbraucherzentrale	☐	☐
Wissenschaft, Labore, Fachexperten		
BfG Bundesanstalt für Gewässerkunde	☐	☐
Wasserchemische Gesellschaft	☐	☐
VDI Verein dt. Ingenieure	☐	☐
BWK Bund der Ingenieure für Wasserwirtschaft	☐	☐
Weitere Akteure		
	☐	☐
	☐	☐
	☐	☐
	☐	☐
	☐	☐

5. Zusammenarbeit mit anderen Akteuren

Im bereits erwähnten Entscheidungsprozess zur Verabschiedung der Oberflächengewässerverordnung (2008 – 2011) haben zahlreiche Akteure zusammengearbeitet. Die nachfolgende Akteursliste ist dieselbe wie in Frage 4.

Bitte kreuzen Sie all diejenigen Akteure an, mit welchen Ihre Organisation im Entscheidungsprozess eng zusammengearbeitet hat, unabhängig davon, ob Sie mit diesen inhaltlich übereinstimmen oder nicht.

Mit enger Zusammenarbeit ist gemeint: das Diskutieren von Erkenntnissen, das Ausarbeiten von Optionen, der Austausch über Positionen, das Bewerten von Alternativen.

Diese Frage ist wichtig, damit wir das Funktionieren von politischen Entscheidungsprozessen verstehen können. Persönliche Namen werden **nicht** veröffentlicht.

Falls Sie Akteure auf der Liste vermissen, können Sie diese in den leeren Zeilen ergänzen und auch hier Ihre Zusammenarbeit angeben.

Akteure	Enge Zusammenarbeit
Bund	
BMU Bundesministerium für Umwelt	❏
BMWI Bundesministerium für Wirtschaft	❏
BMELV Bundesministerium für Landwirtschaft	❏
BMG Bundesministerium für Gesundheit	❏
BMJ Bundesministerium für Justiz	❏
BMF Bundesministerium der Finanzen	❏
UBA Umweltbundesamt	❏
Bundesländer Arbeitsgemeinschaften, BL am Rhein	
LAWA Bund/Länder-Arbeitsgemeinschaft Wasser	❏
BLAK-UQN Bund/Länder-Arbeitskreis Umweltqualitätsnormen	❏
Baden-Württemberg	❏
Bayern	❏
Hessen	❏
Nordrhein-Westfalen	❏
Rheinland-Pfalz	❏
Andere Bundesländer	❏
Dachverbände der Gemeinden, Städte, Kreise, Bezirke	
BVKS Bundesvereinigung der kommunalen Spitzenverbände	❏
VKU Verband kommunaler Unternehmen	❏
Wasser- und Abwasserverbände	
BDEW Bundesverband der Energie- und Wasserwirtschaft	❏
DWA Dt. Vereinigung für Wasserwirtschaft	❏
DVGW Dt. Verein des Gas- und Wasserfaches	❏
AÖW Allianz der öffentlichen Wasserwirtschaft	❏
BDW Bundesverband Dt. Wasserkraftwerke	❏
Umweltverbände	
BUND für Umwelt und Naturschutz	❏
NABU Naturschutzbund	❏
BBU Bundesverband Bürgerinitiativen Umweltschutz	❏
DFV Dt. Fischerei-Verband	❏
VDG Vereinigung dt. Gewässerschutz	❏
DNR Dt. Naturschutzring	❏
Grüne Liga	❏

Akteure	Enge Zusammenarbeit
Wirtschaftsverbände	
VCI Verband der chemischen Industrie	❏
BDI Bundesverband der Dt. Industrie	❏
DIHK Dt. Industrie und- Handelskammertag	❏
WVMetalle Wirtschaftsvereinigung Metalle	❏
BAH Bundesverband der Arzneimittel-Hersteller	❏
VDA Verband der Automobilindustrie	❏
MIRO Bundesverband Mineralische Rohstoffe	❏
MWV Mineralölwirtschaftsverband	❏
Hauptverband der dt. Bauindustrie	❏
VDMA Verband Dt. Maschinen- und Anlagenbau	❏
VGB Technische Vereinigung der Großkraftwerksbetreiber	❏
ZVEI Zentralverband Elektrotechnik- und Elektroindustrie	❏
IVA Industrieverband Agrar	❏
Landwirtschaftsverbände	
DBV Deutscher Bauernverband	❏
Arbeitnehmerverbände, Konsumentenorganisationen	
DGB Deutsche Gewerkschaftsbund	❏
IG Chemie Industriegewerkschaft Chemie	❏
GÖD Gewerkschaft öffentlicher Dienst	❏
vzbv Verbraucherzentrale	❏
Wissenschaft, Labore, Fachexperten	
BfG Bundesanstalt für Gewässerkunde	❏
Wasserchemische Gesellschaft	❏
VDI Verein dt. Ingenieure	❏
BWK Bund der Ingenieure für Wasserwirtschaft	❏
	❏
	❏
	❏
	❏
	❏

6. Diese Frage bezieht sich ebenfalls auf den bereits erwähnten Entscheidungsprozess zur Verabschiedung der Oberflächengewässerverordnung (2008 – 2011). Die nachfolgende Akteursliste ist dieselbe wie in Fragen 4 und 5.

    Bitte kreuzen Sie all diejenigen Akteure an, mit welchen Ihre Organisation mehrheitlich <u>inhaltliche Übereinstimmungen bzw. inhaltliche Divergenzen</u> hatte, unabhängig davon, ob Sie mit diesen zusammengearbeitet haben oder nicht.

    Falls Sie Akteure auf der Liste vermissen, können Sie diese in den leeren Zeilen ergänzen und auch hier das Übereinstimmungs- / Divergenz-Profil evaluieren.

Akteure	Mehrheitlich Überein- stimmungen	Mehrheitlich Divergenzen
Bund		
BMU Bundesministerium für Umwelt	❑	❑
BMWI Bundesministerium für Wirtschaft und Energie	❑	❑
BMEL Bundesministerium für Ernährung und Landwirtschaft	❑	❑
BMG Bundesministerium für Gesundheit	❑	❑
BMJ Bundesministerium der Justiz und für Verbraucherschutz	❑	❑
BMF Bundesministerium der Finanzen	❑	❑
UBA Umweltbundesamt	❑	❑
Bundesländer Arbeitsgemeinschaften, BL am Rhein		
LAWA Bund/Länder-Arbeitsgemeinschaft Wasser	❑	❑
BLAK-UQN Bund/Länder-Arbeitskreis Umweltqualitätsnormen	❑	❑
Baden-Württemberg	❑	❑
Bayern	❑	❑
Hessen	❑	❑
Nordrhein-Westfalen	❑	❑
Rheinland-Pfalz	❑	❑
Andere Bundesländer	❑	❑
Dachverbände der Gemeinden, Städte, Kreise, Bezirke		
BVKS Bundesvereinigung der kommunalen Spitzenverbände	❑	❑
VKU Verband kommunaler Unternehmen	❑	❑
Wasser- und Abwasserverbände		
BDEW Bundesverband der Energie- und Wasserwirtschaft	❑	❑
DWA Dt. Vereinigung für Wasserwirtschaft	❑	❑
DVGW Dt. Verein des Gas- und Wasserfaches	❑	❑
AÖW Allianz der öffentlichen Wasserwirtschaft	❑	❑
BDW Bundesverband Dt. Wasserkraftwerke	❑	❑
Umweltverbände		
BUND für Umwelt und Naturschutz	❑	❑
NABU Naturschutzbund	❑	❑
BBU Bundesverband Bürgerinitiativen Umweltschutz	❑	❑
DFV Dt. Fischerei-Verband	❑	❑
VDG Vereinigung dt. Gewässerschutz	❑	❑
DNR Dt. Naturschutzring	❑	❑
Grüne Liga	❑	❑

Akteure	Mehrheitlich Überein- stimmungen	Mehrheitlich Divergenzen
Wirtschaftsverbände		
VCI Verband der chemischen Industrie	❑	❑
BDI Bundesverband der Dt. Industrie	❑	❑
DIHK Dt. Industrie und- Handelskammertag	❑	❑
WV/Metalle Wirtschaftsvereinigung Metalle	❑	❑
BAH Bundesverband der Arzneimittel-Hersteller	❑	❑
VDA Verband der Automobilindustrie	❑	❑
MIRO Bundesverband Mineralische Rohstoffe	❑	❑
MWV Mineralölwirtschaftsverband	❑	❑
Hauptverband der dt. Bauindustrie	❑	❑
VDMA Verband Dt. Maschinen- und Anlagenbau	❑	❑
VGB Technische Vereinigung der Großkraftwerksbetreiber	❑	❑
ZVEI Zentralverband Elektrotechnik- und Elektroindustrie	❑	❑
IVA Industrieverband Agrar	❑	❑
Landwirtschaftsverbände		
DBV Deutscher Bauernverband	❑	❑
Arbeitnehmerverbände, Konsumentenorganisationen		
DGB Deutsche Gewerkschaftsbund	❑	❑
IG Chemie Industriegewerkschaft Chemie	❑	❑
GÖD Gewerkschaft öffentlicher Dienst	❑	❑
vzbv Verbraucherzentrale	❑	❑
Wissenschaft, Labore, Fachexperten		
BfG Bundesanstalt für Gewässerkunde	❑	❑
Wasserchemische Gesellschaft	❑	❑
VDI Verein dt. Ingenieure	❑	❑
BWK Bund der Ingenieure für Wasserwirtschaft	❑	❑
Weitere Akteure		
	❑	❑
	❑	❑
	❑	❑
	❑	❑
	❑	❑

Teil C:
Positionen Ihrer Organisation

7. Nachfolgend finden Sie eine Liste mit verschiedenen Zielen bezüglich der Verringerung von Mikroschadstoffen in Gewässern.

   Bitte geben Sie den Zustimmungsgrad Ihrer Organisation zu nachfolgenden <u>Zielen</u> an.

Ziele	Meine Organisation			
	Stimmt dem voll und ganz zu	Stimmt dem mehrheitlich zu	Stimmt dem mehrheitlich nicht zu	Stimmt dem ganz und gar nicht zu
Maßnahmen sollten an der Quelle (bei der Verunreinigung) ansetzen.	❏	❏	❏	❏
Maßnahmen sollten end-of-pipe (bei der Abwasserreinigung) ansetzen.	❏	❏	❏	❏
Solange die Auswirkungen von Mikroschadstoffen nicht umfassend erforscht sind, sollten präventive Maßnahmen zur Verringerung des Eintrags von Mikroschadstoffen ergriffen werden (Vorsorgeprinzip).	❏	❏	❏	❏
Solange die Auswirkungen von Mikroschadstoffen nicht umfassend erforscht sind, sollten *keine* Maßnahmen zur Verringerung des Eintrags von Mikroschadstoffen ergriffen werden.	❏	❏	❏	❏
Maßnahmen sollten darauf abzielen, Oberflächengewässer so gut wie gänzlich von Mikroschadstoffen zu befreien.	❏	❏	❏	❏

8. Um Mikroschadstoffe in Gewässern zu verringern, können verschiedene Kompetenzbereiche rechtliche Grundlagen schaffen.

   Bitte geben Sie den Zustimmungsgrad Ihrer Organisation zur Regulierung in nachfolgenden <u>Kompetenzbereichen</u> an.

Die Verringerung des Eintrags von Mikroschadstoffen ist Aufgabe der:	Meine Organisation			
	Stimmt dem voll und ganz zu	Stimmt dem mehrheitlich zu	Stimmt dem mehrheitlich nicht zu	Stimmt dem ganz und gar nicht zu
Gewässerschutzregulierung	❏	❏	❏	❏
Chemikalienregulierung	❏	❏	❏	❏
Landwirtschaftsregulierung	❏	❏	❏	❏
Trinkwasserregulierung	❏	❏	❏	❏
Keine Regulierungsaufgabe, sondern Aufgabe des individuellen Konsumentenverhaltens	❏	❏	❏	❏

9. Nachfolgend finden Sie eine Liste möglicher regulativer Instrumente, die zur Verringerung von Mikroschadstoffen in den Gewässern beitragen können.

   Bitte geben Sie den Zustimmungsgrad Ihrer Organisation zur Regulierung von Mikroschadstoffen durch nachfolgende Instrumente an.

   Falls Sie regulative Instrumente auf der Liste vermissen, können Sie diese in den leeren Zeilen ergänzen und auch hier Ihren Zustimmungsgrad angeben.

Regulative Instrumente	Meine Organisation			
	Stimmt dem voll und ganz zu	Stimmt dem mehrheitlich zu	Stimmt dem mehrheitlich nicht zu	Stimmt dem ganz und gar nicht zu
Zulassungsverbote,- einschränkungen einzelner Mikroschadstoffe	❑	❑	❑	❑
Anwendungseinschränkungen einzelner Mikroschadstoffe	❑	❑	❑	❑
Entsorgungsvorschriften für Produkte, die Mikroschadstoffe enthalten	❑	❑	❑	❑
Festlegung der best-verfügbaren Technik zur Mikroschadstoffelimination (z.B. Aufrüstung der Kläranlagen, Behandlung von Abwasserteilströmen in Betrieben)	❑	❑	❑	❑
Festlegung der besten landwirtschaftlichen Praxis zur Verringerung des Eintrags von Mikroschadstoffen in Gewässer	❑	❑	❑	❑
Festlegung von Umweltqualitätsnormen / Immissionsgrenzwerten für Mikroschadstoffe	❑	❑	❑	❑
Festlegung von Emissionsgrenzwerten für Mikroschadstoffe	❑	❑	❑	❑
Abgabe auf Produkte, die Mikroschadstoffe enthalten	❑	❑	❑	❑
Erhöhung der Abwassergebühr für die Finanzierung von Maßnahmen zur Verringerung von Mikroschadstoffen	❑	❑	❑	❑
Steuerfinanzierte Subventionen (z.B. für Investitionen in Technologie zur Mikroschadstoffelimination, Monitoringtechnologie, optimierte Produktionsprozesse)	❑	❑	❑	❑
Kontrollverfahren (z.B. Erweiterung der Messprogramme, Registrierungspflicht von Mikroschadstoffen)	❑	❑	❑	❑
Freiwillige Maßnahmen durch Betriebe und Zivilgesellschaft (z.B. Technologie zur Mikroschadstoffelimination, optimierte Produktionsprozesse, Labeling, Verzicht)	❑	❑	❑	❑
Staatliche Informationskampagnen, Beratung durch staatliche Behörden	❑	❑	❑	❑
Forschung	❑	❑	❑	❑
Reduktionszielvereinbarungen mit Privatwirtschaft (Branchenvereinbarungen) oder zwischen unterschiedlichen Regierungsebenen (öffentlich-öffentliche Partnerschaft)	❑	❑	❑	❑
Andere:	❑	❑	❑	❑
	❑	❑	❑	❑

Teil D:
Kompetenzen Ihrer Organisation und geographische Eingrenzung

10. Welche Priorität hat die Reduktion von Mikroschadstoffen in Gewässern im Vergleich zu anderen Aufgabenbereichen der Wasserregulierung aus Sicht Ihrer Organisation?

Bitte schätzen Sie ein, in welchen Aufgabenbereichen der Wasserregulierung Maßnahmen mit höherer, gleich hoher oder niedrigerer <u>Priorität</u> im Vergleich zum Bereich Mikroschadstoffe zu treffen sind.

Sollten Sie Aufgaben hinzugefügt haben, bewerten Sie bitte auch hier die Priorität.

Aufgabenbereiche der Wasserregulierung	Höhere Priorität als Mikroschadstoffe	Gleiche hohe Priorität wie Mikroschadstoffe	Niedrigere Priorität als Mikroschadstoffe
Gewässerökologie / Renaturierung	❏	❏	❏
Nährstoffe / Landwirtschaft	❏	❏	❏
Abwasserreinigung (nicht gemeint sind weitergehende Verfahren zur Elimination von Mikroschadstoffen)	❏	❏	❏
Trinkwasser	❏	❏	❏
Wasserkraft	❏	❏	❏
Hochwasser	❏	❏	❏
Gewässerüberwachung,-messungen	❏	❏	❏
Andere:	❏	❏	❏
	❏	❏	❏

11. a) *Bitte geben Sie an, auf welcher Ebene sich Ihre Organisation im Bereich Gewässerschutz* <u>*formell*</u> <u>*zuständig*</u> *sieht, d.h. Regulierungs-, Umsetzungs-, oder Beratungsaufgaben ausübt.*

Mehrere Antworten sind möglich.

Meine Organisation ist zuständig auf folgender/n Ebene/n:	Bitte präzisieren Sie
❑ Lokal	
	.... (Name der Gemeinde/ Stadt/des Kreises/Bezirks)
❑ Bundesland	
	.... (Name des Bundeslandes)
❑ National	
❑ Europa	
❑ Andere	
	.... (Bitte beschreiben Sie)

11. b) Die Ressource Wasser kann Gemeinde-, Länder- und Landesgrenzen überschreiten und somit definierte politische Grenzen. Kann es vorkommen, dass gewässerrelevante Aktivitäten Ihrer Organisation die von Ihnen unter 11a) angegebene Eingrenzung überschreiten (oder auch unterschreiten), um sich der geographischen Ausbreitung der Ressource Wasser anzupassen?

❑   Nein, die politischen Grenzen sind gut auf die Aktivitäten meiner Organisation abgestimmt.

❑   Ja, gewässerrelevante Aktivitäten meiner Organisation können die unter 11a) angegebene Eingrenzung überschreiten bzw. unterschreiten.

Bitte geben Sie an, worin die Diskrepanz besteht und nennen Sie Gründe dafür:

....

....

....

....

12. *Bitte geben Sie an, mit welchen Akteuren Ihre Organisation zum Thema Mikroschadstoffe im Einzugsgebiet des Rheins **international** <u>zusammenarbeitet</u>.*

Meine Organisation arbeitet **international** zusammen mit:	Bitte präzisieren Sie
☐ Arbeitsgruppen:	.... (Namen der Arbeitsgruppen)
☐ Gemeinden / Städten (gemeint ist außerhalb Deutschlands):	.... (Namen der Gemeinden/Städte)
☐ Regionen / Wasseragenturen / Einzugsgebietsorganisationen (außerhalb Deutschlands):	.... (Namen der Regionen/Agenturen)
☐ Anderen Ländern:	.... (Namen der Länder)
☐ Europäischen Akteuren:	.... (Namen der EU Akteure)
☐ Internationalen / ausländischen Verbänden:	.... (Namen der Verbände)
☐ Universitäten / Fachhochschulen aus dem Ausland:	.... (Namen der Universitäten)
☐ Andere:	.... (Bitte präzisieren Sie)

Wir danken Ihnen für Ihre wertvolle Mitarbeit!

Haben Sie noch weitere Anmerkungen oder Ideen zum Thema Mikroschadstoffe oder zum Fragebogen, die Sie mit uns teilen möchten?

....

....

....

....

Bitte senden Sie den ausgefüllten Fragebogen mit beiliegendem frankierten Antwortcouvert an:

Florence Metz
Universität Bern, Institut für Politikwissenschaft
Fabrikstrasse 8
CH-3012 Bern

Annex 8
French Questionnaire

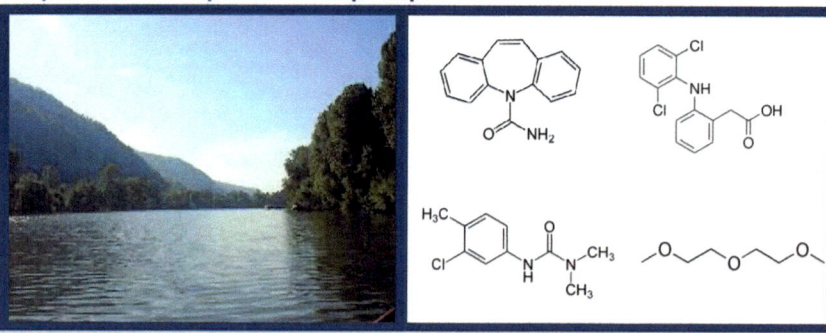

Cette interview fait partie d'un projet de recherche de l'Université de Berne et de l'Institut de Recherche de l'Eau (EAWAG) sur la réduction des micropolluants dans le bassin versant du Rhin. L'objectif est d'**étudier le processus de décision qui a mené à l'adoption du «Plan Micropolluants» et des plans liés (PNRM, Ecophyto).**

L'interview comprend 12 questions et se divise en 4 parties :

Partie A : Participation de votre organisation dans le processus de décision du Plan Micropolluants
Partie B : Collaboration de votre organisation avec d'autres acteurs sur le sujet des micropolluants
Partie C : Position de votre organisation sur la réglementation des micropolluants
Partie D : Compétences de votre organisation dans le domaine de l'eau

Comme votre organisation joue un rôle central dans la politique de l'eau envers les micropolluants, votre participation à l'enquête est particulièrement essentielle.

Nous vous prions de bien vouloir *répondre du point de vue de votre organisation*. L'interview prend entre 30 et 60 minutes. Merci beaucoup d'avance pour votre soutien précieux.

Le projet de recherche est réalisé dans le cadre d'une thèse de doctorat et est financé par le Fonds National Suisse (FNS). Les données sont collectées à des fins de recherche, elles restent confidentielles et ne seront pas transmises à des tiers. Dès que les données seront complètes, nous serions heureux de partager avec vous nos résultats de recherche.

Directrice du projet de recherche:

Prof. Dr. Karin Ingold
Institut d'Etudes Politiques Bern, Eawag

Partenaires et Sponsor:

u^b

UNIVERSITÄT
BERN

eawag
aquatic research ●○○○

FNS NF
SCHWEIZERISCHER NATIONALFONDS
ZUR FÖRDERUNG DER WISSENSCHAFTLICHEN FORSCHUNG

Si vous avez des questions n'hésitez pas à contacter directement :

Florence Metz (Doctorante)
Universität Bern, Institut für Politikwissenschaft, Lerchenweg 36, CH-3000 Bern 9
florence.metz@ipw.unibe.ch
N° mobil: 0041 (0)76 611 71 79, n° bureau: 0041 (0)31 631 48 22

Nom de la personne interviewé:..

Nom de votre organisation et section:...

    Email:...

    N° téléphone:...

Souhaitez-vous recevoir des informations sur les résultats de cette étude ?   Oui ❑ Non ❑

Partie A:
Le processus de décision du Plan Micropolluants

1. La liste suivante regroupe les étapes les plus importantes du processus de décision du Plan Micropolluants. Dans quelles phases décisionnelles du Plan Micropolluants avez-vous participé ?

Veuillez cocher toutes les phases dans lesquelles votre organisation a participé.

Par « participation » nous entendons : apporter une contribution active, participer aux groupes de travail / comités consultatifs / colloques et se prononcer dans des consultations formelles ou informelles.

Phase Date	Étape	Participé
Plan Micropolluants 2010-2013		
Déclencheur Avant 2009	Directive 76/464/CEE Directive Cadre de l'Eau 2000/60/CE PNAR Plan national d'action contre la pollution de milieux aquatiques par certaines substances dangereuses RSDE Action nationale de recherche et de réduction des substances dangereuses pour le milieu aquatique	❑
Conception 2009 – 2010	Réunions pour l'élaboration du plan micropolluants (9.7.2009 et 1.4.2010) Journées programmatiques ONEMA « Micropolluants aquatiques » (10. - 12.3.2010)	❑
Elaboration *Mars 2010*	Avant-projet du Plan Micropolluants	❑
Consultation *Avril 2010*	Avant-projet de plan transmis aux acteurs pour remarques (1.4.2010 - 30.4.2010)	❑
Finalisation *Mai – Juin 2010*	Finalisation du projet du Plan Micropolluants	❑
Approbation *6.7.2010*	Approbation du Plan Micropolluants par le Comité National de l'Eau (6.7.2010)	❑
Adoption *Sept – Nov 2010*	Communication en Conseil des Ministres du Plan Micropolluants (13.10.2010)	❑
Suivi *Depuis 2011*	Rapport « Bilan de présence des micropolluants dans les milieux aquatiques continentaux. Période 2007-2009 » (publié le 17.10.2011) Colloque micropolluants « Plan micropolluants 2010-2013 : quelles avancées en un an ? » (18.10.2011)	❑

J'ai participé à une autre étape du processus de décision / à un autre processus de décision :

❑   Elaboration du Plan National sur les Résidus de Médicaments (PNRM)

❑   Elaboration du Plan Ecophyto 2018

❑   Autre (veuillez préciser) :

....

....

❑   Je n'ai pas participé au processus de décision du Plan Micropolluants, ni aux plans liés.

2. Le plan Micropolluants a pour objectif de définir une stratégie pour réduire la pollution des milieux aquatiques par les micropolluants. L'approche française au plan national s'appuie avant tout sur la surveillance de la présence de micropolluants dans les milieux aquatiques et les rejets. Basées sur ce diagnostic de l'état de contamination, des mesures ciblées seront prises pour réduire les micropolluants.

A quel point est-ce que votre organisation soutient cette approche ?

Mon organisation soutient cette approche			
Tout à fait	Plutôt oui	Plutôt non	Pas du tout
❑	❑	❑	❑

Autres remarques :...

....

....

....

....

3. *A quel point est-ce que les intérêts de votre organisation ont été pris en compte dans le processus de décision du Plan Micropolluants?*

Les intérêts de mon organisation ont été pris en compte			
Tout à fait	Plutôt oui	Plutôt non	Pas du tout
❑	❑	❑	❑

Autres remarques :...

....

....

....

....

Partie B:
Les acteurs

4. Importance des acteurs

Le processus de décision des plans et actions consacrés à la réglementation des micropolluants a associé de nombreux acteurs. Vous trouverez la liste des parties prenantes ci-dessous.

Dans la première colonne, veuillez cocher tous les acteurs qui, selon vous, ont contribué de manière importante dans le processus de décision du Plan Micropolluants.

Parmi les acteurs que vous avez décrits comme importants, qui étaient les trois acteurs les plus importants selon vous? **Dans la deuxième colonne, veuillez cocher exactement trois acteurs parmi toute la liste.**

Veuillez aussi évaluer l'importance des acteurs ajoutés.

Acteurs	Rôle important	Les 3 plus importants
Pouvoirs publics nationaux		
DEB Direction de l'eau et de la biodiversité	❏	❏
DGPR Direction générale de la prévention des risques	❏	❏
ONEMA Office national de l'eau et des milieux aquatiques	❏	❏
Ministère de la santé	❏	❏
Ministère de l'agriculture	❏	❏
ASN Autorité de sûreté nucléaire	❏	❏
CNE Comité national de l'eau	❏	❏
Instituts publics		
AQUAREF	❏	❏
INERIS Institut National de l'Environnement Industriel et des Risques	❏	❏
BRGM Bureau de Recherches Géologiques et Minières	❏	❏
IFREMER Institut Français de Recherche pour l'Exploitation de la Mer	❏	❏
IRSTEA (CEMAGREF) Institut national de recherche en sciences et technologies pour l'environnement et l'agriculture	❏	❏
LNE Laboratoire national de métrologie et d'essais	❏	❏
CNRS Centre national de la recherche scientifique	❏	❏
ANSES Agence nationale de sécurité sanitaire	❏	❏
ADEME Agence de l'Environnement et de la Maîtrise de l'Energie	❏	❏
Services déconcentrés		
Préfet de région	❏	❏
Préfet de département	❏	❏
Préfet coordonnateur de bassin	❏	❏
Police de l'eau et des installations classées	❏	❏
Inspection des installations classées	❏	❏
DREAL Directions Régionales de l'Environnement, de l'Aménagement et du Logement	❏	❏
DDT Directions Départementales des Territoires	❏	❏
Collectivités		
ARF Association des régions de France	❏	❏
ADF Assemblée des départements de France	❏	❏
AMF Association des maires de France	❏	❏

Acteurs	Rôle important	Les 3 plus importants
Bassins versants et sous-unités		
Agences de l'eau	❏	❏
Comité de bassin	❏	❏
CLE Commissions locales de l'eau	❏	❏
Association environnementale		
Robin des bois	❏	❏
FNE France Nature Environnement	❏	❏
WWF	❏	❏
Amis de la terre	❏	❏
Alsace Nature	❏	❏
Associations d'approvisionnement en eau et de l'élimination des eaux usées		
FP2E Fédération Professionnelle des Entreprises de l'Eau	❏	❏
FNCCR Fédération nationale des collectivités concédantes et régies	❏	❏
Associations de l'industrie		
FENARIVE Fédération Nationale des Associations de Riverains et Utilisateurs Industriels de l'Eau	❏	❏
MEDEF Mouvement des entreprises de France	❏	❏
UIC Union des Industries Chimiques	❏	❏
CNIDEP Centre National d'Innovation pour le Développement durable et l'Environnement dans les Petites entreprises	❏	❏
CCI France Chambres de Commerce et d'Industrie	❏	❏
EDF Electricité de France	❏	❏
COPACEL Confédération Française des Industries des Cartons, Papiers et Celluloses	❏	❏
UFIP Union Française des Industries Pétrolières	❏	❏
FIM Fédération des Industries Mécaniques	❏	❏
Agriculture		
Chambres d'agriculture	❏	❏
Associations des consommateurs		
UFC-Que Choisir	❏	❏
CLCV Consommation, logement et cadre de vie	❏	❏
Autres		
	❏	❏
	❏	❏
	❏	❏

5. <u>Collaboration avec d'autres acteurs</u>

Dans le processus de décision des plans et actions consacrés à la réglementation des micropolluants, de nombreux acteurs ont collaboré.

Parmi la même liste d'acteurs, veuillez cocher <u>tous les acteurs</u> avec qui votre organisation a <u>étroitement collaboré</u> lors de la réglementation des micropolluants.

Collaboration ne signifie cependant pas forcément avoir la même opinion ou les mêmes objectifs. Par « collaboration » nous entendons : discuter des résultats scientifiques, élaborer des propositions, échanger les positions, évaluer les alternatives.

Cette question est importante pour comprendre la gouvernance de l'eau. Les noms personnels ne seront pas publiés et leur confidentialité est préservée.

S'il manque des acteurs, vous pouvez compléter la liste dans les lignes vides et indiquer votre collaboration avec les acteurs ajoutés.

Acteurs	Collaboration étroite
Pouvoirs publics nationaux	
DEB Direction de l'eau et de la biodiversité	❑
DGPR Direction générale de la prévention des risques	❑
ONEMA Office national de l'eau et des milieux aquatiques	❑
Ministère de la santé	❑
Ministère de l'agriculture	❑
ASN Autorité de sûreté nucléaire	❑
CNE Comité national de l'eau	❑
Instituts publics	
AQUAREF	❑
INERIS Institut National de l'Environnement Industriel et des Risques	❑
BRGM Bureau de Recherches Géologiques et Minières	❑
IFREMER Institut Français de Recherche pour l'Exploitation de la Mer	❑
IRSTEA (CEMAGREF) Institut national de recherche en sciences et technologies pour l'environnement et l'agriculture	❑
LNE Laboratoire national de métrologie et d'essais	❑
CNRS Centre national de la recherche scientifique	❑
ANSES Agence nationale de sécurité sanitaire	❑
ADEME Agence de l'Environnement et de la Maîtrise de l'Energie	❑
Services déconcentrés	
Préfet de région	❑
Préfet de département	❑
Préfet coordonnateur de bassin	❑
Police de l'eau et des installations classées	❑
Inspection des installations classées	❑
DREAL Directions Régionales de l'Environnement, de l'Aménagement et du Logement	❑
DDT Directions Départementales des Territoires	❑
Collectivités	
ARF Association des régions de France	❑
ADF Assemblée des départements de France	❑
AMF Association des maires de France	❑

Acteurs	Collaboration étroite
Bassins versants et sous-unités	
Agences de l'eau	❑
Comité de bassin	❑
CLE Commissions locales de l'eau	❑
Association environnementale	
Robin des bois	❑
FNE France Nature Environnement	❑
WWF	❑
Amis de la terre	❑
Alsace Nature	❑
Associations d'approvisionnement en eau et de l'élimination des eaux usées	
FP2E Fédération Professionnelle des Entreprises de l'Eau	❑
FNCCR Fédération nationale des collectivités concédantes et régies	❑
Associations de l'industrie	
FENARIVE Fédération Nationale des Associations de Riverains et Utilisateurs Industriels de l'Eau	❑
MEDEF Mouvement des entreprises de France	❑
UIC Union des Industries Chimiques	❑
CNIDEP Centre National d'Innovation pour le Développement durable et l'Environnement dans les Petites entreprises	❑
CCI France Chambres de Commerce et d'Industrie	❑
EDF Electricité de France	❑
COPACEL Confédération Française des Industries des Cartons, Papiers et Celluloses	❑
UFIP Union Française des Industries Pétrolières	❑
FIM Fédération des Industries Mécaniques	❑
Agriculture	
Chambres d'agriculture	❑ ❑
Associations des consommateurs	
UFC-Que Choisir	❑
CLCV Consommation, logement et cadre de vie	❑
Autres	
	❑ ❑

6. *Veuillez cocher tous les acteurs avec qui votre organisation a eu principalement des <u>convergences ou des divergences</u> sur le contenu des plans et actions consacrés à la réglementation des micropolluants. Les deux sont possibles simultanément.*

Par convergence ou divergence nous entendons des positions concordantes ou, au contraire, des positions discordantes, indépendamment du fait que vous ayez collaboré ou non avec les acteurs en question.

S'il manque des acteurs, vous pouvez compléter la liste dans les lignes vides et évaluer le profil de convergence et divergence.

Acteurs	Plutôt convergences	Plutôt divergences
Pouvoirs publics nationaux		
DEB Direction de l'eau et de la biodiversité	❑	❑
DGPR Direction générale de la prévention des risques	❑	❑
ONEMA Office national de l'eau et des milieux aquatiques	❑	❑
Ministère de la santé	❑	❑
Ministère de l'agriculture	❑	❑
ASN Autorité de sûreté nucléaire	❑	❑
CNE Comité national de l'eau	❑	❑
Instituts publics		
AQUAREF	❑	❑
INERIS Institut National de l'Environnement Industriel et des Risques	❑	❑
BRGM Bureau de Recherches Géologiques et Minières	❑	❑
IFREMER Institut Français de Recherche pour l'Exploitation de la Mer	❑	❑
IRSTEA (CEMAGREF) Institut national de recherche en sciences et technologies pour l'environnement et l'agriculture	❑	❑
LNE Laboratoire national de métrologie et d'essais	❑	❑
CNRS Centre national de la recherche scientifique	❑	❑
ANSES Agence nationale de sécurité sanitaire	❑	❑
ADEME Agence de l'Environnement et de la Maîtrise de l'Energie	❑	❑
Services déconcentrés		
Préfet de région	❑	❑
Préfet de département	❑	❑
Préfet coordonnateur de bassin	❑	❑
Police de l'eau et des installations classées	❑	❑
Inspection des installations classées	❑	❑
DREAL Directions Régionales de l'Environnement, de l'Aménagement et du Logement	❑	❑
DDT Directions Départementales des Territoires	❑	❑
Collectivités		
ARF Association des régions de France	❑	❑
ADF Assemblée des départements de France	❑	❑
AMF Association des maires de France	❑	❑

Acteurs	Plutôt convergences	Plutôt divergences
Bassins versants et sous-unités		
Agences de l'eau	❑	❑
Comité de bassin	❑	❑
CLE Commissions locales de l'eau	❑	❑
Association environnementale		
Robin des bois	❑	❑
FNE France Nature Environnement	❑	❑
WWF	❑	❑
Amis de la terre	❑	❑
Alsace Nature	❑	❑
Associations d'approvisionnement en eau et de l'élimination des eaux usées		
FP2E Fédération Professionnelle des Entreprises de l'Eau	❑	❑
FNCCR Fédération nationale des collectivités concédantes et régies	❑	❑
Associations de l'industrie		
FENARIVE Fédération Nationale des Associations de Riverains et Utilisateurs Industriels de l'Eau	❑	❑
MEDEF Mouvement des entreprises de France	❑	❑
UIC Union des Industries Chimiques	❑	❑
CNIDEP Centre National d'Innovation pour le Développement durable et l'Environnement dans les Petites entreprises	❑	❑
CCI France Chambres de Commerce et d'Industrie	❑	❑
EDF Electricité de France	❑	❑
COPACEL Confédération Française des Industries des Cartons, Papiers et Celluloses	❑	❑
UFIP Union Française des Industries Pétrolières	❑	❑
FIM Fédération des Industries Mécaniques	❑	❑
Agriculture		
Chambres d'agriculture	❑	❑
Associations des consommateurs		
UFC-Que Choisir	❑	❑
CLCV Consommation, logement et cadre de vie	❑	❑
Autres		
	❑	❑
	❑	❑
	❑	❑

Partie C:
Positions de votre organisation

7. Ci-dessous vous trouverez une liste avec différents objectifs concernant la réduction de micropolluants dans les milieux aquatiques.

Veuillez indiquer dans quelle mesure votre organisation est d'accord avec les <u>objectifs</u> suivants.

Objectifs	Mon organisation est			
	Tout à fait d'accord	Plutôt d'accord	Plutôt pas d'accord	Pas du tout d'accord
Les mesures doivent agir à la source de la contamination.	❏	❏	❏	❏
Les mesures doivent agir « end-of-pipe » (traitement des eaux usées.	❏	❏	❏	❏
Même si les conséquences des micropolluants ne sont pas étudiées en détail, des mesures préventives doivent être prises.	❏	❏	❏	❏
Tant que les conséquences des micropolluants ne sont pas étudiées en détail, *aucune* mesure ne doit être prise.	❏	❏	❏	❏
L'objectif des mesures doit être d'éliminer aussi complètement que possible les micropolluants dans les milieux aquatiques.	❏	❏	❏	❏

8. Pour réduire les micropolluants dans les milieux aquatiques, des mesures peuvent être prises dans différents domaines de compétence.

Dans quel <u>domaine</u> de compétence des mesures pour réduire les micropolluants dans les milieux aquatiques devraient être prises selon vous ?

La réduction des micropolluants doit être réglementée dans le domaine	Mon organisation est			
	Tout à fait d'accord	Plutôt d'accord	Plutôt pas d'accord	Pas du tout d'accord
Protection de l'eau / des milieux aquatiques	❏	❏	❏	❏
Substances chimiques	❏	❏	❏	❏
Agriculture	❏	❏	❏	❏
Eau potable	❏	❏	❏	❏
Comportement des consommateurs	❏	❏	❏	❏

9. La liste suivante vous propose des instruments politiques qui peuvent contribuer à la réduction des micropolluants dans les milieux aquatiques et les rejets.

Lesquels des <u>instruments politiques</u> suivants seraient d'après votre organisation les <u>plus favorisés</u> pour réglementer les micropolluants ?

Instruments politiques	Mon organisation est			
	Tout à fait d'accord	Plutôt d'accord	Plutôt pas d'accord	Pas du tout d'accord
Restrictions pour la mise sur le marché de substances problématiques	❏	❏	❏	❏
Restrictions d'utilisation de substances problématiques	❏	❏	❏	❏
Règles pour le tri des déchets contenant des substances problématiques	❏	❏	❏	❏
Etablir des «Meilleures Techniques Disponibles» (MTD) pour l'élimination des micropolluants dans les eaux usées	❏	❏	❏	❏
Etablir des «Bonnes Pratiques Agricoles» (BPA) pour réduire l'émission aux milieux aquatiques	❏	❏	❏	❏
Etablir des Normes de Qualité Environnementale (NQE) pour micropolluants	❏	❏	❏	❏
Etablir des Valeurs Limites pour micropolluants	❏	❏	❏	❏
Taxe sur les produits qui contiennent des substances problématiques	❏	❏	❏	❏
Augmenter la redevance d'assainissement / de l'eau pour financer les mesures de réduction	❏	❏	❏	❏
Subventions provenant des impôts pour financer les mesures de réduction	❏	❏	❏	❏
Surveillance des milieux aquatiques	❏	❏	❏	❏
Mesures volontaires par les entreprises et la société	❏	❏	❏	❏
Campagnes publiques d'information et consultation	❏	❏	❏	❏
Recherche	❏	❏	❏	❏
Contrat d'objectifs de réductions avec le secteur privé ou les Agences de l'Eau	❏	❏	❏	❏
Autres:	❏	❏	❏	❏
	❏	❏	❏	❏

Partie D:
Compétences et collaboration internationale

10. Dans lesquels des domaines suivants, selon vous, des mesures doivent être prises avec une priorité supérieure, égale ou inférieure à la réduction des micropolluants.

Veuillez évaluer les <u>priorités</u> que votre organisation accorde aux domaines suivants par rapport à l'importance accordée à la réduction des micropolluants.

Si vous avez ajouté un domaine de responsabilité, veuillez aussi évaluer sa priorité s'il vous plaît.

Domaines de responsabilité:	Priorité supérieure aux micropolluants	Priorité égale aux micropolluants	Priorité inférieure aux micropolluants
Bon état écologique des eaux / revitalisation	❑	❑	❑
Eutrophisation / nitrates / macropolluants	❑	❑	❑
Assainissement (type secondaire)	❑	❑	❑
Eau potable	❑	❑	❑
Hydroélectricité	❑	❑	❑
Inondations	❑	❑	❑
Surveillance d'autres substances que les micropolluants	❑	❑	❑
Autres (veuillez préciser):			

11.1 Veuillez indiquer à quel niveau votre organisation est <u>formellement compétente</u> pour décider, mettre en œuvre ou conseiller en matière de la protection des eaux. Plusieurs réponses possibles.

Mon organisation est compétente au niveau / aux niveaux:	Veuillez préciser:
❑ Local	....(Nom de la/les commune/s)
❑ Départemental	....(Nom du département)
❑ Régional	....(Nom de la région)
❑ De l'agence / du bassin hydrographique	....(Nom de l'agence / du bassin)
❑ National	
❑ Européen	
❑ Autres:	

11.2 En matière de la ressource de l'eau, les compétences formelles peuvent diverger du champ d'activité effectif. *Si cela est le cas, veuillez indiquer l'espace géographique dans lequel votre organisation est effectivement active en matière de la protection des eaux.*

Mon organisation est surtout active dans l'espace géographique:	Veuillez préciser:
❑ Local	....(Nom de la/les commune/s)
❑ Départemental	....(Nom du/des départements)
❑ Régional	....(Nom de la/des régions)
❑ De (sous)bassin hydrographique	....(Nom du (sous)bassin)
❑ National	
❑ Européen	
❑ Autres:	

❑ Le champ d'activité ne diverge pas de l'espace défini par les compétences formelles.

12. *Veuillez indiquer si votre organisation maintient des collaborations <u>internationales</u> au sujet des micropolluants.*

	Nom des collaborations internationales:
❑ Groupes de travail ou comités	....(Nom du groupe de travail ou comité)
❑ Communes ou régions (à l'extérieur du territoire français)	....(Nom des communes ou régions étrangère/s)
❑ Agences de bassin ou bassins hydrographiques (à l'extérieur du territoire français)	....(Nom des agences de bassin ou bassins hydrographiques)
❑ Autres pays	....(Nom des pays)
❑ Acteurs européens	....(Nom des acteurs européens)
❑ Associations internationales / étrangères	....(Nom des associations)
❑ Universités étrangères	....(Nom des universités)
❑ Autres	....(Veuillez préciser)

Nous vous remercions pour votre collaboration !

Avez-vous d'autres remarques ou idées concernant les micropolluants ou concernant le questionnaire ?
N'hésitez pas à les partager avec nous?

Si vous avez des questions n'hésitez pas à contacter directement :

Florence Metz
Universität Bern, Institut für Politikwissenschaft
Lerchenweg 36
CH-3000 Bern 9

florence.metz@ipw.unibe.ch
Mobil: 0041 (0)76 611 71 79
Bureau: 0041 (0)31 631 4822

Annex 9
Dutch Questionnaire

Your Views Regarding and Participation in Aquatic Micropollution Policies
2014 Survey of Policy Actors

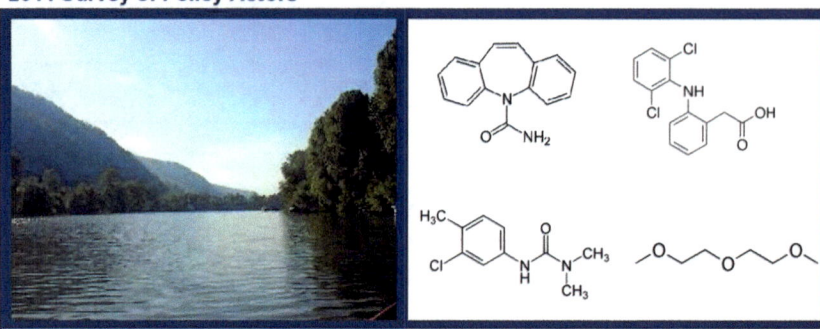

General Instructions

As mentioned in the cover letter, this questionnaire is part of a research project of the University of Twente in collaboration with the University of Berne and the Swiss Water Research Institute Eawag. The aim is to understand the process of developing policy measures for the reduction of pharmaceutical micropollution in Dutch waters that began in 1997. As your organization has a central position in the policy process under study, we need your help and practitioner's expertise to succeed in our research.

This questionnaire contains 12 questions in total and will allow you to express the views and experiences of your organization on:

      Part A: Participation of your organization in the policy process on pharmaceutical micropollution
      Part B: Collaboration of your organization with other actors in the policy process
      Part C: Views of your organization on the various policy options to reduce pharmaceutical
            micropollution
      Part D: Responsibilities of your organization in water protection

We kindly ask you to participate in the survey and *respond to the questions from the perspective of your organization*. **It will take no longer than 30 minutes** to complete the questionnaire. In return for sharing your views we will be happy to inform you of the results of the study.

Please return the completed questionnaire by mail in the enclosed pre-paid envelope **by 9 May 2014**. Thank you in advance for sharing your thoughts and for your support.

The research project is part of a PhD dissertation supported by the Swiss National Science Foundations. The information will be used exclusively for research purposes and will not be shared with any third parties.

Project Director:

Prof. Dr. Karin Ingold
Institute for Political Science University of Berne, Eawag

Project Partner and Sponsors:

u^b

UNIVERSITÄT
BERN

eawag
aquatic research ○○○

FNS NF
SCHWEIZERISCHER NATIONALFONDS
ZUR FÖRDERUNG DER WISSENSCHAFTLICHEN FORSCHUNG

If you have questions, please don't hesitate to contact:

Florence Metz (PhD)
University of Berne, Institute for Political Science, Fabrikstrasse 8, CH-3012 Berne
florence.metz@ipw.unibe.ch
Tel. 0049 (0)163 60 77 689

Name of the person completing the questionnaire:..

Name of your organization:...

Phone:...Email: ....

Would you like to receive information about the results of the study?   Yes ❑   No ❑

Part A:
The Policy Process on Measures for the Reduction of Pharmaceutical Micropollution

1. The following table summarizes the ongoing policy process since 1997 during which measures for the reduction of pharmaceutical micropollution have been developed. In which phases of the process did your organization participate?
   Please check all phases in which your organization participated.

   Participation is defined as: being actively involved in and contributing to research or implementation; participating in working groups, workshops or informal consultations.

Date	Phase	Participation
Reduction of pharmaceutical micropollution		
Trigger 1997 - 2002	1997 – 2001: Tweede Kamer inquiry to Minister of Environment about risks arising from endocrine substances in waters resulting in "Strategienota Omgaan Met Stoffen - SOMS"	
	Report "**Milieurisico's van geneesmiddelen**", Gezondheidsraad (no. 2001/17)	❑
	RIZA reports „Vergeten stoffen in Nederlands oppervlaktewater" (no. 2001-020) and „Estrogens in the aquatic environment" (no. 2002-001)	
Concept phase 2002 - 2013	Set up of "**Interdepartementale werkgroep (dier)geneesmiddelen en het watermilieu**" to disucss policy options (Kamerstuk nr. 28808-35)	❑
	Implementation of pilot measures (e.g. monitoring, wastewater treatment, elektronisch patiënten dossier) (30.9.2009 nr.30535 / 27625-19)	
Parliamentary involvement 2007 - 2014	Environmental Ministry informs the Tweede Kamer of policy options and pilot measures (21.2.2007 nr. 28808-39; 30.9.2009 nr.30535; 4.9.2012 no. 27625-281; 25.6.2013 no. 27625-305)	❑
	Inquiry by Groenlinks about state of improved wastewater treatment (25.3.2010 no. 27625-281)	
	Meeting of Vaste Commissie voor Infrastructuur en Milieu (27.6.2013) to discuss results of pilot measures	
	Motion 27625-299 and 27625-300 calling for regulation of pharmaceutical micropollution in waters	
	Parliamentary round-table conference "Geneesmiddelen en waterkwaliteit" (30.1.2014)	
Further research 2013	Report "Evaluatie screening RWS 2011-2012.Rapportage screeningsonderzoek van microverontreinigingen in de Nederlandse oppervlaktewateren van Rijkswaterstaat" (20.9.2013)	❑
	BTO/KWR report "Vóórkomen en voorkómen van geneesmiddelen in bronnen van drinkwater" (Nov 2013)	

❑ *Have you been involved in measures for the reduction of pharmaceutical micropollution in waters in another way? If yes, please indicate with an "x" and describe your involvement below.*

Further remarks: ....

❑ My organization did not participate in the policy process on pharmaceutical micropollution in waters.

2. *From the perspective of your organization, are you satisfied with the degree to which the positions, ideas or responsibilities of your organization have been taken into consideration in the policy process on pharmaceutical micropollution (1997 – present)?*

The positions of my organization were			
Strongly taken into consideration	Somewhat taken into consideration	Not taken into consideration enough	Not taken into consideration at all
❏	❏	❏	❏

Further remarks: ....
....

3. a) The policy process on pharmaceutical micropollution (1997 – present) aims to develop measures for the reduction of pharmaceutical micropollution in Dutch waters. Measures adopted so far include research about concentration levels of pharmaceuticals in waters as well as pilot measures to separately treat wastewater from hospitals or other healthcare facilities.

To what degree does your organization support this policy approach?

My organization supports this policy approach			
Strongly	Somewhat	Rather not	Not at all
❏	❏	❏	❏

Further remarks:...
....

b) From the perspective of your organization are the policy measures adopted so far comprehensive enough?

Policy measures taken so far are			
Far too comprehensive	Somewhat too comprehensive	Rather not comprehensive enough	Not at all comprehensive enough
❏	❏	❏	❏

Further remarks:...
....

Part B:
Actors in the Policy Process

4. Importance of actors

A number of actors have been involved in the policy process on pharmaceutical micropollution (1997 – present). The following table attempts to comprehensively enumerate the involved actors.

In the first column, please check all the actors that have been <u>particularly important</u> in the policy process from the point of view of your organization.

*In the second column, please make **exactly three crosses for the whole list** to indicate which actors are the <u>three most important</u> in the policy process.*

If there are actors missing, please add them to the bottom of the list and evaluate their importance.

Actors	Particularly important	3 most important
National		
IenM Ministerie van Infrastructuur en Milieu	☐	☐
RWS Rijkswaterstaat	☐	☐
VWS Ministerie van Volksgezondheid, Welzijn en Sport	☐	☐
RIVM Rijksinstituut voor Volksgezondheid en Milieu	☐	☐
EZ Ministerie van Economische Zaken	☐	☐
RVO Rijksdienst voor Ondernemend Nederland	☐	☐
Voormalig LNV Ministerie van Landbouw, Natuurbeheer en Viedselkwaliteit	☐	☐
Voormalig V&W Ministerie van Verkeer en Waterstaat	☐	☐
Voormalig VROM Volkshuisvesting, Ruimtelijke Ordening en Milieubeheer	☐	☐
Gezondheidsraad	☐	☐
CBG / BD College ter Beoordeling van Geneesmiddelen / Bureau diergeneesmiddelen	☐	☐
Bestuurlijk Overleg Water	☐	☐
Adviescommissie Water	☐	☐
Voormalig RIKZ Rijksinstituut voor Kust en Zee	☐	☐
Provinces, municipalities and water boards		
IPO Interprovinciaal Overleg	☐	☐
UvW Unie van Waterschappen	☐	☐
VNG Vereniging van Nederlandse Gemeenten	☐	☐
Parliament and political parties		
Eerste Kamer der Staten-Generaal	☐	☐
Tweede Kamer der Staten-General	☐	☐
Vaste commissie voor Infrastructuur en Milieu, Tweede Kamer	☐	☐
Vaste commissie voor Volksgezondheid, Welzijn en Sport, Tweede Kamer	☐	☐
Vaste commissie voor Economische Zaken, Tweede Kamer	☐	☐
Partij ChristenUnie	☐	☐
Partij GroenLinks	☐	☐
Agricultural associations		
LTO Land- en Tuinbouw Organisatie	☐	☐

Actors	Particularly important	3 most important
Economic associations		
Vereniging VNO-NCW	☐	☐
VNCI Vereniging van de Nederlandse Chemische Industrie	☐	☐
VEMW Vereniging voor Energie, Milieu en Water	☐	☐
Pharmaceutical and health sector		
BOGIN Bond van de Generieke Geneesmiddelenindustrie Nederland	☐	☐
Nefarma Vereniging Innovatieve Geneesmiddelen Nederland	☐	☐
KNMP Koninklijke Nederlandse Maatschappij ter Bevordering der Pharmacie	☐	☐
SFK Stichting Farmaceutische Kengetallen	☐	☐
SWAB Stichting Werkgroep Antibioticabeleid	☐	☐
Reinier de Graaf ziekenhuis	☐	☐
Water associations		
VEWIN Vereniging van waterbedrijven in Nederland	☐	☐
Vitens	☐	☐
RIWA Vereniging van Rivierwaterbedrijven	☐	☐
Stichting Rioned	☐	☐
Environmental associations		
Coöperatieve Visserij Organisatie	☐	☐
Stichting Huize Aarde	☐	☐
Vereniging tot Behoud van Natuurmonumenten	☐	☐
WWF/WNF Wereld Natuur Fonds	☐	☐
Consumer associations		
Consumentenbond	☐	☐
Research and consultancy		
KWR Watercycle Research Institute	☐	☐
Deltares / voormalig RIZA	☐	☐
STOWA Stichting Toegepast Onderzoek Waterbeheer	☐	☐
Other		
	☐	☐
	☐	☐
	☐	☐
	☐	☐

5. Your collaboration with other actors

The following table shows the same actor list as previously shown in question 4.

Please check all the actors with whom your organization has closely collaborated during the policy process on pharmaceutical micropollution (1997 – present). Collaboration does not necessarily imply that you share the same views.
Close collaboration is defined as: discussing new findings, developing policy options, exchanging positions, evaluating alternatives.

This question is important for understanding how policy processes work. Your name will be kept confidential.

If there are actors missing, please add them to the bottom of the list and indicate with an "x" if you closely collaborate.

Actors	Close collaboration
National	
IenM Ministerie van Infrastructuur en Milieu	❏
RWS Rijkswaterstaat	❏
VWS Ministerie van Volksgezondheid, Welzijn en Sport	❏
RIVM Rijksinstituut voor Volksgezondheid en Milieu	❏
EZ Ministerie van Economische Zaken	❏
RVO Rijksdienst voor Ondernemend Nederland	❏
Voormalig LNV Ministerie van Landbouw, Natuurbeheer en Viedselkwaliteit	❏
Voormalig V&W Ministerie van Verkeer en Waterstaat	❏
Voormalig VROM Volkshuisvesting, Ruimtelijke Ordening en Milieubeheer	❏
Gezondheidsraad	❏
CBG / BD College ter Beoordeling van Geneesmiddelen / Bureau diergeneesmiddelen	❏
Bestuurlijk Overleg Water	❏
Adviescommissie Water	❏
Voormalig RIKZ Rijksinstituut voor Kust en Zee	❏
Provinces, municipalities and water boards	
IPO Interprovinciaal Overleg	❏
UvW Unie van Waterschappen	❏
VNG Vereniging van Nederlandse Gemeenten	❏
Parliament and political parties	
Eerste Kamer der Staten-Generaal	❏
Tweede Kamer der Staten-General	❏
Vaste commissie voor Infrastructuur en Milieu, Tweede Kamer	❏
Vaste commissie voor Volksgezondheid, Welzijn en Sport, Tweede Kamer	❏
Vaste commissie voor Economische Zaken, Tweede Kamer	❏
Partij ChristenUnie	❏
Partij GroenLinks	❏
Agricultural associations	
LTO Land- en Tuinbouw Organisatie	❏

Actors	Close collaboration
Economic associations	
Vereniging VNO-NCW	❏
VNCI Vereniging van de Nederlandse Chemische Industrie	❏
VEMW Vereniging voor Energie, Milieu en Water	❏
Pharmaceutical and health sector	
BOGIN Bond van de Generieke Geneesmiddelenindustrie Nederland	❏
Nefarma Vereniging Innovatieve Geneesmiddelen Nederland	
KNMP Koninklijke Nederlandse Maatschappij ter Bevordering der Pharmacie	❏
SFK Stichting Farmaceutische Kengetallen	❏
SWAB Stichting Werkgroep Antibioticabeleid	❏
Reinier de Graaf ziekenhuis	❏
Water associations	
VEWIN Vereniging van waterbedrijven in Nederland	❏
Vitens	❏
RIWA Vereniging van Rivierwaterbedrijven	❏
Stichting Rioned	❏
Environmental associations	
Coöperatieve Visserij Organisatie	❏
Stichting Huize Aarde	❏
Vereniging tot Behoud van Natuurmonumenten	❏
WWF/WNF Wereld Natuur Fonds	❏
Consumer associations	
Consumentenbond	❏
Research and consultancy	
KWR Watercycle Research Institute	❏
Deltares / voormalig RIZA	❏
STOWA Stichting Toegepast Onderzoek Waterbeheer	❏
Other	
	❏
	❏
	❏
	❏

6. Convergences and/or divergences with other actors

The following table shows the same actor list as previously shown in question 4 and 5.

Please check all the actors with whom your organization had convergences and/or divergences about policy content during the policy process on pharmaceutical micropollution (1997 – present). Converging or diverging about policy content does not necessarily imply collaboration.
Convergence is defined as agreement on policy content; divergence as disagreement

If there are actors missing, please add them to the bottom of the list and indicate your convergences and divergences.

Actors	Convergence	Divergence
National		
IenM Ministerie van Infrastructuur en Milieu	☐	☐
RWS Rijkswaterstaat	☐	☐
VWS Ministerie van Volksgezondheid, Welzijn en Sport	☐	☐
RIVM Rijksinstituut voor Volksgezondheid en Milieu	☐	☐
EZ Ministerie van Economische Zaken	☐	☐
RVO Rijksdienst voor Ondernemend Nederland	☐	☐
Voormalig LNV Ministerie van Landbouw, Natuurbeheer en Viedselkwaliteit	☐	☐
Voormalig V&W Ministerie van Verkeer en Waterstaat	☐	☐
Voormalig VROM Volkshuisvesting, Ruimtelijke Ordening en Milieubeheer	☐	☐
Gezondheidsraad	☐	☐
CBG / BD College ter Beoordeling van Geneesmiddelen / Bureau diergeneesmiddelen	☐	☐
Bestuurlijk Overleg Water	☐	☐
Adviescommissie Water	☐	☐
Voormalig RIKZ Rijksinstituut voor Kust en Zee	☐	☐
Provinces, municipalities and water boards		
IPO Interprovinciaal Overleg	☐	☐
UvW Unie van Waterschappen	☐	☐
VNG Vereniging van Nederlandse Gemeenten	☐	☐
Parliament and political parties		
Eerste Kamer der Staten-Generaal	☐	☐
Tweede Kamer der Staten-Generaal	☐	☐
Vaste commissie voor Infrastructuur en Milieu, Tweede Kamer	☐	☐
Vaste commissie voor Volksgezondheid, Welzijn en Sport, Tweede Kamer	☐	☐
Vaste commissie voor Economische Zaken, Tweede Kamer	☐	☐
Partij ChristenUnie	☐	☐
Partij GroenLinks	☐	☐
Agricultural associations		
LTO Land- en Tuinbouw Organisatie	☐	☐

Actors	Convergence	Divergence
Economic associations		
Vereniging VNO-NCW	☐	☐
VNCI Vereniging van de Nederlandse Chemische Industrie	☐	☐
VEMW Vereniging voor Energie, Milieu en Water	☐	☐
Pharmaceutical and health sector		
BOGIN Bond van de Generieke Geneesmiddelenindustrie Nederland	☐	☐
Nefarma Vereniging Innovatieve Geneesmiddelen Nederland	☐	☐
KNMP Koninklijke Nederlandse Maatschappij ter Bevordering der Pharmacie	☐	☐
SFK Stichting Farmaceutische Kengetallen	☐	☐
SWAB Stichting Werkgroep Antibioticabeleid	☐	☐
Reinier de Graaf ziekenhuis	☐	☐
Water associations		
VEWIN Vereniging van waterbedrijven in Nederland	☐	☐
Vitens	☐	☐
RIWA Vereniging van Rivierwaterbedrijven	☐	☐
Stichting Rioned	☐	☐
Environmental associations		
Coöperatieve Visserij Organisatie	☐	☐
Stichting Huize Aarde	☐	☐
Vereniging tot Behoud van Natuurmonumenten	☐	☐
WWF/WNF Wereld Natuur Fonds	☐	☐
Consumer associations		
Consumentenbond	☐	☐
Research and consultancy		
KWR Watercycle Research Institute	☐	☐
Deltares / voormalig RIZA	☐	☐
STOWA Stichting Toegepast Onderzoek Waterbeheer	☐	☐
Other		
	☐	☐
	☐	☐
	☐	☐
	☐	☐

Part C:
Positions of your Organization

7. Hereafter you find a list with different statements regarding the reduction of pharmaceutical micropollution in waters.

   Please indicate your organization's level of agreement with the following <u>*statements*</u>.

Statements	My Organisation			
	Strongly agrees	Agrees somewhat	Disagrees somewhat	Strongly disagrees
Measures should address the sources of pollution.	❑	❑	❑	❑
Measures should be end-of-pipe (waste-water treatment).	❑	❑	❑	❑
Preventive measures should be taken to reduce potential risks for humans and the environment (precautionary principle).	❑	❑	❑	❑
It is reasonable to wait with policy measures until the impact of micropollution is fully understood.	❑	❑	❑	❑
Policy measures should aim at *completely* eliminating pharmaceutical micropollution in waters.	❑ (completely)	❑ (largely)	❑ (only a few substances)	❑ (not at all)
The financial burden for adopting measures to reduce pharmaceutical micropollution in waters is too high.	❑	❑	❑	❑

8. Pharmaceutical micropollution can be reduced by adopting measures in diverse areas of policy responsibility and diverse levels (European to local level).

   In which <u>*area(s) of responsibility and at which level(s)*</u> *should measures for the reduction of micropollution mainly be adopted according to your organization?*

Micropollution is the responsibility of:	My organization			
	Strongly agrees	Agrees somewhat	Disagrees somewhat	Strongly disagrees
Water Protection Policy	❑	❑	❑	❑
Chemical Policy	❑	❑	❑	❑
Agricultural Policy	❑	❑	❑	❑
Drinking Water Policy	❑	❑	❑	❑
Health Policy	❑	❑	❑	❑
Reducing pharmaceutical micropollution is a consumer responsibility.	❑	❑	❑	❑
European level	❑	❑	❑	❑
National level	❑	❑	❑	❑
Provincial level	❑	❑	❑	❑
Water agency level	❑	❑	❑	❑
Municipal level	❑	❑	❑	❑

9. Below is a list of policy instruments which may contribute to the reduction of pharmaceutical micropollution in waters.

Please indicate your organization's level of agreement with adopting each of the following policy instruments for the reduction of pharmaceutical micropollution, independently of what has been done in the Netherlands thus far.

If there are policy instruments missing, please add them to the bottom of the list and indicate your level of agreement.

Policy Instruments	My organization			
	Strongly agrees	Agrees somewhat	Disagrees somewhat	Strongly disagrees
Bans or authorization restrictions of single pharmaceutical substances	❏	❏	❏	❏
Use restrictions of single pharmaceutical substances	❏	❏	❏	❏
Discharge requirements for products containing pharmaceutical substances	❏	❏	❏	❏
Use of best available technique (BAT) for the elimination of pharmaceutical micropollution (e.g. technically upgrading wastewater treatment plants, treatment of wastewater partial flows in companies or hospitals)	❏	❏	❏	❏
Use of best environmental practice (BEP) for the reduction of micropollution inputs into waters	❏	❏	❏	❏
Establishment of environmental quality norms = immission limit for pharmaceutical micropollutants	❏	❏	❏	❏
Definition of emission limits for micropollutants	❏	❏	❏	❏
Product charge for pharmaceuticals	❏	❏	❏	❏
Increase of the wastewater charge to fund measures for the reduction of pharmaceutical micropollution	❏	❏	❏	❏
Subsidies (e.g. for investments in filtering technology or monitoring technology, optimization of production processes)	❏	❏	❏	❏
Control measures (e.g. expanding monitoring programs, obligatory registries for pharmaceuticals)	❏	❏	❏	❏
Voluntary measures of companies and civil society (e.g. investments in filtering technology, optimize production processes, labeling, abdication)	❏	❏	❏	❏
Information campaigns, consulting	❏	❏	❏	❏
Research	❏	❏	❏	❏
Private-public partnerships Public-public partnerships	❏	❏	❏	❏
Other:	❏	❏	❏	❏
	❏	❏	❏	❏

Part D: Responsibilities of your Organization

10. To what extent does your organization prioritize the reduction of micropollution in waters?

    Please indicate if the reduction of micropollution has a higher <u>priority</u>, equal or lower priority in comparison to other water-related responsibilities.

    If you add further responsibilities, please indicate its level of priority in comparison to the reduction of micropollution.

Water-related responsibilities	Higher priority to micropollution	Equal priority to micropollution	Lower priority to micropollution
Ecological status (when considering micropollution as part of the chemical status)	❑	❑	❑
"Macro"pollution such as nutrients, fertilizers, inorganic pollution	❑	❑	❑
Sewage system/infrastructure	❑	❑	❑
Drinking water	❑	❑	❑
Hydropower	❑	❑	❑
Flood protection	❑	❑	❑
Waterways	❑	❑	❑
Other:	❑	❑	❑
	❑	❑	❑

11. *Please indicate with whom your organization <u>collaborates</u> **internationally** on micropollution in the Rhine catchment area.*

My organization collaborates **internationally** with:	Please specify:
❑ Working groups:	-- (Name of the working groups)
❑ Municipalities / cities outside of the Netherlands	-- (Name of the municipalities / cities)
❑ Regions or water agencies outside of the Netherlands	-- (Name of the regions / water agencies)
❑ Other countries:	-- (Name of the countries)
❑ European actors	-- (Name of the EU actors)
❑ International / foreign associations:	-- (Name of the associations)
❑ Foreign universities / research institutes / consultancies	-- (Name of the universities / institutes)
❑ Other:	-- (Please specify)

12. a) *Please indicate at which level your organization is* <u>*formally responsible*</u> *for water protection.* Formally responsible is defined as: adopting or implementing policies, having a mandate to carry out research or other water-related responsibilities, etc.

Indicate as many levels as are applicable.

My organization is responsible for the following levels:	Please indicate
❏ Local	
	(Name of the municipality/city)
❏ Provincial	
	(Name of the province)
❏ Water Basin	
	(Name of the water agency)
❏ National	
❏ European	
	(Please describe)
❏ Other	
	(Please describe)

b) Water can flow across politically-defined borders, such as municipal, provincial, regional or state frontiers. Do the water-related activities of your organization extend beyond (or not encompass the entirety of) the area for which you are responsible as indicated under 11a)?

❏ No, the politically-defined frontiers are well-adapted to the water-related activities of my organization.

❏ Yes, water-related activities of my organization can extend beyond (or do not encompass the entirety of) the area indicated under 11a).

Please describe any discrepancies indicated above and explain their cause(s):

....

....

Thank you for your valuable cooperation!

If you have further remarks or ideas about the topic of micropollution or about the questionnaire, please share them below.

....

....

Please return your completed questionnaire in the enclosed pre-paid envelope
by 9 May 2014 to:

Florence Metz
Universität Bern, Institut für Politikwissenschaft
Fabrikstrasse 8
CH-3012 Bern

Annex 10
Actor List of Swiss Policy Actors

Actor code	Full actor name in English	Full actor name in German
AGIDEA	Swiss Association for Agricultural Development and Rural Areas	Agridea
ARPEA	Western Swiss Association for Water and Air Protection	Association romande pour la protection des eaux et de l'air
BAFU-CHEM	Federal Office for the Environment, Department of Air Protection and Chemicals	Bundesamt für Umwelt, Abteilung Luftreinhaltung und Chemikalien
BAFU-W/UVEK	Federal Office for the Environment, Department for Water	Bundesamt für Umwelt, Abteilung Wasser, Sektion Oberflächengewässer Qualität/Eidgenössisches Departement für Umwelt, Verkehr, Energie und Kommunikation
BAG	Federal Office for Health	Bundesamt für Gesundheit
BDP	Civil Democratic Party	Bürgerlich - Demokratische Partei
BFE	Federal Office for Energy	Bundesamt für Energie
BJ	Federal Office for Justice	Bundesamt für Justiz
BLW	Federal Office for Agriculture	Bundesamt für Landwirtschaft
BMG	BMG Engineering AG	BMG Engineering AG
BPUK	Conference of Cantonal Directors of Construction, Planning, and Environmental Protection	Bau-, Planungs- und Umweltdirektoren-Konferenz
BR	Federal Council	Bundesrat
CERCL	Cercl'eau	Cercl'eau
CVP	Christian Democratic People's Party	Christlich demokratische Volkspartei
EAWAG	Swiss Federal Institute of Aquatic Science and Technology	Eidgenössische Anstalt für Wasserversorgung, Abwasserreinigung und Gewässerschutz
ECON/SAV	Economiesuisse/Swiss Employers' Association	Economiesuisse, Verband der Schweizer Unternehmen/Schweizerische Arbeitgeberverband
EFV	Swiss Finance Administration	Eidgenössisches Finanzverwaltung
EPA	Swiss Office of Personnel	Eidgenössisches Personalamt
EPFL	Swiss Federal Institute for Technology Lausanne	École polytechnique fédérale de Lausanne
ERFA	Sewage Treatment Plants in Large Cities Initiative	ERFA Klärwerke Grossstädte CH

(continued)

(continued)

Actor code	Full actor name in English	Full actor name in German
FDP	Free Democratic Party. The Liberals	FDP. Die Liberalen
FHNW	University of Applied Sciences of North-West Switzerland	Fachhochschule Nordwestschweiz
FISCH	Swiss Fishery Association	Schweizerische Fischereiverband
GLP	Green Liberal Party of Switzerland	Grünliberale Partei Schweiz
GPS	Swiss Green Party	Grüne Partei der Schweiz/Grünes Bündnis
GRESE	Western Swiss Group of Sewage Treatment Plants Operators	Groupement Romand des Exploitants de Stations d'Epuration
HKBB	Basel Chamber of Commerce	Handelskammer beider Basel
HOLINGER	Holinger Engineering	Holinger AG
HUNZIKER	Hunziker-Betatech	Hunziker-Betatech
INDUS	Scienceindustries	Scienceindustries
KF	Consumer Forum	Konsumentenforum
KI/SSV/SGV	Communal Infrastructure/Swiss Cities Association/Swiss Municipalities Association	Kommunale Infrastruktur/Schweizerischer Städteverband/Schweizerischer Gemeindeverband
KVU	Conference of Heads of Cantonal Offices for Environmental Protection	Konferenz der Vorsteher der Umweltschutzämter
LABEAUX	Competence Network of Cantonal Laboratories for Water and Environmental Protection	Lab'Eaux
OEKOTOX	Ecotox Centre	Oekotoxzentrum
PRONA	Pro Natura	Pro Natura
SBV	Swiss Farmers' Association	Schweiz. Bauernverband
SECO	State Secretariat for Economic Affairs	Staatssekretariat für Wirtschaft
SGEWV	Swiss Trade Association	Schweizerischer Gewerbeverband
SKW	Swiss Cosmetics and Detergent Association	Schweiz. Kosmetik- und Waschmittelverband
SMEDIC	Swissmedic	Swissmedic
SMEM	Swiss Mechanical and Electrical Engineering Industry Association	Swissmem
SP	Swiss Social Democratic Party	Sozialdemokratische Partei der Schweiz
SVGW	Swiss Gas and Water Industry Association	Schweiz. Verein des Gas- und Wasserfaches
SVP	Swiss People's Party	Schweizerische Volkspartei

(continued)

(continued)

Actor code	Full actor name in English	Full actor name in German
UBAS	University of Basel	Universität Basel
UNIL	University of Lausanne	Université de Lausanne
UREKN	National Council's Committee on the Environment, Spatial Planning, and Energy	Kommissionen für Umwelt, Raumplanung und Energie des Nationalrates
UREKS	Council of State's Committee on the Environment, Spatial Planning, and Energy	Kommissionen für Umwelt, Raumplanung und Energie des Ständerates
VKCS	Association of Cantonal Chemists of Switzerland	Verband der Kantonschemiker der Schweiz
VSA	Swiss Water Association	Verband Schweizer Abwasser- und Gewässerschutzfachleute
WWF	World Wide Fund For Nature, Switzerland	World Wide Fund For Nature, Schweiz

Annex 11
Actor List of German Policy Actors

	Actor code	Full actor name in English	Full actor name in German
1	AÖW	Alliance of Public Water Management	Allianz der öffentlichen Wasserwirtschaft
2	BAH	Federal Association of Medicine Manufacturers	Bundesverband der Arzneimittel-Hersteller
3	BauInd	German Association of Construction Industry	Hauptverband der Deutschen Bauindustrie
4	BaWü	State of Baden-Württemberg	Baden-Württemberg
5	BAY	State of Bavaria	Bayern
6	BBU	Federal Association of Citizens' Initiative on Environmental Protection	Bundesverband Bürgerinitiativen Umweltschutz
7	BDEW	Federal Association of Energy and Water Industry	Bundesverband der Energie- und Wasserwirtschaft
8	BDI	Federation of German Industries	Bundesverband der Deutschen Industrie
9	BfG	German Federal Institute of Hydrology	Bundesanstalt für Gewässerkunde, Referat Gewässerkunde
10	BMF	Federal Ministry of Finance	Bundesministerium der Finanzen
11	BMG	Federal Ministry of Health	Bundesministerium für Gesundheit, Referat Gesundheit und Umwelt (TrinkW)

(continued)

(continued)

	Actor code	Full actor name in English	Full actor name in German
12	BMJ	Federal Ministry of Justice	Bundesministerium der Justiz
13	BML	Federal Ministry of Agriculture	Bundesministeriums für Landwirtschaft
14	BMU	Federal Ministry for the Environment	Bundesministerium für Umwelt
15	BMWi	Federal Ministry of Economics	Bundeswirtschaftsministerium
16	BUND	Friends of the Earth Germany	BUND für Umwelt und Naturschutz
17	BV	Federation of Municipal Organizations	Bundesvereinigung der kommunalen Spitzenverbände
18	BWK	Association of Engineers for Water Management, Waste Management and Environmental Construction	Bund der Ingenieure für Wasserwirtschaft, Abfallwirtschaft und Kulturbau
19	DAFV	German Fishing Association	Deutsche Angelfischer-Verband/ Mitglied Deutscher Fischerei-Verband
20	DBV	German Farmers' Association	Deutscher Bauernverband
21	DIHK	Association of German Chambers of Commerce and Industry	Deutscher Industrie- und Handelskammertag
22	DNR	German League for Nature, Animal Protection and Environment	Deutscher Naturschutzring
23	DVGW	German Technical and Scientific Association for Gas and Water	Deutscher Verein des Gas- und Wasserfaches
24	DWA	German Association for Water, Wastewater and Waste	Deutsche Vereinigung für Wasserwirtschaft, Abwasser und Abfall
25	GÖD	Union of Public Services	Gewerkschaft öffentlicher Dienst und Dienstleistungen
26	GrLi	Green League	Grüne Liga
27	HES	State of Hesse	Hessen
28	IVA	Agrochemical Association	Industrieverband Agrar
29	LAWA/BLAK	Common Working Group on Water of the Federal Government and State Governments	Bund/Länder-Arbeitsgemeinschaft Wasser/Bund/Länder-Arbeitskreis Wasser
30	MIRO	Federal Association of Mineral Raw Materials	Bundesverband Mineralische Rohstoffe
31	MWV	Association of German Petroleum Industry	Mineralölwirtschaftsverband

(continued)

(continued)

	Actor code	Full actor name in English	Full actor name in German
32	NRW	State of North Rhine-Westphalia	Nordrhein-Westfalen
33	RLP	State of Rhineland-Palatinate	Rheinland-Pfalz
34	UBA	Federal Environmental Agency	Umweltbundesamt
35	VCI	German Chemical Industry Association	Verband der Chemischen Industrie
36	VDA	German Automotive Industry Association	Verband der Automobilindustrie
37	VDMA	German Engineering Federation	Verband Deutscher Maschinen- und Anlagenbau
38	VKU	Association of Municipal Companies	Verband kommunaler Unternehmen
39	WVM	German Metal Industry Association	WirtschaftsVereinigung Metalle

Annex 12
Actor List of French Policy Actors

	Actor code	Full actor name in English	Full actor name in French
1	ADF	Assembly of the French Departments	Assemblée des Départements de France
2	AGENCE	Water Agency	Agence de l'Eau
3	ALSACENATURE	Alsace Nature	Alsace Nature
4	AMF	Association of French Mayors	Association des Maires de France
5	ANSES	French Agency for Food, Environmental and Occupational Health & Safety	Agence Nationale de Sécurité Sanitaire
6	AQUAREF/INERIS	National Reference Laboratory for Water Monitoring/National Competence Centre for Industrial Safety and Environmental Protection	Laboratoire National de Référence pour la Surveillance des Milieux Aquatiques/Institut National de l'Environnement Industriel et des Risques
7	BRGM	Geology and Mining Research Institute	Bureau de Recherches Géologiques et Minières
8	CHAMBREAGRI	Farmers' Association	Chambres d'Agriculture

(continued)

(continued)

	Actor code	Full actor name in English	Full actor name in French
9	CLCV	Consumers, Housing, and Well-being Association	Association Consommation, Logement et Cadre de Vie
10	CLE	Local Water Commissions	Commissions Locales de l'Eau
11	CNIDEP	National Innovation Centre for Sustainable Development and Environment in Small Companies	Centre National d'Innovation pour le Développement Durable et l'Environnement dans les Petites Entreprises
12	DDT	Local Territorial Authority	Directions Départementales des Territoires
13	DEB	Department of Water and Biodiversity/Ministry of Ecology	Direction de l'Eau et de la Biodiversité/Ministère de l'Ecologie, du Développement Durable et de l'Energie
14	DELEGBASSIN	Water Basin Delegations/Prefect of Water Basin	Délégation de Bassin/Préfet Coordonnateur de Bassin
15	DGPR	General Directorate for Risk Prevention	Direction Générale de la Prévention des Risques
16	DREAL	Regional Authority for the Environment, Spatial Planning, and Housing	Directions Régionales de l'Environnement, de l'Aménagement et du Logement
17	FENARIVE	National Federation of Industrial Water Users	Fédération Nationales des Associations de Riverains et utilisateurs industriels de l'Eau
18	FNCCR	National Federation of Public Water Services	Fédération Nationale des Collectivités Concédantes et Régies
19	FNE	France Nature Environment	France Nature Environnement
20	FP2E	French Professional Federation of Water Companies	Fédération Professionnelle des Entreprises de l'Eau
21	IFREMER	French Research Institute for Exploitation of the Sea	Institut Français de Recherche pour l'Exploitation de la Mer
22	IRSTEA	National Research Institute of Science and Technology for Environment and Agriculture	Institut National de Recherche en Sciences et Technologies pour l'Environnement et l'Agriculture
23	LNE	French National Metrology and Testing Laboratory	Laboratoire National de Métrologie et d'Essais

(continued)

(continued)

	Actor code	Full actor name in English	Full actor name in French
24	MEDEF	French Business Confederation	Mouvement des entreprises de France
25	MINISTAGRI	Ministry of Agriculture	Ministère de l'Agriculture
26	MINISTSANTE	Ministry of Health	Ministère de la Santé
27	ONEMA	National Agency for Water and Aquatic Environments	Office National de l'Eau et des Milieux Aquatiques
28	POLICEEAU	Water Police	Police de l'Eau
29	ROBINDESBOIS	Robin des Bois	Robin des Bois
30	UFC	Federal Union of Consumers	Union Fédérale des Consommateurs
31	UIC	Chemical Industries Union	Union des Industries Chimiques
32	WWF	World Wide Fund For Nature, France	World Wide Fund For Nature, France

Annex 13
Actor List of Dutch Policy Actors

Actor code	Full actor name in English	Full actor name in Dutch
BOGIN	Association of the Dutch Generic Medicines Industry	Bond van de Generieke Geneesmiddelenindustrie Nederland
CBG	Medicines Evaluation Board	College ter Beoordeling van Geneesmiddelen
CHRISTENUNIE	Christian Union Party	ChristenUnie
GEZONDHEIDSRAAD	Health Council of the Netherlands	Gezondheidsraad
GROENLINKS	Green Left Party	GroenLinks
IenM	Ministry for Infrastructure and Environment	Ministerie van Infrastructuur en Milieu
KNMP	Royal Dutch Society for the Advancement of Pharmacy	Koninklijke Nederlandse Maatschappij ter bevordering der Pharmacie
KWR	Watercycle Research Institute	Watercycle Research Institute
LNV-ex	Former Ministry of Agriculture, Nature, and Food Quality	Voormalig Ministerie van Landbouw, Natuur en Voedselkwaliteit

(continued)

(continued)

Actor code	Full actor name in English	Full actor name in Dutch
NEFARMA	Association for Innovative Medicines in the Netherlands	Vereniging Innovatieve Geneesmiddelen Nederland
RDGG	Reinier de Graaf Hospital	Reinier de Graaf Ziekenhuis
RIVM	National Institute for Public Health and the Environment	Rijksinstituut voor Volksgezondheid en Milieu
RIWA	Association for River Waterworks	Vereniging van Rivierwaterbedrijven
RIZA/RIKZ-ex	Former National Institute for Integrated Freshwater and Wastewater Management/Former National Institute for Coast and Sea	Voormalig Rijksinstituut voor Integraal Zoetwaterbeheer en Afvalwaterbehandeling/Voormalig Rijksinstituut voor Kust en Zee
RWS	Department for Water Management	Rijkswaterstaat
STICHTINGHA	Foundation House of the Earth	Stichting Huize Aarde
STOWA	Foundation for Applied Water Research	Stichting Toegepast Onderzoek Waterbeheer
TWEEDEKAMER	Second Chamber	Tweede Kamer der Staten-General
UvW	Union of Water Boards	Unie van Waterschappen
V&W-ex	Ministry of Transport and Water Management	Ministerie van Verkeer en Waterstaat
VCIenM	Parliamentary Committee for Infrastructure and Environment of the Second Chamber	Vaste commissie voor Infrastructuur en Milieu, Tweede Kamer
VEWIN	Association of Dutch Drinking Water Companies	Vereniging van Waterbedrijven in Nederland
VITENS	Vitens Water Supply Company	Vitens
VROM-ex	Former Ministry of Housing, Spatial Planning, and the Environment	Voormalig Ministerie Volkshuisvesting, Ruimtelijke Ordening en Milieubeheer
VWS	Ministry of Health, Well-being, and Sport	Ministerie van Volksgezondheid, Welzijn en Sport
WWF	World Wide Fund For Nature, NL	Wereld Natuur Fonds - Nederland

Annex 14
Swiss Actors' Scores for Reputational Power, Degree Centrality, Betweenness Centrality, and Eigenvector Centrality

Swiss actors	Indegree	Degree	Normalized betweenness centrality	Normalized eigenvector centrality
Type of tie	Reputational power	Collaboration	Collaboration	Collaboration
AGIDEA	1	2	0	5.18
ARPEA	8	8	0.08	20.76
BAFU-CHEM	12	8	0.62	14.50
BAFU-W	35	43	49.54	58.75
BAG	14	9	0.97	16.44
BDP	2	2	0.01	4.96
BFE	2	2	0	5.98
BJ	3	1	0	4.67
BLW	13	4	2.27	6.47
BMG	5	9	0.40	21.07
BPUK	20	16	2.50	30.82
BR	23	2	0	6.56
CERCL	9	6	0.09	14.93
CVP	4	8	0.66	17.84
EAWAG	29	20	3.85	39.12
ECON/SAV	8	4	0.14	3.09
EFV	3	3	0	6.33
EPA	2	3	0	6.33
EPFL	15	7	0.03	18.70
ERFA	20	22	7.86	38.69
FDP	6	5	0.03	12.66
FHNW	4	5	0	16.19
FISCH	16	6	0.15	11.15
GLP	4	1	0	1.89
GPS	5	5	0.06	8.96
GRESE	13	11	0.38	26.44
HKBB	1	4	0.95	9.22
HOLINGER	4	6	0.02	16.35
HUNZIKER	5	7	0.08	18.80
INDUS	15	10	2.71	15.41
KF	3	0	0	−3.5 (isolate)

(continued)

(continued)

Swiss actors	Indegree	Degree	Normalized betweenness centrality	Normalized eigenvector centrality
Type of tie	Reputational power	Collaboration	Collaboration	Collaboration
KI/SSV/SGV	17	8	0.25	19.35
KVU	23	13	0.91	28.03
LABEAUX	8	7	0.09	17.78
OEKOTOX	14	12	0.67	27.51
PRONA	10	7	0.62	13.85
SBV	5	2	0.12	1.08
SECO	1	3	0.07	7.20
SGEWV	2	7	0.72	13.75
SKW	9	4	0.61	7.19
SMEDIC	6	1	0	4.67
SMEM	5	6	0.64	13.17
SP	6	6	0.06	15.34
SVGW	8	7	0.12	16.98
SVP	3	7	2.55	7.06
UBAS	5	13	1.52	28.46
UNIL	4	11	0.54	23.94
UREKN	19	10	1.46	18.76
UREKS	22	17	9.06	23.74
VKCS	18	6	0.03	13.42
VSA	30	20	3.63	38.59
WWF	12	8	0.62	17.71
Network mean	10.21	7.96	1.88	16.07

Annex 15
German Actors' Scores for Reputational Power, Degree Centrality, Betweenness Centrality, and Eigenvector Centrality

German actors	Indegree	Degree	Normalized betweenness centrality	Normalized eigenvector centrality
Type of tie	Reputational power	Collaboration	Collaboration	Collaboration
AOEW	2	7	1.69	17.75
BAH	2	0	0	0
BauInd	2	1	0	1.44
BaWue	9	6	0.02	22.29
Bay	5	3	0	10.84
BBU	4	2	0	3.96
BDEW	12	14	4.14	40.61
BDI	13	8	2.30	23.20
BfG	6	0	0	−6.24 (isolate)
BMF	0	1	0	4.52
BMG	3	5	0	19.80
BMJ	2	1	0	4.59
BML	11	3	0	12.74
BMU	22	22	27.26	52.95
BMWi	12	8	1.93	25.37
BUND	6	9	12.98	23.35
BV	3	4	0.06	12.17
BWK	2	1	0	4.01
DAFV	0	1	0	2.25
DBV	7	2	0	8.15
DIHK	0	3	0	9.75
DNR	2	1	0	2.25
DVGW	6	10	1.36	32.54
DWA	9	11	2.81	34.11
GOED	0	0	0	−7.38 (isolate)
GrLi	4	4	0.27	15.31
HES	4	7	0.62	23.23
IVA	4	11	2.02	31.68

(continued)

(continued)

German actors	Indegree	Degree	Normalized betweenness centrality	Normalized eigenvector centrality
Type of tie	Reputational power	Collaboration	Collaboration	Collaboration
LAWA	23	10	1.33	32.57
MIRO	2	1	0	5.10
MWV	1	3	0	6.88
NRW	13	10	1.31	33.34
RLP	3	15	9.34	41.58
UBA	20	19	16.13	47.66
VCI	17	12	6.13	33.214
VDA	1	1	0	1.44
VDMA	1	1	0	1.44
VKU	8	6	0.45	19.05
WVM	5	9	14.68	15.00
Network mean	6.31	5.95	2.74	17.34

Annex 16
French Actors' Scores for Reputational Power, Degree Centrality, Betweenness Centrality, and Eigenvector Centrality

French actors	Indegree	Degree	Normalized betweenness centrality	Normalized eigenvector centrality
Type of tie	Reputational power	Collaboration	Collaboration	Collaboration
ADF	0	2	0.11	5.53
AGENCE	17	18	15.36	47.69
ALSACENATURE	1	16	13.35	39.73
AMF	2	4	1.55	9.43
ANSES	8	7	0.23	25.64
AQUAREF/INERIS	11	11	2.32	33.44
BRGM	3	8	0.35	29.34
CHAMBREAGRI	3	2	0.03	6.34
CLCV	1	2	0	9.04

(continued)

(continued)

French actors	Indegree	Degree	Normalized betweenness centrality	Normalized eigenvector centrality
Type of tie	Reputational power	Collaboration	Collaboration	Collaboration
CLE	6	2	0	8.64
CNIDEP	1	4	0.33	12.53
DDT	4	5	0	18.16
DEB	15	24	33.85	53.53
DELEGBASSIN	4	10	2.21	31.81
DGPR	12	13	5.54	37.15
DREAL	10	14	5.98	41.45
FENARIVE	7	4	0.03	13.25
FNCCR	4	7	1.31	20.36
FNE	9	9	3.20	26.76
FP2E	4	11	6.44	25.70
IFREMER	4	3	0	11.02
IRSTEA	2	6	0.39	21.12
LNE	2	8	0.77	26.78
MEDEF	4	3	0.03	11.28
MINISTAGRI	7	4	0	16.74
MINISTSANTE	8	2	0	7.16
ONEMA	15	12	2.27	38.78
POLICEEAU	8	4	0.06	13.96
ROBINDESBOIS	2	2	0	8.27
UFC	2	3	0	11.63
UIC	3	2	0	8.79
WWF	3	2	0	6.44
Network mean	5.69	7	2.99	21.17

Annex 17
Dutch Actors' Scores for Reputational Power, Degree Centrality, Betweenness Centrality, and Eigenvector Centrality

Dutch actors	Indegree	Degree	Normalized betweenness centrality	Normalized eigenvector centrality
Type of tie	Reputational power	Collaboration	Collaboration	Collaboration
BOGIN	5	8	0.25	25.51
CBG	5	5	0	16.20
CHRISTENUNIE	4	4	0.06	13.09
GEZONDHEIDSRAAD	7	2	0	7.35
GROENLINKS	1	3	0	9.97
IenM	12	22	18.86	48.03
KNMP	4	3	0	10.32
KWR	11	16	5.49	40.67
LNV-ex	5	5	0	16.97
NEFARMA	9	16	8.26	38.46
RDGG	8	4	0	14.64
RIVM	11	18	6.69	44.49
RIWA	7	9	0.78	26.99
RIZA/RIKZ-ex	7	7	0.04	23.67
RWS	8	12	1.55	34.58
STICHTINGHA	3	6	0.25	18.52
STOWA	10	10	0.43	31.43
TWEEDEKAMER	9	4	0	14.78
UvW	10	19	8.83	45.18
V&W-ex	7	5	0	18.75
VCIenM	8	4	0	14.16
VEWIN	9	18	9.53	42.32
VITENS	7	6	0	21.04
VROM-ex	9	8	0.19	26.50
VWS	10	12	1.45	34.20
WWF	0	0	0	0
Network mean	7.15	8.69	2.41	24.53

Printed by Printforce, the Netherlands